P9-AOU-937

The
Art
of
Strategy

**A GAME THEORIST'S
GUIDE TO SUCCESS
IN BUSINESS & LIFE**

Avinash K. Dixit

Barry J. Nalebuff

W. W. NORTON & COMPANY

New York • London

Copyright © 2008 by Avinash K. Dixit and Barry J. Nalebuff

All rights reserved
Printed in the United States of America
First Edition

Doonesbury cartoon: © 1993 G. B. Trudeau. Reprinted with permission of
Universal Press Syndicate. All rights reserved.

Peanuts cartoon: © United Features Syndicate, Inc.

For information about permission to reproduce selections from this book,
write to Permissions, W. W. Norton & Company, Inc.
500 Fifth Avenue, New York, NY 10110

For information about special discounts for bulk purchases,
please contact W. W. Norton Special Sales at
specialsales@wwnorton.com or 800-233-4830

Manufacturing by RR Donnelley, Harrisonburg
Book design by Charlotte Staub
Production manager: Anna Oler

Library of Congress Cataloging-in-Publication Data

Dixit, Avinash K.
The art of strategy : a game theorist's guide to success in
business & life / Avinash K. Dixit, Barry J. Nalebuff. — 1st ed.
p. cm.
Includes bibliographical references and index.
ISBN 978-0-393-06243-4 (hardcover)
1. Strategic planning. 2. Strategy. 3. Game theory.
4. Decision making. I. Nalebuff, Barry, 1958– II. Title.
HD30.28.D587 2008
658.4'012—dc22

2008021347

W. W. Norton & Company, Inc.
500 Fifth Avenue, New York, N.Y. 10110
www.wwnorton.com

W. W. Norton & Company Ltd.
Castle House, 75/76 Wells Street, London W1T 3QT

1 2 3 4 5 6 7 8 9 0

To all our students,
from whom we have learned so much

*(especially Seth—*BJN*)*

CONTENTS

Preface

We didn't set out to write a new book. The plan was simply to revise our 1991 book, *Thinking Strategically*. But it didn't quite turn out that way.

One model for writing a revision comes from Borges's character Pierre Menard, who decides to rewrite Cervantes's *Don Quixote*. After great effort, Menard's revision ends up being word-for-word identical to the original. However, 300 years of history and literature have passed since *Quixote*, including *Quixote* itself. Although Menard's words are the same, his meaning is now entirely different.

Alas, our original text wasn't *Don Quixote*, and so the revision did require changing a few words. In fact, most of the book is entirely new. There are new applications, new developments in the theory, and a new perspective. So much is new that we decided a new title was called for as well. Although the words are new, our meaning remains the same. We intend to change the way you see the world, to help you think strategically by introducing the concepts and logic of game theory.

Like Menard, we have a new perspective. When we wrote *Thinking Strategically*, we were younger, and the zeitgeist was one of self-centered competition. We have since come to the full realization of the important part that cooperation plays in strategic situations, and how good strategy must appropriately mix competition and cooperation.*

We started the original preface with: "Strategic thinking is the art of outdoing an adversary, knowing that the adversary is trying to do the same to you." To this we now add: It is also the art of finding ways to cooperate, even when others are motivated by self-interest, not benevolence. It is the art of convincing others, and even yourself, to do what you say. It is the art of interpreting and revealing information. It is the art of putting yourself in others' shoes so as to predict and influence what they will do.

We like to think that *The Art of Strategy* includes this older, wiser perspective. But there is also continuity. Even though we offer more real-life stories, our purpose remains to help you develop your own ways of thinking about the strategic situations you will face; this is not an airport book offering "seven steps for sure strategic success." The situations you face will be so diverse that you will succeed better by knowing some general principles and adapting them to the strategic games you are playing.

Businessmen and corporations must develop good competitive strategies to survive, and find cooperative opportunities to grow the pie. Politicians have to devise campaign strategies to get elected and legislative strategies to implement their visions. Football coaches plan strategies for players to execute on the field. Parents trying to elicit good behavior from children must become amateur strategists—the children are pros.

Good strategic thinking in such numerous diverse contexts remains an art. But its foundations consist of some simple basic principles—an emerging science of strategy, namely game theory. Our premise is that readers from a variety of backgrounds and occupations can become better strategists if they know these principles.

Some people question how we can apply logic and science to a

---

* Pursuing this line led one of us to write a book on this idea; see Adam Brandenburger and Barry J. Nalebuff, *Co-opetition* (New York: Doubleday, 1996).

world where people act irrationally. It turns out that there is often method to the madness. Indeed, some of the most exciting new insights have come from recent advances in behavioral game theory, which incorporates human psychology and biases into the mix and thus adds a social element to the theory. As a result, game theory now does a much better job dealing with people as they are, rather than as we might like them to be. We incorporate these insights into our discussions.

While game theory is a relatively young science—just over seventy years old—it has already provided many useful insights for practical strategists. But, like all sciences, it has become shrouded in jargon and mathematics. These are essential research tools, but they prevent all but the specialists from understanding the basic ideas. Our main motive for writing *Thinking Strategically* was the belief that game theory is too interesting and important to leave to the academic journals. The insights prove useful in many endeavors—business, politics, sports, and everyday social interactions. Thus we translated the important insights back into English and replaced theoretical arguments with illustrative examples and case studies.

We are delighted to find our view becoming mainstream. Game theory courses are some of the most popular electives at Princeton and Yale, and most other schools where they are offered. Game theory permeates strategy courses in MBA programs, and a Google search for game theory produces more than 6 million pages. You'll find game theory in newspaper stories, op-eds, and public policy debates.

Of course, much of the credit for these developments belongs to others: to the Economics Nobel Prize Committee, which has awarded two prizes in game theory—in 1994, to John Harsanyi, John Nash, and Reinhard Selten and in 2005, to Robert Aumann and Thomas Schelling;* to Sylvia Nasar, who wrote *A Beautiful*

---

* There have also been three Nobel Prizes awarded for work in mechanism design and information economics, both of which are closely related to game theory: in 1996, to William Vickrey and James Mirrlees; in 2001, to George Akerlof, Michael Spence, and Joseph Stiglitz; and in 2007, to Leonid Hurwicz, Eric Maskin, and Roger Myerson.

Mind, the best-selling biography of John Nash; to those who made the award-winning movie of the same name; and to all those who have written books popularizing the subject. We might even share a bit of the credit. Since publication, *Thinking Strategically* has sold 250,000 copies. It has been translated into numerous languages, and the Japanese and Hebrew translations have been best sellers.

We owe a special debt to Tom Schelling. His writings on nuclear strategies, particularly *The Strategy of Conflict* and *Arms and Influence*, are justly famous. In fact, Schelling pioneered a lot of game theory in the process of applying it to nuclear conflict. Michael Porter's *Competitive Strategy*, drawing on the lessons of game theory for business strategy, is equally important and influential. An annotated guide to the works of Schelling, Porter, and many others is provided in our Further Reading section.

In this book we do not confine the ideas to any particular context. Instead, we offer a wide range of illustrations for each basic principle. Thus readers from different backgrounds will all find something familiar here. They will also see how the same principles bear on strategies in less familiar circumstances; we hope this will give them a new perspective on many events in news as well as history. We also draw on the shared experience of our readers, with illustrations from, for example, literature, movies, and sports. Serious scientists may think this trivializes strategy, but we believe that familiar examples are an effective vehicle for conveying the important ideas.

The idea of writing a book at a more popular level than that of a course text came from Hal Varian, now at Google and the University of California, Berkeley. He also gave us many useful ideas and comments on earlier drafts. Drake McFeely at W. W. Norton was an excellent if exacting editor for *Thinking Strategically.* He made extraordinary efforts to fashion our academic writing into a lively text. Many readers of *Thinking Strategically* gave us encouragement, advice, and criticism, all of which have helped us when writing *The Art of Strategy.* At the grave risk of omitting some, we must mention ones to whom we owe special thanks. Our coauthors on other related and unrelated book projects, Ian Ayres, Adam Brandenburger, Robert Pindyck, David Reiley, and Susan Skeath, generously gave us much useful input. Others whose influence continues in this new book include David Austen-Smith, Alan Blinder, Peter Grant, Seth Masters, Benjamin Polak, Carl Shapiro, Terry Vaughn, and Robert Willig. Jack Repcheck at W. W. Norton has been a constantly supportive, understanding, and perceptive editor for *The Art of Strategy*. Our manuscript editors, Janet Byrne and Catherine Pichotta, were generous to our faults. Every time you don't find a mistake, you should thank them.

We owe special thanks to Andrew St. George, book critic for the *Financial Times*. In choosing *Thinking Strategically* as a book he enjoyed reading most in the year 1991, he said: "it is a trip to the gym for the reasoning facilities" (*FT Weekend,* December 7/8, 1991). This gave us the idea of labeling some intriguing challenges we pose to the readers in this edition "Trips to the Gym." Finally, John Morgan, of the University of California, Berkeley, gave us a powerful incentive with the threat, "If you don't write a revision, I will write a competing book." And after we saved him the trouble, he helped us out with many ideas and suggestions.

AVINASH DIXIT

BARRY J. NALEBUFF

October 2007

How Should
People Behave
in Society?

Our answer does not deal with ethics or etiquette. Nor do we aim to compete with philosophers, preachers, or parents. Our theme, although less lofty, affects the lives of all of us just as much as do morality and manners. This book is about strategic behavior. All of us are strategists, whether we like it or not. It is better to be a good strategist than a bad one, and this book aims to help you improve your skills at discovering and using effective strategies.

Work, even social life, is a constant stream of decisions. What career to follow, how to manage a business, whom to marry, how to bring up children, and whether to run for president are just some examples of such fateful choices. The common element in these situations is that you do not act in a vacuum. Instead, you are surrounded by active decision makers whose choices interact with yours. This interaction has an important effect on your thinking and actions.

To illustrate the point, think of the difference between the decisions of a lumberjack and those of a general. When the lumberjack decides how to chop wood, he does not expect the wood to fight back: his environment is neutral. But when the general tries to cut down the enemy's army, he must anticipate and overcome resistance to his plans. Like the general, you must recognize that your business rivals, prospective spouse, and even your children are strategic. Their aims often conflict with yours, but they may well coincide. Your own choice must allow for the conflict and utilize the cooperation. This book aims to help you think strategically, and then translate these thoughts into action.

The branch of social science that studies strategic decision making is called *game theory*. The games in this theory range from chess to child rearing, from tennis to takeovers, and from advertising to arms control. As the Hungarian humorist George Mikes expressed it, "Many continentals think life is a game; the English think cricket is a game." We think both are right.

Playing these games requires many different kinds of skills. Basic skills, such as shooting ability in basketball, knowledge of precedents in law, or a blank face in poker, are one kind; strategic thinking is another. Strategic thinking starts with your basic skills and considers how best to use them. Knowing the law, you must decide the strategy for defending your client. Knowing *how well* your football team can pass or run and how well the other team can defend against each choice, your decision as the coach is *whether* to pass or to run. Sometimes, as in the case of nuclear brinkmanship, strategic thinking also means knowing when not to play.

The science of game theory is far from being complete, and much of strategic thinking remains an art. Our ultimate aim is to make you better practitioners of that art, but this requires a good foundation in some elementary concepts and methods of the science. Therefore we mix the two approaches. Chapter 1 begins with examples of the art, showing how strategic issues arise in a variety of decisions. We point out some effective strategies, some less effective ones, and even some downright bad ones that were used by players in these real-life games. These examples begin to suggest a conceptual framework. In the next set of chapters, 2–4, we build this basis for the science using examples, each of which is designed to bring out one principle. Then we turn our attention to more specific concepts and strategies for dealing with particular situations—

how to mix moves when any systematic action can be exploited by the other player, how to change a game to your advantage, and how to manipulate information in strategic interaction. Finally, we take up several broad classes of strategic situations—bargaining, auctions, voting, and the design of incentives—where you can see these principles and strategies in action.

Science and art, by their very nature, differ in that science can be learned in a systematic and logical way, whereas expertise in art has to be acquired by example, experience, and practice. Our exposition of the basic science generates some principles and broad rules—for example, the idea and method of backward reasoning that is developed in chapter 2, and the concept of Nash equilibrium in chapter 4. On the other hand, the art of strategy, in all the varied situations where you may need it, requires you to do more work. Each such situation will have some unique features that you must take into account and combine with the general principles of the science. The only way to improve your skill at this art is the inductive way—by seeing how it has been done before in similar situations. That is exactly how we aim to improve your strategic IQ: by giving numerous examples, including a case study, in each chapter and in a collection of case studies in the final chapter.

The examples range from the familiar, trivial, or amusing—usually drawn from literature, sports, or movies—to the frightening—nuclear confrontation. The former are merely a nice and palatable vehicle for the game-theoretic ideas. As to the latter, at one point in time many readers would have thought the subject of nuclear war too horrible to permit rational analysis. But with the cold war now long over, we hope that the game-theoretic aspects of the arms race and the Cuban missile crisis can be examined for their strategic logic with some detachment from their emotional content.

The case studies are similar to ones you might come across in a business-school class. Each case sets out a particular set of circumstances and invites you to apply the principles discussed in that chapter to find the right strategy for that situation. Some cases are open-ended; but that is also a feature of life. At times there is no clearly correct solution, only imperfect ways to cope with the problem. A serious effort to think each case through before reading our discussion is a better way to understand the ideas than any amount of reading of the text alone. For more practice, the final chapter is a collection of cases, in roughly increasing order of difficulty.

By the end of the book, we hope that you will emerge a more effective manager, negotiator, athlete, politician, or parent. We warn you that some of the strategies that are good for achieving these goals may not earn you the love of your rivals. If you want to be fair, tell them about our book.

Part I

CHAPTER 1

Ten Tales
of Strategy

WE BEGIN WITH ten tales of strategy from different aspects of life and offer preliminary thoughts on how best to play. Many of you will have faced similar problems in everyday life and will have reached the correct solution after some thought or trial and error. For others, some of the answers may be surprising, but surprise is not the primary purpose of the examples. Our aim is to show that such situations are pervasive, that they amount to a coherent set of questions, and that methodical thinking about them is likely to be fruitful.

In later chapters, we develop these systems of thought into prescriptions for effective strategy. Think of these tales as a taste of dessert before the main course. They are designed to whet your appetite, not fill you up.

#1. PICK A NUMBER

Believe it or not, we are going to ask you to play a game against us. We've picked a number between 1 and 100, and your goal is to guess the number. If you guess correctly on the first try, we'll pay you $100.

Actually, we aren't really going to pay you $100. It would be rather costly for us, especially since we want to give you some help along the way. But, as you play the game, we'd like you to imagine that we really are going to give you money, and we'll play the same way.

The chance of getting the number right on the first shot is quite low, only one in a hundred. So to improve your chances, we'll give you five guesses, and after each wrong guess, we'll also tell you if you are too high or too low. Of course, there's a bigger reward for getting the right answer quickly. If you guess correctly on the second try, you'll get $80. On the third try, the payment is down to $60, then $40 for the fourth guess, and $20 if you get the number on the fifth try. If it takes more than five guesses, the game is over and you get nothing.

Are you ready to play? We are, too. If you are wondering how to play a game with a book, it is a bit of a challenge, but not impossible. You can go to the artofstrategy.info web site and play the game interactively. Or, we can anticipate how you might be playing the game and respond accordingly.

Is your first guess 50? That's the most common first guess and, alas for you, it's too high.

Might your second guess be 25? Following 50, that is what most folks do. Sorry, that's too low. The next step for many is 37. We're afraid that 37 is too low. What about 42? Too low, again.

Let's pause, take a step back, and analyze the situation. This is your fifth guess coming up and your last chance to take our money. You know the number is above 42 and less than 50. There are seven options: 43, 44, 45, 46, 47, 48, and 49. Which of those seven do you think it will be?

So far, you have been guessing in a way that divides the interval into two equal parts and picking the midpoint. This is the ideal strategy in a game where the number has been chosen at random.* You are getting the most information possible from each guess and therefore will converge to the number as quickly as possible.

---

* The technical term for this search strategy is *minimizing the entropy*.

Indeed, Microsoft CEO Steven Ballmer is said to have used this game as a job interview question. For Ballmer the correct answer was 50, 25, 37, 42, . . . He was interested in seeing if the candidate approached the search problem in the most logical and efficient manner.

Our answer is a bit different. In Ballmer's problem, the number was picked at random, and so the engineer's strategy of "divide the set in two and conquer" was just right. Getting the most information from each guess minimizes the expected number of guesses and therefore leads you to get the most money. In our case, however, the number was *not* picked at random. Remember that we said that we were playing this game as if we actually had to pay you the money. Well, no one is reimbursing us for money that, hypothetically, we would have to pay you. And as much as we like you for having bought our book, we like ourselves even more. We'd rather keep the money than give it to you. So we deliberately picked a number that would be hard for you to find. Think about it for a moment—would it have made any sense for us to have picked 50 as the number? That would have cost us a fortune.

The key lesson of game theory is to put yourself in the other player's shoes. We put ourselves in your shoes and anticipated that you would guess 50, then 25, then 37, then 42. Understanding how you would play the game allowed us to greatly decrease the chance that you would guess our number and thus reduce how much we'd have to pay out.

In explaining all of this to you before the game is over, we've given you a big hint. So now that you understand the real game you are playing, you've got one last guess, for $20. What number do you pick?

49?

Congratulations. To us, not you. You've fallen right into our trap again. The number we picked was 48. Indeed, the whole speech about picking a number that was hard to find according to the split-the-interval rule was further designed to mislead you. We wanted you to pick 49 so that our choice of 48 would remain safe. Remember our objective is not to give you money.

To beat us at that game, you had to be one step ahead of us. You would have had to think, "They want us to pick 49, so I'm going to pick 48." Of course, if we had thought you would have been so clever, then we would have picked 47 or even 49.

The larger point of this game is not that we are selfish professors or cunning tricksters. Rather, the point is to illustrate as cleanly as possible what makes something a game: you have to take into account the objectives and strategies of the other players. When guessing a number picked at random, the number isn't trying to hide. You can take the engineer's mindset and divide the interval in two and do the best possible. But if you are playing a game, then you have to consider how the other player will be acting and how those decisions will influence your strategy.

#2. WINNING BY LOSING

We admit it: we watched *Survivor*. We would never have made it on the island. If we hadn't starved first, the others would surely have voted us off for being eggheads. For us, the challenge was trying to predict how the game would play out. We weren't surprised when the pudgy nudist Richard Hatch outwitted, outplayed, and outlasted his rivals to become the first champion of the CBS series and earn the million-dollar prize. He was gifted in his ability to act strategically without appearing to be strategic.

Hatch's most cunning ploy was in the last episode. The game was down to three players. Richard's two remaining rivals were 72-year-old retired Navy SEAL Rudy Boesch and 23-year-old river guide Kelly Wiglesworth. For their final challenge, the three of them had to stand on a pole with one hand on the immunity idol. The last one standing would go into the finals. And just as important, the winner would get to choose his or her opponent in the finals.

Your first impression might be that this was just a physical endurance contest. Think again. All three players understood that Rudy was the most popular of the three. If Rudy made it to the finals, Rudy would likely win. Richard's best hope was to go against Kelly in the finals.

There were two ways that could happen. One is that Kelly would win the pole-standing competition and pick Richard. The other is that Richard would win and pick Kelly. Richard could count on Kelly picking him. She was also aware of Rudy's popularity. Her best hope of winning was to get to the finals against Richard.

It would seem that if either Richard or Kelly won the final challenge, each would pick the other as his or her opponent. Hence Richard should try to stay in the game, at least until Rudy had fallen off. The only problem is that Richard and Rudy had a long-standing alliance. If Richard won the challenge and didn't pick Rudy, that would have turned Rudy (and all Rudy's friends) against Richard, and this could have cost him the victory. One of the great twists of *Survivor* is that the ousted contestants vote to determine the final winner. One has to be very careful how one disposes of rivals.

From Richard's perspective, the final challenge could go one of three ways:

i. Rudy wins. Rudy then picks Richard, but Rudy would be the likely victor.

ii. Kelly wins. Kelly would be smart enough to know her best hope was to eliminate Rudy and go against Richard.

iii. Richard wins. If he picks Rudy to go on, Rudy beats him in the finals. If he picks Kelly to go on, Kelly might beat him because Richard would lose the support of Rudy and his many friends.

Comparing these options, Richard does best by losing. He wants Rudy eliminated, but it is better if Kelly does the dirty work for him. The smart money was on Kelly winning the challenge. She had won three of the previous four and as an outdoors guide was in the best shape of the three.

As a bonus, throwing the game saved Richard the trouble of standing on a pole under a hot sun. Early in the competition, host Jeff Probst offered a slice of orange to anyone willing to call it quits. Richard stepped off the pole and took the orange.

Throughout the book, you'll find these asides, which contain what we call a "Trip to the Gym." These trips take a look at more advanced elements of the game that we glossed over. For example, Richard could have tried to wait and see who dropped out first. If Kelly fell early, Richard might have preferred to beat Rudy and choose Kelly than to let Rudy win and have to go against Rudy in the finals. He might also have been concerned that Kelly would be savvy enough to do the same calculation and drop out early. The next chapters will show you how to use a more systematic approach to solve a game. The end goal is to help change the way you approach strategic situations, recognizing that you won't always have time to analyze every possible option.

After 4 hours and 11 minutes, Rudy fumbled when shifting his stance, let go of the immunity idol, and lost. Kelly picked Richard to go on to the finals. Rudy cast the swing vote in his favor, and Richard Hatch became the first *Survivor* champion.

With the benefit of hindsight it may all seem easy. What makes Richard's play so impressive is that he was able to anticipate all the different moves before they happened.* In chapter 2, we'll provide some tools to help you anticipate the way a game will play out and even give you a chance to have a go at another *Survivor* game.

#3. THE HOT HAND

Do athletes ever have a "hot hand"? Sometimes it seems that Yao Ming cannot miss a basket or that Sachin Tendulkar cannot fail to score a century in cricket. Sports announcers see these long streaks of consecutive successes and proclaim that the athlete has a hot hand. Yet according to psychology professors Thomas Gilovich, Robert Vallone, and Amos Tversky, this is a misperception of reality.[1]

They point out that if you flip a coin long enough, you will find some very long series of consecutive heads. The psychologists suspect that sports commentators, short on insightful things to say, are just finding patterns in what amounts to a long series of coin

---

* Richard would have done well to anticipate the consequences of not paying taxes on his $1 million winnings. On May 16, 2006, he was sentenced to 51 months in prison for tax evasion.

tosses over a long playing season. They propose a more rigorous test. In basketball, they look at all the instances of a player's baskets and observe the percentage of times that player's next shot is also a basket. A similar calculation is made for the shots immediately following misses. If a basket is more likely to follow a basket than to follow a miss, then there really is something to the theory of the hot hand.

They conducted this test on the Philadelphia 76ers basketball team. The results contradicted the hot hand view. When a player made his last shot, he was less likely to make his next; when he missed his previous attempt, he was more likely to make his next. This was true even for Andrew Toney, a player with the reputation for being a streak shooter. Does this mean we should be talking of the "stroboscopic hand," like the strobe light that alternates between on and off?

Game theory suggests a different interpretation. While the statistical evidence denies the presence of streak shooting, it does not refute the possibility that a hot player might warm up the game in some other way. The difference between streak shooting and a hot hand arises because of the interaction between the offensive and defensive strategies. Suppose Andrew Toney does have a truly hot hand. Surely the other side would start to crowd him. This could easily lower his shooting percentage.

That is not all. When the defense focuses on Toney, one of his teammates is left unguarded and is more likely to shoot successfully. In other words, Toney's hot hand leads to an improvement in the 76ers' *team* performance, although there may be a deterioration in Toney's *individual* performance. Thus we might test for hot hands by looking for streaks in team success.

Similar phenomena are observed in many other team sports. A brilliant running back on a football team improves the team's passing game and a great receiver helps the running game, as the opposition is forced to allocate more of its defensive resources to guard the stars. In the 1986 soccer World Cup final, the Argentine star Diego Maradona did not score a goal, but his passes through a ring of West German defenders led to two Argentine goals. The value of a star cannot be assessed by looking only at his scoring performance; his contribution to his teammates' performance is crucial, and assist statistics help measure this contribution. In ice hockey, assists and goals are given equal weight for ranking individual performance.

A player may even assist himself when one hot hand warms up the other. The Cleveland Cavaliers star LeBron James eats and writes with his left hand but prefers shooting with his right (though his left hand is still better than most). The defense knows that LeBron is right-handed, so they concentrate on defending against right-handed shots. But they do not do so exclusively, since LeBron's left-handed shots are too effective to be left unguarded.

What happens when LeBron spends his off season working to improve his left-handed shooting? The defense responds by spending more time covering his left-handed shots. The result is that this frees his right hand more often. A better left-handed shot results in a more effective right-handed shot. In this case, not only does the left hand know what the right hand is doing, it's helping it out.

Going one step further, in chapter 5 we show that when the left hand is stronger it may even be used *less* often. Many of you will have experienced this seemingly strange phenomenon when playing tennis. If your backhand is much weaker than your forehand, your opponents will learn to play to your backhand. Eventually, as a result of all this backhand practice, your backhand will improve. As your two strokes become more equal, opponents can no longer exploit your weak backhand. They will play more evenly between forehand and backhand. You get to use your better forehand more often; this could be the real advantage of improving your backhand.

#4. TO LEAD OR NOT TO LEAD

After the first four races in the 1983 America's Cup finals, Dennis Conner's *Liberty* led 3–1 in a best-of-seven series. On the morning of the fifth race, "cases of champagne had been delivered to *Liberty*'s dock. And on their spectator yacht, the wives of the crew were wearing red-white-and-blue tops and shorts, in anticipation of having their picture taken after their husbands had prolonged the United States' winning streak to 132 years.[2] It was not to be.

At the start, *Liberty* got off to a 37-second lead when *Australia II* jumped the gun and had to recross the starting line. The Australian skipper, John Bertrand, tried to catch up by sailing way over to the left of the course in the hopes of catching a wind shift. Dennis Conner chose to keep *Liberty* on the right hand side of the course. Bertrand's gamble paid off. The wind shifted five degrees in *Australia II*'s favor and she won the race by one minute and forty-seven seconds. Conner was criticized for his strategic failure to follow *Australia II*'s path. Two races later, *Australia II* won the series.

Sailboat racing offers the chance to observe an interesting reversal of a "follow the leader" strategy. The leading sailboat usually copies the strategy of the trailing boat. When the follower tacks, so does the leader. The leader imitates the follower even when the follower is clearly pursuing a poor strategy. Why? Because in sailboat racing (unlike ballroom dancing) close doesn't count; only winning matters. If you have the lead, the surest way to stay ahead is to play monkey see, monkey do.*

Stock-market analysts and economic forecasters are not immune to this copycat strategy. The leading forecasters have an incentive to follow the pack and produce predictions similar to everyone else's. This way people are unlikely to change their perception of these forecasters' abilities. On the other hand, newcomers take the risky strategies; they tend to predict boom or doom. Usually they are wrong and are never heard of again, but now and again they are proven correct and move to the ranks of the famous.

---

* This strategy no longer applies once there are more than two competitors. Even with three boats, if one boat tacks right and the other tacks left, the leader has to choose which (if either) to follow.

Industrial and technological competitions offer further evidence. In the personal-computer market, Dell is less known for its innovation than for its ability to bring standardized technology to the mass market. More new ideas have come from Apple, Sun, and other start-up companies. Risky innovations are their best and perhaps only chance of gaining market share. This is true not just of high-technology goods. Procter & Gamble, the Dell of diapers, followed Kimberly-Clark's innovation of resealable diaper tape and recaptured its commanding market position.

There are two ways to move second. You can imitate as soon as the other has revealed his approach (as in sailboat racing) or wait longer until the success or failure of the approach is known (as in computers). The longer wait is more advantageous in business because, unlike in sports, the competition is usually not winner-take-all. As a result, market leaders will not follow the upstarts unless they also believe in the merits of their course.

#5. HERE I STAND

When the Catholic Church demanded that Martin Luther repudiate his attack on the authority of popes and councils, he refused to recant: "I will not recant anything, for to go against conscience is neither right nor safe." Nor would he compromise: "Here I stand, I cannot do otherwise."[3] Luther's intransigence was based on the divinity of his positions. When defining what was right, there was no room for compromise. His firmness had profound long-term consequences; his attacks led to the Protestant Reformation and substantially altered the medieval Catholic Church.

Similarly, Charles de Gaulle used the power of intransigence to become a powerful player in the arena of international relations. As his biographer Don Cook expressed it, "[de Gaulle] could create power for himself with nothing but his own rectitude, intelligence, personality and sense of destiny."[4] But above all, his was "the power of intransigence." During the Second World War, as the self-proclaimed leader in exile of a defeated and occupied nation, he held his own in negotiations with Roosevelt and Churchill. In the

1960s, his presidential "Non!" swung several decisions France's way in the European Economic Community (EEC).

In what way did his intransigence give him power in bargaining? When de Gaulle took a truly irrevocable position, the other parties in the negotiation were left with just two options—to take it or to leave it. For example, he single-handedly kept England out of the European Economic Community, once in 1963 and again in 1968; the other countries were forced either to accept de Gaulle's veto or to break up the EEC. De Gaulle judged his position carefully to ensure that it would be accepted. But that often left the larger (and unfair) division of the spoils to France. De Gaulle's intransigence denied the other party an opportunity to come back with a counteroffer that was acceptable.

In practice, this is easier said than done, for two kinds of reasons. The first kind stems from the fact that bargaining usually involves considerations other than the pie on today's table. The perception that you have been excessively greedy may make others less willing to negotiate with you in the future. Or, next time they may be more firm bargainers as they try to recapture some of their perceived losses. On a personal level, an unfair win may spoil business relations, or even personal relations. Indeed, biographer David Schoenbrun faulted de Gaulle's chauvinism: "In human relations, those who do not love are rarely loved: those who will not be friends end up by having none. De Gaulle's rejection of friendship thus hurt France."[5] A compromise in the short term may prove a better strategy over the long haul.

The second kind of problem lies in achieving the necessary degree of intransigence. Luther and de Gaulle achieved this through their personalities, but this entails a cost. An inflexible personality is not something you can just turn on and off. Although being inflexible can sometimes wear down an opponent and force him to make concessions, it can equally well allow small losses to grow into major disasters.

Ferdinand de Lesseps was a mildly competent engineer with extraordinary vision and determination. He is famous for building the Suez Canal in what seemed almost impossible conditions. He did not recognize the impossible and thereby accomplished it. Later, he tried using the same technique to build the Panama Canal. It ended in disaster.* Whereas the sands of the Nile yielded to his will, tropical malaria did not. The problem for de Lesseps was that his inflexible personality could not admit defeat even when the battle was lost.

How can one achieve selective inflexibility? Although there is no ideal solution, there are various means by which commitment can be achieved and sustained; this is the topic for chapter 7.

#6. THINNING STRATEGICALLY

Cindy Nacson-Schechter wanted to lose weight. She knew just what to do: eat less and exercise more. She knew all about the food pyramid and the hidden calories in soft drinks. Still, nothing had worked. She had gained forty pounds with the birth of her second child and it just wasn't coming off.

That's why she accepted ABC's offer to help her lose weight. On December 9, 2005, she came into a photographer's studio on Manhattan's West Side, where she found herself changing into a bikini. She hadn't worn a bikini since she was nine, and this wasn't the time to start again.

The setup felt like backstage at the *Sports Illustrated* swimsuit issue shoot. There were lights and cameras everywhere, and all she had on was a tiny lime-green bikini. The producers had thoughtfully placed a hidden space heater to keep her warm. Snap. Smile. Snap. Smile. What in the world was she thinking? Snap.

If things worked out as she hoped, no one would ever see these pictures. The deal she made with ABC *Primetime* was that they would destroy the pictures if she lost 15 pounds over the next two months. They wouldn't help her in any way. No coach, no trainer,

---

* The Suez Canal is a sea-level passage. The digging was relatively easy since the land was already low-lying and desert. Panama involved much higher elevations, lakes along the way, and dense jungle. Lesseps's attempt to dig down to sea level failed. Much later, the U.S. Army Corps of Engineers succeeded using a very different method—a sequence of locks, using the lakes along the way.

no special diets. She already knew what she had to do. All she needed was some extra motivation and a reason to start today rather than tomorrow.

Now she had that extra motivation. If she didn't lose the promised weight, ABC would show the photos and the videos on prime-time television. She had already signed the release giving them permission.

Fifteen pounds in two months was a safe amount to lose, but it wouldn't be a cakewalk. There was a series of holiday parties and Christmas dinners. She couldn't risk waiting until the New Year. She had to start now.

Cindy knew all about the dangers of being overweight—the increased risk of diabetes, heart attack, and death. And yet that wasn't enough to scare her into action. What she feared more than anything was the possibility that her ex-boyfriend would see her hanging out of a bikini on national TV. And there was little doubt that he would watch the show. Her best friend was going to tell him if she failed.

Laurie Edwards didn't like the way she looked or how she felt. It didn't help that she worked part-time tending bar, surrounded by hot twenty-somethings. She had tried Weight Watchers, South Beach, Slim-Fast, you name it. She was headed in the wrong direction and needed something to help her change course. When she told her girlfriends about the show, they thought it was the stupidest thing she'd ever done. The cameras captured that "what am I doing?" look on her face and a lot more.

Ray needed to lose weight, too. He was a newlywed in his twenties but looked closer to forty. As he walked the red carpet in his racing swimsuit, it wasn't a pretty picture. Click. Smile. Click.

He wasn't taking any chances. His wife wanted him to lose weight and was willing to help. She offered to diet with him. Then she took the plunge. She changed into a bikini, too. She wasn't as overweight as Ray, but she wasn't bikini-ready, either.

Her deal was different from Cindy's. She didn't have to weigh in. She didn't even have to lose weight. The pictures of her in a bikini would only be shown if *Ray* didn't lose the weight.

For Ray, the stakes had been raised even higher. He was either going to lose the weight or his wife.

All together, four women and one couple bared their soles and much more in front of the cameras. What were they doing? They weren't exhibitionists. The ABC producers had carefully screened them out. None of the five wanted to see these photos appear on TV, and none of them expected they ever would.

They were playing a game against their future selves. Today's self wants the future self to diet and exercise. The future self wants the ice cream and the television. Most of the time, the future self wins because it gets to move last. The trick is to change the incentives for the future self so as to change its behavior.

In Greek mythology, Odysseus wanted to hear the Sirens' songs. He knew that if he allowed his future self to listen to their song, that future self would sail his ship into the rocks. So he tied his hands—literally. He had his crew bind his hands to the mast (while plugging their own ears). In dieting, this is known as the empty-fridge strategy.

Cindy, Laurie, and Ray went one step further. They put themselves in a bind that only dieting would get them out of. You might think that having more options is always a good thing. But thinking strategically, you can often do better by cutting off options. Thomas Schelling describes how the Athenian General Xenophon fought with his back against an impassable ravine. He purposefully set himself up so that his soldiers had no option of retreat.[6] Backs stiffened, they won.

Similarly, Cortés scuttled his ships upon arrival in Mexico. This decision was made with the support of his troops. Vastly outnumbered, his six hundred soldiers decided that they would either defeat the Aztecs or perish trying. The Aztecs could retreat inland, but for Cortés's soldiers there was no possibility of desertion or retreat. By making defeat worse, Cortés increased his chance of victory and indeed conquered.*

---

* Cortés was also helped by the Aztecs' misconception that he was Quetzalcoatl, a fair-skinned god.

What worked for Cortés and Xenophon worked for Cindy, Laurie, and Ray. Two months later, just in time for Valentine's Day, Cindy had lost 17 pounds. Ray was down 22 pounds and two belt loops. While the threat was the motivator to get them started, once they got going, they were doing it for themselves. Laurie lost the required 15 pounds in the first month. She kept on going and lost another 13 in month two. Laurie's 28 pounds translated into two dress sizes and over 14 percent of her body weight. Her friends no longer think the ABC show was a stupid idea.

At this point, you shouldn't be surprised to know that one of us was behind the show's design.[7] Perhaps we should have called this book *Thinning Strategically* and sold many more copies. Alas, not, and we return to study these types of strategic moves in chapter 6.

#7. BUFFETT'S DILEMMA

In an op-ed promoting campaign finance reform, the Oracle of Omaha, Warren Buffett, proposed raising the limit on individual contributions from $1,000 to $5,000 and banning all other contributions. No corporate money, no union money, no soft money. It sounds great, except that it would never pass.

Campaign finance reform is so hard to pass because the incumbent legislators who have to approve it are the ones who have the most to lose. Their advantage in fundraising is what gives them job security.* How do you get people to do something that is against their interest? Put them in what is known as the prisoners' dilemma.† According to Buffett:

Well, just suppose some eccentric billionaire (not me, not me!) made the following offer: If the bill was defeated, this person—

---

* Between 1992 and 2000, Dan Rostenkowski was the only case of an incumbent congressman losing a contest for reelection. The incumbent success rate was 604 out of 605, or 99.8 percent. When Rostenkowski lost, he was under indictment on seventeen counts of extortion, obstruction of justice, and misuse of funds.

† While prisoner's dilemma is the more common usage, we prefer the plural, because unless there are two or more prisoners involved, there is no dilemma.

the E.B.—would donate $1 billion in an allowable manner (soft money makes all possible) to the political party that had delivered the most votes to getting it passed. Given this diabolical application of game theory, the bill would sail through Congress and thus cost our E.B. nothing (establishing him as not so eccentric after all).[8]

Consider your options as a Democratic legislator. If you think that the Republicans will support the bill and you work to defeat it, then if you are successful, you will have delivered $1 billion to the Republicans, thereby handing them the resources to dominate for the next decade. Thus there is no gain in opposing the bill if the Republicans are supporting it. Now, if the Republicans are against it and you support it, then you have the chance of making $1 billion.

Thus whatever the Republicans do, the Democrats should support the bill. Of course, the same logic applies to the Republicans. They should support the bill no matter what the Democrats do. In the end, both parties support the bill, and our billionaire gets his proposal for free. As a bonus, Buffett notes that the very effectiveness of his plan "would highlight the absurdity of claims that money doesn't influence Congressional votes."

This situation is called a prisoners' dilemma because both sides are led to take an action that is against their mutual interest.* In the classic version of the prisoners' dilemma, the police are separately interrogating two suspects. Each is given an incentive to be the first to confess and a much harsher sentence if he holds out while the other confesses. Thus each finds it advantageous to confess, though they would both do better if each kept quiet.

Truman Capote's *In Cold Blood* provides a vivid illustration. Richard "Dick" Hickock and Perry Edward Smith have been arrested for the senseless murder of the Clutter family. While there were no witnesses to the crime, a jailhouse snitch had given their

---

* The active players in the game are the losers, but outsiders can benefit. While incumbent politicians might be unhappy with campaign finance reform, the rest of us would be better off.

names to the police. During the interrogation, the police play one against the other. Capote takes us into Perry's mind:

> . . . that it was just another way of getting under his skin, like that phony business about a witness—"a living witness." There couldn't be. Or did they mean— If only he could talk to Dick! But he and Dick were being kept apart; Dick was locked in a cell on another floor. . . . And Dick? Presumably they'd pulled the same stunt on him. Dick was smart, a convincing performer, but his "guts" were unreliable, he panicked too easily. . . . "And before you left that house you killed all the people in it." It wouldn't amaze him if every Old Grad in Kansas had heard that line. They must have questioned hundreds of men, and no doubt accused dozens; he and Dick were merely two more. . . .
>
> And Dick, awake in a cell on the floor below, was (he later recalled) equally eager to converse with Perry—find out what the punk had told them.[9]

Eventually Dick confessed and then Perry.* That's the nature of the game.

The problem of collective action is a variant of the prisoners' dilemma, albeit one with many more than two prisoners. In the children's story about belling the cat, the mice decide that life would be much safer if the cat were stuck with a bell around its neck. The problem is, who will risk his life to bell the cat?

This is a problem for both mice and men. How can unpopular tyrants control large populations for long periods? Why can a lone bully terrorize a schoolyard? In both cases, a simultaneous move by the masses stands a very good chance of success.

But the communication and coordination required for such action is difficult, and the oppressors, knowing the power of the masses, take special steps to keep it difficult. When the people must act individually and hope that the momentum will build up, the question arises, "Who is going to be the first?" Such a leader

---

* Although each of the two thought that confession would bring more favorable treatment, in this instance that did not happen—both were sentenced to death.

will pay a high cost—a broken nose or possibly his life. His reward may be posthumous glory or gratitude. There are people who are moved by considerations of duty or honor, but most find the costs exceed the benefits.

Khrushchev first denounced Stalin's purges at the Soviet Communist Party's 20th Congress. After his dramatic speech, someone in the audience shouted out, asking what Khrushchev had been doing at the time. Khrushchev responded by asking the questioner to please stand up and identify himself. The audience remained silent. Khrushchev replied, "That is what I did, too."

Each person acts in his or her self-interest, and the result is a disaster for the group. The prisoners' dilemma is perhaps the most famous and troubling game in game theory, and we return to the topic in chapter 3 to discuss what can be done. We should emphasize right from the start that we have no presumption that the outcome of a game will be good for the players. Many economists, ourselves included, tout the advantages of the free market. The theory behind this conclusion relies on a price system that guides individual behavior. In most strategic interactions, there is no invisible hand of prices to guide the baker or the butcher or anyone else. Thus there is no reason to expect that the outcome of a game will be good for the players or society. It may not be enough to play a game well—you must also be sure you are playing the right game.

#8. MIX YOUR PLAYS

Apparently Takashi Hashiyama has trouble making decisions. Both Sotheby's and Christie's had made attractive offers to be the auction house for the sale of his company's $18 million art collection. Rather than choose one over the other, he suggested the two of them play a game of Rock Paper Scissors to determine the winner. Yes, Rock Paper Scissors. Rock breaks scissors, scissors cuts paper, and paper covers rock.

Christie's chose scissors and Sotheby's chose paper. Scissors cut paper and so Christie's won the assignment and a nearly $3 million commission. With the stakes so high, could game theory have helped?

The obvious point is that in this type of game, one can't be predictable. If Sotheby's had known that Christie's would be playing scissors, then they would have chosen rock. No matter what you choose, there is something else that beats it. Hence it is important that the other side can't predict your play.

As part of their preparation, Christie's turned to local experts, namely the kids of their employees who play the game regularly. According to eleven-year-old Alice, "Everybody knows you always start with scissors." Alice's twin sister, Flora, added her perspective: "Rock is way too obvious, and scissors beats paper. Since they were beginners, scissors was definitely the safest."[10]

Sotheby's took a different tack. They thought this was simply a game of chance and hence there was no room for strategy. Paper was as good as anything else.

What is interesting here is that both sides were half right. If Sotheby's picked its strategy at random—with an equal chance of rock, scissors, or paper—then whatever Christie's did would have been equally good. Each option has a one-third chance of winning, a one-third chance of losing, and a one-third chance of a tie.

But Christie's didn't pick at random. Thus Sotheby's would have done better to think about the advice Christie's would likely get and then play to beat it. If it's true that everyone knows you start with scissors, Sotheby's should have started with Bart Simpson's favorite, good old rock.

In that sense, both players got it half wrong. Given Sotheby's lack of strategy, there was no point in Christie's efforts. But given Christie's efforts, there would have been a point to Sotheby's thinking strategically.

In a single play of a game, it isn't hard to choose randomly. But when games get repeated, the approach is trickier. Mixing your plays does not mean rotating your strategies in a predictable manner. Your opponent can observe and exploit any systematic pattern almost as easily as he can exploit an unchanging repetition of a single strategy. It is *unpredictability* that is important when mixing.

It turns out most people fall into predictable patterns. You can test this yourself online where computer programs are able to find the pattern and beat you.[11] In an effort to mix things up, players often rotate their strategies too much. This leads to the surprise success of the "avalanche" strategy: rock, rock, rock.

People are also too influenced by what the other side did last time. If both Sotheby's and Christie's had opened with scissors, there would have been a tie and a rematch. According to Flora, Sotheby's would expect Christie's to play rock (to beat their scissors). That should lead Sotheby's to play paper and so Christie's should stick with scissors. Of course, that formulaic approach can't be right, either. If it were, Sotheby's could then play rock and win.

Imagine what would happen if there were some known formula that determined who would be audited by the IRS. Before you submitted a tax return, you could apply the formula to see if you would be audited. If an audit was predicted, but you could see a way to "amend" your return until the formula no longer predicted an audit, you probably would do so. If an audit was unavoidable, you would choose to tell the truth. The result of the IRS being completely predictable is that it would audit exactly the wrong people. All those audited would have anticipated their fate and chosen to act honestly, while those spared an audit would have only their consciences to watch over them. When the IRS audit formula is somewhat fuzzy, everyone stands some risk of an audit; this gives an added incentive for honesty.

The importance of randomized strategies was one of the early insights of game theory. The idea is simple and intuitive but needs refinement to be useful in practice. It is not enough for a tennis player to know that he should mix his shots between the opponent's forehand and backhand. He needs some idea of whether he should go to the forehand 30 percent or 64 percent of the time and how the answer depends on the relative strengths of the two sides. In chapter 5 we develop methods to answer such questions.

We'd like to leave you with one last commentary. The biggest loser in the Rock Paper Scissors game wasn't Sotheby's; it was Mr.

Hashiyama. His decision to deploy Rock Paper Scissors gave each of the two auction houses a 50 percent chance of winning the commission. Instead of letting the two contestants effectively agree to split the commission, he could have run his own auction. Both firms were willing, even eager, to lead the sale with a 12 percent commission.* The winning house would be the one willing to take the lowest fee. Do I hear 11 percent? Going once, going twice, . . .

#9. NEVER GIVE A SUCKER AN EVEN BET

In *Guys and Dolls*, gambler Sky Masterson relates this valuable advice from his father:

> One of these days in your travels, a guy is going to show you a brand-new deck of cards on which the seal is not yet broken. Then this guy is going to offer to bet you that he can make the jack of spades jump out of this brand-new deck of cards and squirt cider in your ear. But, son, you do not accept this bet because, as sure as you stand there, you're going to wind up with an ear full of cider.

The context of the story is that Nathan Detroit has offered Sky Masterson a bet about whether Mindy's sells more strudel or cheesecake. Nathan had just discovered the answer (strudel) and is willing to bet if Sky will bet on cheesecake.†

This example may sound somewhat extreme. Of course no one would take such a sucker bet. Or would they? Look at the market for futures contracts on the Chicago Board of Exchange. If

---

* The standard commission is 20 percent on the first $800,000 and 12 percent thereafter. Mr. Hashiyama's four paintings sold for a combined $17.8 million, suggesting a total commission of $2.84 million.

† We should add that Sky never quite learned his father's lesson. A minute later, he offers to bet that Nathan does not know the color of his own bow tie. Sky can't win. If Nathan knows the color, he takes the bet and wins. As it turns out, Nathan doesn't know the color and thus doesn't bet. Of course, that was the real gamble. Sky is betting that Nathan won't take the offer.

another speculator offers to sell you a futures contract, he will make money only if you lose money.*

If you happen to be a farmer with soy beans to sell in the future, then the contract can provide a hedge against future price movements. Similarly, if you sell soy milk and hence need to buy soy beans in the future, this contract is insurance, not a gamble.

But the volume of the contracts on the exchange suggests that most people buying and selling are traders, not farmers and manufacturers. For them, the deal is a *zero-sum* game. When both sides agree to trade, each one thinks it will make money. One of them must be wrong. That's the nature of a zero-sum game. Both sides can't win.

This is a paradox. How can both sides think that they can outsmart the other? Someone must be wrong. Why do you think the other person is wrong, not you? Let us assume that you don't have any insider information. If someone is willing to sell you a futures contract, any money you make is money they lose. Why do you think that you are smarter than they are? Remember that their willingness to trade means that they think they are smarter than you.

In poker, players battle this paradox when it comes to raising the stakes. If a player bets only when he has a strong hand, the other players will soon figure this out. In response to a raise, most other players will fold, and he'll never win a big pot. Those who raise back will have even stronger hands, and so our poor player will end up a big loser. To get others to bet against a strong hand, they have to think you might be bluffing. To convince them of this possibility, it helps to bet often enough that you must be bluffing some of the time. This leads to an interesting dilemma. You'd like others to fold against your bluffs and thereby win with weak hands. But that won't lead to high-pot victories. To convince others to raise your bets, you also need to get caught bluffing.

As the players get even more sophisticated, persuading others to take big bets against you becomes harder and harder. Consider the

---

* Buying stocks is not the same as betting on a futures contract. In the case of stocks, the capital you provide to the firm allows it to grow faster, and thus you and the firm can both win.

following high-stakes game of wits between Erick Lindgren and Daniel Negreanu, two of poker's top-ranked players.

> . . . Negreanu, sensing a weak hand, raised him two hundred thousand [dollars]. "I put two hundred and seventy thousand in, so I have two hundred thousand left," Negreanu said. "And Erick looks over my chips and says, 'How much you got left?' And he moves all in"—wagering all he had. Under the special betting rules governing the tournament, Negreanu had only ninety seconds to decide whether to call the bet, and risk losing all his money if Lindgren wasn't bluffing, or to fold, and give up the hefty sum he had already put into the pot.

> "I didn't think he could be so stupid," Negreanu said. "But it wasn't stupid. It was like a step above. He knows that I know that he wouldn't do something so stupid, so by doing something so quote-unquote stupid it actually became a great play."[12]

While it is obvious that you shouldn't bet against these poker champions, when should you take a gamble? Groucho Marx famously said that he didn't care to belong to any club that would accept him as a member. For similar reasons, you might not want to take any bet that others offer. You should even be worried when you win an auction. The very fact that you were the highest bidder implies that everyone else thought the item was worth less than you did. The result of winning an auction and discovering you've overpaid is called the winner's curse.

Every action someone takes tells us something about what he knows, and you should use these inferences along with what you already know to guide your actions. How to bid so that you won't be cursed when you win is something we discuss in chapter 10.

There are some rules of the game that can help put you on more equal footing. One way to allow trading with lopsided information is to let the less informed party pick which side of the bet to take. If Nathan Detroit agreed in advance to take the bet whatever side Sky picked, then Nathan's inside information would be of no help. In stock markets, foreign exchange markets, and other financial markets, people are free to take either side of the bet. Indeed, in some exchanges, including the London stock market, when you ask for a quote on a stock the market maker is required to state both the buying and selling prices *before* he knows which side of the transaction you want. Without such a safeguard, market makers could stand to profit from private information, and the outside investors' fear of being suckered might cause the entire market to fold. The buy and sell prices are not quite the same; the difference is called the bid-ask spread. In liquid markets the spread is quite small, indicating that little information is contained in any buy or sell order. We return to the role of information in chapter 8.

#10. GAME THEORY CAN BE DANGEROUS TO YOUR HEALTH

Late one night, after a conference in Jerusalem, two American economists—one of whom is this book's coauthor—found a taxi-cab and gave the driver directions to the hotel. Immediately recognizing us as American tourists, the driver refused to turn on his meter; instead, he proclaimed his love for Americans and promised us a lower fare than the meter. Naturally, we were somewhat skeptical of this promise. Why should this stranger offer to charge less than the meter when we were willing to pay the metered fare? How would we even know whether or not we were being overcharged?

On the other hand, we had not promised to pay the driver anything more than what would be on the meter. We put on our game-theory hats. If we were to start bargaining and the negotiations broke down, we would have to find another taxi. But if we waited until we arrived at the hotel, our bargaining position would be much stronger. And taxis were hard to find.

We arrived. The driver demanded 2,500 Israeli shekels ($2.75). Who knew what fare was fair? Because people generally bargain in Israel, Barry protested and counteroffered 2,200 shekels. The driver was outraged. He claimed that it would be impossible to get from there to here for that amount. Before negotiations could continue, he locked all the doors automatically and retraced the route at breakneck speed, ignoring traffic lights and pedestrians. Were we being kidnapped to Beirut? No. He returned to the original position and ungraciously kicked us out of his cab, yelling, "See how far your 2,200 shekels will get you now."

We found another cab. This driver turned on his meter, and 2,200 shekels later we were home.

Certainly the extra time was not worth the 300 shekels. On the other hand, the story was well worth it. It illustrates the dangers of bargaining with those who have not yet read our book. More generally, pride and irrationality cannot be ignored. Sometimes, it may be better to be taken for a ride when it costs only two dimes.

There is a second lesson to the story. We didn't really think far enough ahead. Think of how much stronger our bargaining position would have been if we had begun to discuss the price *after* getting out of the taxi. (Of course, for hiring a taxi, this logic should be reversed. If you tell the driver where you want to go before getting in, you may find your taxi chasing after some other customer. Get in first, then say where you want to go.)

Some years after this story was first published, we received the following letter:

Dear Professors,

You certainly don't know my name, but I think you will remember my story. I was a student in Jerusalem moonlighting as a taxi driver. Now I am a consultant and chanced upon your book when it was translated into Hebrew. What you might find interesting is that I too have been sharing the story with my clients. Yes, it was indeed a late night in Jerusalem. As for the rest, well, I recall things a bit differently.

Between classes and working nights as a taxi driver, there was almost no time for me to spend with my new bride. My solution was to have her ride with me in the front seat. Although she was silent, it was a big mistake for you to have left her out of the story.

My meter was broken, but you didn't seem to believe me. I was too tired to argue. When we arrived, I asked for 2,500 shekels, a fair price. I was even hoping you would round the fare up to 3,000. You rich Americans could well afford a 50¢ tip.

I couldn't believe you tried to cheat me. Your refusal to pay a fair price dishonored me in front of my wife. As poor as I was, I did not need to take your meager offer.

Americans think that we should be happy to take whatever crumbs you offer. I say that we should teach you a lesson in the game of life. My wife and I are now married twenty years. We still laugh about those stupid Americans who spent a half an hour riding back and forth in taxis to save twenty cents.

Sincerely,
(name withheld)

Truth be told, we never received such a letter. Our point in creating it was to illustrate a critical lesson in game theory: you need to understand the other player's perspective. You need to consider what they know, what motivates them, and even how they think about you. George Bernard Shaw's quip on the golden rule was to not do unto others as you would have them do unto you— their tastes may be different. When thinking strategically, you have to work extra hard to understand the perspective and interactions of all the other players in the game, including ones who may be silent.

That brings us to one last point. You may be thinking you are playing one game, but it is only part of a larger game. There is always a larger game.

THE SHAPE OF THINGS TO COME

These examples have given us glimpses of principles that guide strategic decisions. We can summarize these principles with a few "morals" from our tales.

Think 48 when you are wondering what the other player is trying to achieve. Recall Richard Hatch's ability to play out all the future moves to figure out what he should do. The story of the hot hand told us that in strategy, no less than in physics, "For every action we take, there is a reaction." We do not live and act in a vacuum. Therefore, we cannot assume that when we change our behavior everything else will remain unchanged. De Gaulle's success in negotiations suggests that "the stuck wheel gets the grease."* But being stubborn is not always easy, especially when one has to be more stubborn than an obstinate adversary. That stubborn adversary might well be your future self, especially when it comes to dieting. Fighting or dieting with your back up against the wall can help strengthen your resolve.

In Cold Blood and the story of belling the cat demonstrate the difficulty of obtaining outcomes that require coordination and individual sacrifice. In technology races, no less than in sailboat races, those who trail tend to employ more innovative strategies; the leaders tend to imitate the followers.

Rock Paper Scissors points out the strategic advantage of being unpredictable. Such behavior may also have the added advantage that it makes life just a little more interesting. Our taxi rides make it clear that the other players in games are people, not machines. Pride, spite, and other emotions may color their decisions. When you put yourself in others' shoes, you have to take them as they are, not as you are.

We could go on offering more examples and drawing morals from them, but this is not the best way to think methodically about strategic games. That is better done by approaching the subject from a different angle. We pick up the principles—for example, commitment, cooperation, and mixing—one at a time. In each instance, we explore examples that bear centrally on that issue, until the principle is clear. Then you will have a chance to apply that principle in the case study that ends each chapter.

CASE STUDY: MULTIPLE CHOICE

We think almost everything in life is a game, even things that might not seem that way at first. Consider the following question from the GMAT (the test given to MBA applicants).

---

* You may have heard this expression as the "squeaky wheel"—a stuck wheel needs even more grease. Of course, sometimes it gets replaced.

Unfortunately, issues of copyright clearance have prevented us from reproducing the question, but that shouldn't stop us. Which of the following is the correct answer?

a. 4π sq. inches   c. 16 sq. inches   e. 32π sq. inches
b. 8π sq. inches   d. 16π sq. inches

Okay, we recognize that you're at a bit of a disadvantage not having the question. Still, we think that by putting on your game-theory hat you can still figure it out.

Case Discussion

The odd answer in the series is c. Since it is so different from the other answers, it is probably not right. The fact that the units are in square inches suggests an answer that has a perfect square in it, such as 4π or 16π.

This is a fine start and demonstrates good test-taking skills, but we haven't really started to use game theory. Think of the game being played by the person writing the question. What is that person's objective?

He or she wants people who understand the problem to get the answer right and those who don't to get it wrong. Thus wrong answers have to be chosen carefully so as to be appealing to folks who don't quite know the answer. For example, in response to the question: How many feet are in a mile, an answer of "Giraffe," or even 16π, is unlikely to attract any takers.

Turning this around, imagine that 16 square inches really is the right answer. What kind of question might have 16 square inches as the answer but would lead someone to think 32π is right? Not many. People don't often go around adding π to answers for the fun of it. "Did you see my new car—it gets 10π miles to the gallon." We think not. Hence we can truly rule out 16 as being the correct solution.

Let's now turn to the two perfect squares, 4π and 16π. Assume for a moment that 16π square inches is the correct solution. The problem might have been what is the area of a circle with a radius of 4? The correct formula for the area of a circle is πr^2. However, the person who didn't quite remember the formula might have mixed it up with the formula for the circumference of a circle, $2\pi r$. (Yes, we know that the circumference is in inches, not square inches, but the person making this mistake would be unlikely to recognize this issue.)

Note that if $r = 4$, then $2\pi r$ is 8π, and that would lead the person to the wrong answer of b. The person could also mix and match and use the formula $2\pi r^2$ and hence believe that 32π or e was the right answer. The person could leave off the π and come up with 16 or c, or the person could forget to square the radius and simply use πr as the area, leading to 4π or a. In summary, if 16π is the correct answer, then we can tell a plausible story about how each of the other answers might be chosen. They are all good wrong answers for the test maker.

What if 4π is the correct solution (so that $r = 2$)? Think now about the most common mistake, mixing up circumference with area. If the student used the wrong formula, $2\pi r$, he or she would still get 4π, albeit with incorrect units. There is nothing worse, from a test maker's perspective, than allowing the person to get the right answer for the wrong reason. Hence 4π would be a terrible right answer, as it would allow too many people who didn't know what they were doing to get full credit.

At this point, we are done. We are confident that the right answer is 16π. And we are right. By thinking about the objective of the person writing the test, we can suss out the right answer, often without even seeing the question.

Now, we don't recommend that you go about taking the GMAT and other tests without bothering to even look at the questions. We appreciate that if you are smart enough to go through this logic, you most likely know the formula for the area of a circle. But you never know. There will be cases where you don't know the meaning of one of the answers or the material for the question wasn't covered in your course. In those cases, thinking about the testing game may lead you to the right answer.

CHAPTER 2

Games Solvable by Backward Reasoning

IT'S YOUR MOVE, CHARLIE BROWN

In a recurring theme in the comic strip *Peanuts*, Lucy holds a football on the ground and invites Charlie Brown to run up and kick it. At the last moment, Lucy pulls the ball away. Charlie Brown, kicking only air, lands on his back, and this gives Lucy great perverse pleasure.

Anyone could have told Charlie that he should refuse to play Lucy's game. Even if Lucy had not played this particular trick on him last year (and the year before and the year before that), he knows her character from other contexts and should be able to predict her action.

At the time when Charlie is deciding whether or not to accept Lucy's invitation, her action lies in the future. However, just because it lies in the future does not mean Charlie should regard it as uncertain. He should know that of the two possible outcomes—letting him kick and seeing him fall—Lucy's preference is for the latter. Therefore he should forecast that when the time comes, she is going to pull the ball away. The logical possibility that Lucy will let him kick the ball is realistically irrelevant. Reliance on it would be, to borrow Dr. Johnson's characterization of remarriage, a triumph of hope over experience. Charlie should disregard it, and forecast that acceptance will inevitably land him on his back. He should decline Lucy's invitation.

TWO KINDS OF STRATEGIC INTERACTIONS

The essence of a game of strategy is the interdependence of the players' decisions. These interactions arise in two ways. The first is *sequential,* as in the Charlie Brown story. The players make alternating moves. Charlie, when it is his turn, must look ahead to how his current actions will affect the future actions of Lucy, and his own future actions in turn.

The second kind of interaction is *simultaneous*, as in the prisoners' dilemma tale of chapter 1. The players act at the same time, in ignorance of the others' current actions. However, each must be aware that there are other active players, who in turn are similarly aware, and so on. Therefore each must figuratively put himself in the shoes of all and try to calculate the outcome. His own best action is an integral part of this overall calculation.

When you find yourself playing a strategic game, you must determine whether the interaction is simultaneous or sequential. Some games, such as football, have elements of both, in which case you must fit your strategy to the context. In this chapter, we develop, in a preliminary way, the ideas and rules that will help you play sequential games; simultaneous-move games are the subject of chapter 3. We begin with really simple, sometimes contrived, examples, such as the Charlie Brown story. This is deliberate; the stories are not of great importance in themselves, and the right strategies are usually easy to see by simple intuition, allowing the underlying ideas to stand out much more clearly. The examples get increasingly realistic and more complex in the case studies and in the later chapters.

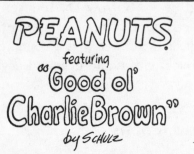

The First Rule of Strategy

The general principle for sequential-move games is that each player should figure out the other players' future responses and use them in calculating his own best current move. This idea is so important that it is worth codifying into a basic rule of strategic behavior:

RULE 1: Look forward and reason backward.

Anticipate where your initial decisions will ultimately lead and use this information to calculate your best choice.

In the Charlie Brown story, this was easy to do for anyone (except Charlie Brown). He had just two alternatives, and one of them led to Lucy's decision between two possible actions. Most strategic situations involve a longer sequence of decisions with several alternatives at each. A tree diagram of the choices in the game sometimes serves as a visual aid for correct reasoning in such games. Let us show you how to use these trees.

DECISION TREES AND GAME TREES

A sequence of decisions, with the need to look forward and reason backward, can arise even for a solitary decision maker not involved in a game of strategy with others. For Robert Frost in the yellow wood:

> Two roads diverged in a wood, and I—
> I took the road less traveled by,
> And that has made all the difference.[1]

We can show this schematically.

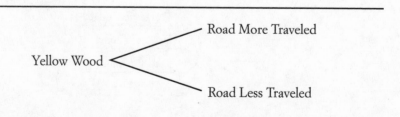

This need not be the end of the choice. Each road might in turn have further branches. The road map becomes correspondingly complex. Here is an example from our own experience.

Travelers from Princeton to New York have several choices. The first decision point involves selecting the mode of travel: bus, train, or car. Those who drive then have to choose among the Verrazano-Narrows Bridge, the Holland Tunnel, the Lincoln Tunnel, and the George Washington Bridge. Rail commuters must decide whether to switch to the PATH train at Newark or continue to Penn Station. Once in New York, rail and bus commuters must choose among going by foot, subway (local or express), bus, or taxi to get to their final destination. The best choices depend on many factors, including price, speed, expected congestion, the final destination in New York, and one's aversion to breathing the air on the New Jersey Turnpike.

This road map, which describes one's options at each junction, looks like a tree with its successively emerging branches—hence the term. The right way to use such a map or tree is not to take the

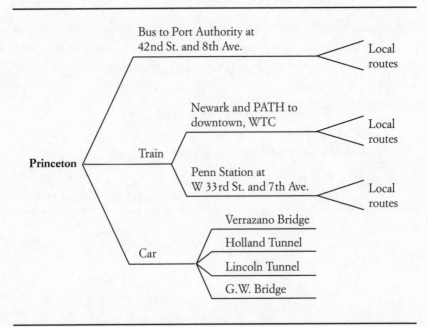

route whose first branch looks best—for example, because you would prefer driving to taking the train when all other things are equal—and then "cross the Verrazano Bridge when you get to it." Instead, you anticipate the future decisions and use them to make your earlier choices. For example, if you want to go downtown, the PATH train would be superior to driving because it offers a direct connection from Newark.

We can use just such a tree to depict the choices in a game of strategy, but one new element enters the picture. A game has two or more players. At various branching points along the tree, it may be the turn of different players to make the decision. A person making a choice at an earlier point must look ahead, not just to his own future choices but to those of others. He must forecast what the others will do, by putting himself figuratively in their shoes, and thinking as they would think. To remind you of the difference, we will call a tree showing the decision sequence in a game of strategy a *game tree,* reserving *decision tree* for situations in which just one person is involved.

Charlie Brown in Football and in Business

The story of Charlie Brown that opened this chapter is absurdly simple, but you can become familiar with game trees by casting that story in such a picture. Start the game when Lucy has issued her invitation, and Charlie faces the decision of whether to accept. If Charlie refuses, that is the end of the game. If he accepts, Lucy has the choice between letting Charlie kick and pulling the ball away. We can show this by adding another fork along this road.

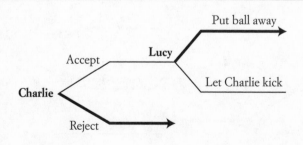

As we said earlier, Charlie should forecast that Lucy will choose the upper branch. Therefore he should figuratively prune the lower branch of her choice from the tree. Now if he chooses his own upper branch, it leads straight to a nasty fall. Therefore his better choice is to follow his own lower branch. We show these selections by making the branches thicker and marking them with arrowheads.

Are you thinking that this game is too frivolous? Here is a business version of it. Imagine the following scenario. Charlie, now an adult, is vacationing in the newly reformed formerly Marxist country of Freedonia. He gets into a conversation with a local businessman named Fredo, who talks about the wonderful profitable opportunities that he could develop given enough capital, and then makes a pitch: "Invest $100,000 with me, and in a year I will turn it into $500,000, which I will share equally with you. So you will more than double your money in a year." The opportunity Fredo describes is indeed attractive, and he is willing to write up a proper contract under Freedonian law. But how secure is that law? If at the end of the year Fredo absconds with all the money, can Charlie, back in the United States, enforce the contract in Freedonian courts? They may be biased in favor of their national, or too slow, or bribed by Fredo. So Charlie is playing a game with Fredo, and the tree is as shown here. (Note that if Fredo honors the contract, he pays Charlie $250,000; therefore Charlie's profit is that minus the initial investment of $100,000—that is, $150,000.)

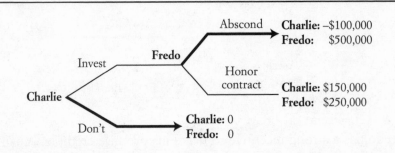

What do you think Fredo is going to do? In the absence of a clear and strong reason to believe his promise, Charlie should predict that Fredo will abscond, just as young Charlie should have been sure that Lucy would pull the ball away. In fact the trees of the two games are identical in all essential respects. But how many Charlies have failed to do the correct reasoning in such games?

What reasons can there be for believing Fredo's promise? Perhaps he is engaged in many other enterprises that require financing from the United States or export goods to the United States. Then Charlie may be able to retaliate by ruining his reputation in the United States or seizing his goods. So this game may be part of a larger game, perhaps an ongoing interaction, that ensures Fredo's honesty. But in the one-time version we showed above, the logic of backward reasoning is clear.

We would like to use this game to make three remarks. First, different games may have identical or very similar mathematical forms (trees, or the tables used for depictions in later chapters). Thinking about them using such formalisms highlights the parallels and makes it easy to transfer your knowledge about a game in one situation to that in another. This is an important function of the "theory" of any subject: it distills the essential similarities in apparently dissimilar contexts and enables one to think about them in a unified and therefore simplified manner. Many people have an instinctive aversion to theory of any kind. But we think this is a mistaken reaction. Of course, theories have their limitations. Specific contexts and experiences can often add to or modify the prescriptions of theory in substantial ways. But to abandon theory altogether would be to abandon a valuable starting point for thought, which may be a beachhead for conquering the problem. You should make game theory your friend, and not a bugbear, in your strategic thinking.

The second remark is that Fredo should recognize that a strategic Charlie would be suspicious of his pitch and not invest at all, depriving Fredo of the opportunity to make $250,000. Therefore Fredo has a strong incentive to make his promise credible. As an individual businessman, he has little influence over Freedonia's weak legal system and cannot allay the investor's suspicion that way. What other methods may be at his disposal? We will examine the general issue of credibility, and devices for achieving it, in chapters 6 and 7.

The third, and perhaps most important, remark concerns comparisons of the different outcomes that could result based on the different choices the players could make. It is not always the case that more for one player means less for the other. The situation where Charlie invests and Fredo honors the contract is better for both than the one where Charlie does not invest at all. Unlike sports or contests, games don't have to have winners and losers; in the jargon of game theory, they don't have to be zero-sum. Games can have win-win or lose-lose outcomes. In fact, some combination of commonality of interest (as when Charlie and Fredo can both gain if there is a way for Fredo to commit credibly to honoring the contract) and some conflict (as when Fredo can gain at Charlie's expense by absconding after Charlie has invested) coexist in most games in business, politics, and social interactions. And that is precisely what makes the analysis of these games so interesting and challenging.

More Complex Trees

We turn to politics for an example of a slightly more complex game tree. A caricature of American politics says that Congress likes pork-barrel expenditures and presidents try to cut down the bloated budgets that Congress passes. Of course presidents have their own likes and dislikes among such expenditures and would like to cut only the ones they dislike. To do so, they would like to have the power to cut out specific items from the budget, or a line-item veto. Ronald Reagan in his State of the Union address in January 1987 said this eloquently: "Give us the same tool that 43 governors have—a line-item veto, so we can carve out the boondoggles and pork, those items that would never survive on their own."

At first sight, it would seem that having the freedom to veto parts of a bill can only increase the president's power and never yield him any worse outcomes. Yet it is possible that the president may be better off without this tool. The point is that the existence of a line-item veto will influence the Congress's strategies in passing bills. A simple game shows how.

For this purpose, the essence of the situation in 1987 was as follows. Suppose there were two items of expenditure under consideration: urban renewal (U) and an antiballistic missile system (M). Congress liked the former and the president liked the latter. But both preferred a package of the two to the status quo. The following table shows the ratings of the possible scenarios by the two players, in each case 4 being best and 1, worst.

Outcomes	Congress	President
Both U and M	3	3
U only	4	1
M only	1	4
Neither	2	2

The tree for the game when the president does not have a line-item veto is shown on the following page. The president will sign a bill containing the package of U and M, or one with M alone, but will veto one with U alone. Knowing this, the Congress chooses the package. Once again we show the selections at each point by thickening the chosen branches and giving them arrowheads. Note that we have to do this for all the points where the president might conceivably be called upon to choose, even though some of these are rendered moot by Congress's previous choice. The reason is that Congress's actual choice is crucially affected by its calculation of what the president would have done if Congress had counterfactually made a different choice; to show this logic we must show the president's actions in all logically conceivable situations.

Our analysis of the game yields an outcome in which both sides get their second best preference (rating 3).

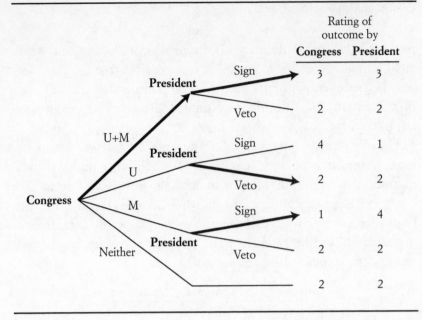

| | Rating of outcome by | |
	Congress	President
Sign	3	3
Veto	2	2
Sign	4	1
Veto	2	2
Sign	1	4
Veto	2	2
	2	2

Next, suppose the president has a line-item veto. The game changes to the following:

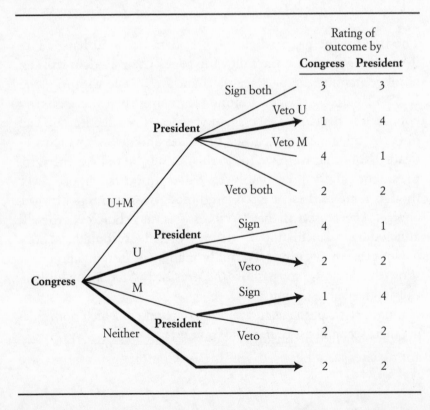

| | Rating of outcome by | |
	Congress	President
Sign both	3	3
Veto U	1	4
Veto M	4	1
Veto both	2	2
Sign	4	1
Veto	2	2
Sign	1	4
Veto	2	2
	2	2

Now Congress foresees that if it passes the package, the president will selectively veto U, leaving only M. Therefore Congress's best action is now either to pass U only to see it vetoed, or pass nothing. Perhaps it may have a preference for the former, if it can score political points from a presidential veto, but perhaps the president may equally score political points by this show of budgetary discipline. Let us suppose the two offset each other, and Congress is indifferent with respect to the two choices. But either gives each party only their third-best outcome (rating 2). Even the president is left worse-off by his extra freedom of choice.[2]

This game illustrates an important general conceptual point. In single-person decisions, greater freedom of action can never hurt. But in games, it can hurt because its existence can influence other players' actions. Conversely, tying your own hands can help. We will explore this "advantage of commitment" in chapters 6 and 7.

We have applied the method of backward reasoning in a game tree to a very trivial game (Charlie Brown), and extended it to a slightly more complicated game (the line-item veto). The general principle remains applicable, no matter how complicated the game may be. But trees for games where each player has several choices available at any point, and where each player gets several turns to move, can quickly get too complicated to draw or use. In chess, for example, 20 branches emerge from the root—the player with the white pieces can move any of his/her eight pawns forward one square or two, or move one of his two knights in one of two ways. For each of these, the player with the black pieces has 20 moves, so we are up to 400 distinct paths already. The number of branches emerging from later nodes in chess can be even larger. Solving chess fully using the tree method is beyond the ability of the most powerful computer that exists or might be developed during the next several decades, and other methods of partial analysis must be sought. We will discuss later in the chapter how chess experts have tackled this problem.

Between the two extremes lie many moderately complex games that are played in business, politics, and everyday life. Two approaches can be used for these. Computer programs are available to construct trees and compute solutions.[3] Alternatively, many games of moderate complexity can be solved by the logic of tree analysis, without drawing the tree explicitly. We illustrate this using a game that was played in a TV show that is all about games, where each player tries to "outplay, outwit, and outlast" the others.

STRATEGIES FOR "SURVIVORS"

CBS's *Survivor* features many interesting games of strategy. In the sixth episode of *Survivor: Thailand*, the two teams or tribes played a game that provides an excellent example of thinking forward and reasoning backward in theory and in practice.[4] Twenty-one flags were planted in the field of play between the tribes, who took turns removing the flags. Each tribe at its turn could choose to remove 1 or 2 or 3 flags. (Thus zero—passing up one's turn—was not permitted; nor was it within the rules to remove four or more at one turn.) The team to take the last flag, whether standing alone or as a part of a group of 2 or 3 flags, won the game.[5] The losing tribe had to vote out one of its own members, thus weakening it in future contests. In fact the loss proved crucial in this instance, and a member of the other tribe went on to win the ultimate prize of a million dollars. Thus the ability to figure out the correct strategy for this game would prove to be of great value.

The two tribes were named Sook Jai and Chuay Gahn, and Sook Jai had the first move. They started by taking 2 flags and leaving 19. Before reading on, pause a minute and think. If you were in their place, how many would you have chosen?

Write down your choice somewhere, and read on. To understand how the game should be played, and compare the correct strategy with how the two tribes actually played, it helps to focus on two very revealing incidents. First, each tribe had a few minutes to discuss the game among its own members before the play started. During this discussion within Chuay Gahn, one of the members, Ted Rogers, an African American software developer, pointed out, "At the end, we must leave them with four flags." This is correct: if Sook Jai faces 4 flags, it must take 1 or 2 or 3, leaving Chuay Gahn to take the remaining 3 or 2 or 1, respectively, at its next turn and win the game. Chuay Gahn did in fact get and exploit this opportunity correctly; facing 6 flags, they took 2.

But here is the other revealing incident. At the previous turn, just as Sook Jai returned from having taken 3 flags out of the 9 facing them, the realization hit one of their members, Shii Ann, a feisty and articulate competitor who took considerable pride in her analytical skills: "If Chuay Gahn now takes two, we are sunk." So Sook Jai's just-completed move was wrong. What should they have done?

Shii Ann or one of her Sook Jai colleagues should have reasoned as Ted Rogers did but carried the logic of leaving the other tribe with 4 flags to its next step. How do you ensure leaving the other tribe with 4 flags at its next turn? By leaving it with 8 flags at its previous turn. When it takes 1 or 2 or 3 out of eight, you take 3 or 2 or 1 at your next turn, leaving them with 4 as planned. Therefore Sook Jai should have turned the tables on Chuay Gahn and taken just 1 flag out of the 9. Shii Ann's analytical skill kicked into high gear one move too late! Ted Rogers perhaps had the better analytical insights. But did he?

How did Sook Jai come to face 9 flags at its previous move? Because Chuay Gahn had taken 2 from 11 at *its* previous turn. Ted Rogers should have carried his own reasoning one step further. Chuay Gahn should have taken 3, leaving Sook Jai with 8, which would be a losing position.

The same reasoning can be carried even farther back. To leave the other tribe with 8 flags, you must leave them with 12 at their previous turn; for that you must leave them with 16 at the turn before that and 20 at the turn before that. So Sook Jai should have started the game by taking just 1 flag, not 2 as it actually did. Then it could have had a sure win by leaving Chuay Gahn with 20, 16, . . . 4 at their successive turns.*

---

* Does the first mover always have a sure win in all games? No. If the flags game started with 20 flags instead of 21, the second mover would have a sure win. And in some games, for example the simple 3-by-3 tic-tac-toe, either player can ensure a tie with correct play.

Now think of Chuay Gahn's very first turn. It faced 19 flags. If it had carried its own logic back far enough, it would have taken 3, leaving Sook Jai with 16 and already on the way to certain defeat. Starting from any point in the middle of the game where the opponent has played incorrectly, the team with the turn to move can seize the initiative and win. But Chuay Gahn did not play the game perfectly either.*

The table below shows the comparison between the actual and the correct moves at each point in the game. (The entry "No move" means that all moves are losing moves if the opponent plays correctly.) You can see that almost all the choices were wrong, except Chuay Gahn's move when facing 13 flags, and that must have been accidental, because at their next turn they faced 11 and took 2 when they should have taken 3.

Tribe	No. of flags before move	No. of flags taken	Move to put team on path to sure victory
Sook Jai	21	2	1
Chuay Gahn	19	2	3
Sook Jai	17	2	1
Chuay Gahn	15	1	3
Sook Jai	14	1	2
Chuay Gahn	13	1	1
Sook Jai	12	1	No move
Chuay Gahn	11	2	3
Sook Jai	9	3	1
Chuay Gahn	6	2	2
Sook Jai	4	3	No move
Chuay Gahn	1	1	1

* The fates of the two key people were also interesting. Shii Ann made another key miscalculation in the next episode and was voted out, at number 10 among the 16 contestants who started the game. Ted, more quiet but perhaps somewhat more skillful, made it to the last five.

Before you judge the tribes harshly, you should recognize that it takes time and some experience to learn how to play even very simple games. We have played this game between pairs or teams of students in our classes and found that it takes Ivy League freshmen three or even four plays before they figure out the complete reasoning and play correctly all the way through from the first move. (By the way, what number did you choose when we asked you to initially, and what was your reasoning?) Incidentally, people seem to learn faster by watching others play than by playing themselves; perhaps the perspective of an observer is more conducive to seeing the game as a whole and reasoning about it coolly than that of a participant.

TRIP TO THE GYM NO. 1

Let us turn the flag game into hot potato: now you win by forcing the other team to take the last flag. It's your move and there are 21 flags. How many do you take?

To fix your understanding of the logic of the reasoning, we offer you the first of our Trips to the Gym—questions on which you can exercise and hone your developing skills in strategic thinking. The answers are in the Workouts section in the end of the book.

Now that you are invigorated by this exercise, let us proceed to think about some general issues of strategy in this whole class of games.

WHAT MAKES A GAME FULLY SOLVABLE BY BACKWARD REASONING?

The 21-flags game had a special property that made it fully solvable, namely the absence of uncertainty of any kind: whether about some natural chance elements, the other players' motives and capabilities, or their actual actions. This seems a simple point to make, but it needs some elaboration and clarification.

First, at any point in the game when one tribe had the move, it knew exactly what the situation was, namely how many flags remained. In many games there are elements of pure chance, thrown up by nature or by the gods of probability. For example, in many card games, when a player makes a choice, he/she does not know for sure what cards the other players hold, although their previous actions may give some basis for drawing some inferences about that. In many subsequent chapters, our examples and analysis will involve games that have this natural element of chance.

Secondly, the tribe making its choice also knew the other tribe's objective, namely to win. And Charlie Brown should have known that Lucy enjoyed seeing him fall flat on his back. Players have such perfect knowledge of the other player's or players' objectives in many simple games and sports, but that is not necessarily the case in games people play in business, politics, and social interactions. Motives in such games are complex combinations of selfishness and altruism, concern for justice or fairness, short-run and long-run considerations, and so on. To figure out what the other players will choose at future points in the game, you need to know what their objectives are and, in the case of multiple objectives, how they will trade one off against the other. You can almost never know this for sure and must make educated guesses. You must not assume that other people will have the same preferences as you do, or as a hypothetical "rational person" does, but must genuinely think about their situation. Putting yourself in the other person's shoes is a difficult task, often made more complicated by your emotional involvement in your own aims and pursuits. We will have more to say about this kind of uncertainty later in this chapter and at various points throughout the book. Here we merely point out that the uncertainty about other players' motives is an issue for which it may be useful to seek advice from an objective third party—a strategic consultant.

Finally, players in many games must face uncertainty about other players' choices; this is sometimes called strategic uncertainty to distinguish it from the natural aspects of chance, such as a distribution of cards or the bounce of a ball from an uneven surface. In 21-flags there was no strategic uncertainty, because each tribe saw and knew exactly what the other had done previously. But in many games, players take their actions simultaneously or in such rapid sequence that one cannot see what the other has done and react to it. A soccer goalie facing a penalty kick must decide whether to move to his/her own right or left without knowing which direction the shooter will aim for; a good shooter will conceal his/her own intentions up to the last microsecond, by which time it is too late for the goalie to react. The same is true for serves and passing shots in tennis and many other sports. Each participant in a sealed-bid auction must make his/her own choice without knowing what the other bidders are choosing. In other words, in many games the players make their moves simultaneously, and not in a preassigned sequence. The kind of thinking that is needed for choosing one's action in such games is different from, and in some respects harder than, the pure backward reasoning of sequential-move games like 21-flags; each player must be aware of the fact that others are making conscious choices and are in turn thinking about what he himself is thinking, and so on. The games we consider in the next several chapters will elucidate the reasoning and solution tools for simultaneous-move games. In this tools chapter, however, we focus solely on sequential-move games, as exemplified by 21-flags and, at a much higher level of complexity, chess.

Do People Actually Solve Games by Backward Reasoning?

Backward reasoning along a tree is the correct way to analyze and solve games where the players move sequentially. Those who fail to do so either explicitly or intuitively are harming their own objectives; they should read our book or hire a strategic consultant. But that is an advisory or normative use of the theory of backward reasoning. Does the theory have the usual explanatory or positive value that most scientific theories do? In other words, do we observe the correct outcomes from the play of actual games? Researchers in the new and exciting fields of behavioral economics and behavioral game theory have conducted experiments that yield mixed evidence.

What seems to be the most damaging criticism comes from the ultimatum game. This is the simplest possible negotiation game: there is just one take-it-or-leave-it offer. The ultimatum game has two players, a "proposer," say A, a "responder," say B, and a sum

A QUICK TRIP TO THE GYM: REVERSE ULTIMATUM GAME

In this variant of the ultimatum game, A makes an offer to B about how to divide up the 100 dollars. If B says yes, the money is divided up and the game is over. But if B says no, then A must decide whether to make another offer or not. Each subsequent offer from A must be more generous to B. The game ends when either B says yes or A stops making offers. How do you predict this game will end up?

In this case, we can suppose that A will keep on making offers until he has proposed 99 to B and 1 for himself. Thus, according to tree logic, B should get almost all of the pie. If you were B, would you hold out for 99:1? We'd advise against it.

of money, say 100 dollars. Player A begins the game by proposing a division of the 100 dollars between the two. Then B decides whether to agree to A's proposal. If B agrees, the proposal is implemented; each player gets what A proposed and the game ends. If B refuses, then neither player gets anything, and the game ends.

Pause a minute and think. If you were playing this game in the A role, what division would you propose?

Now think how this game would be played by two people who are "rational" from the point of view of conventional economic theory—

that is, each is concerned only with his or her self-interest and can calculate perfectly the optimal strategies to pursue that interest. The proposer (A) would think as follows. "No matter what split I propose, B is left with the choice between that and nothing. (The game is played only once, so B has no reason to develop a reputation for toughness, or to engage in any tit-for-tat response to A's actions.) So B will accept whatever I offer. I can do best for myself by offering B as little as possible—for example, just one cent, if that is the minimum permissible under the rules of the game." Therefore A would offer this minimum and B would accept.*

Pause and think again. If you were playing this game in the B role, would you accept one cent?

Numerous experiments have been conducted on this game.[6] Typically, two dozen or so subjects are brought together and are matched randomly in pairs. In each pair, the roles of proposer and

---

* This argument is another example of tree logic without drawing a tree.

responder are assigned, and the game is played once. New pairs are formed at random, and the game played again. Usually the players do not know with whom they are matched in any one play of the game. Thus the experimenter gets several observations from the same pool in the same session, but there is no possibility of forming ongoing relationships that can affect behavior. Within this general framework, many variations of conditions are attempted, to study their effects on the outcomes.

Your own introspection of how you would act as proposer and as responder has probably led you to believe that the results of actual play of this game should differ from the theoretical prediction above. And indeed they differ, often dramatically so. The amounts offered to the responder differ across proposers, but one cent or one dollar, or in fact anything below 10 percent of the total sum at stake, is very rare. The median offer (half of the proposers offer less than that and half offer more) is in the 40–50 percent range; in many experiments a 50:50 split is the single most frequent proposal. Proposals that would give the responder less than 20 percent are rejected about half the time.

IRRATIONALITY VERSUS OTHER-REGARDING RATIONALITY

Why do proposers offer substantial shares to the responders? Three reasons suggest themselves. First, the proposers may be unable to do the correct backward reasoning. Second, the proposers may have motives other than the pure selfish desire to get as much as they can; they act altruistically or care about fairness. Third, they may fear that responders would reject low offers.

The first is unlikely because the logic of backward reasoning is so simple in this game. In more complex situations, players may fail to do the necessary calculations fully or correctly, especially if they are novices to the game being played, as we saw in 21-flags. But the ultimatum game is surely simple enough, even for novices. The explanation must be the second, the third, or a combination thereof.

Early results from ultimatum experiments favored the third. In fact, Harvard's Al Roth and his coauthors found that, given the pattern of rejection thresholds that prevailed in their subject pool, the proposers were choosing their offers to achieve an optimal balance between the prospect of obtaining a greater share for themselves against the risk of rejection. This suggests a remarkable conventional rationality on part of the proposers.

However, later work to distinguish the second and the third possibilities led to a different idea. To distinguish between altruism and strategy, experiments were done using a variant called the dictator game. Here the proposer dictates how the available total is to be split; the other player has no say in the matter at all. Proposers in the dictator game give away significantly smaller sums on average than they offer in the ultimatum game, but they give away substantially more than zero. Thus there is something to both of those explanations; proposers' behavior in the ultimatum game has both generous and strategic aspects.

Is the generosity driven by altruism or by a concern for fairness? Both explanations are different aspects of what might be called a regard for others in people's preferences. Another variation of the experiment helps tell these two possibilities apart. In the usual setup, after the pairs are formed, the roles of proposer and responder are assigned by a random mechanism like a coin toss. This may build in a notion of equality or fairness in the players' minds. To remove this, a variant assigns the roles by holding a preliminary contest, such as a test of general knowledge, and making its winner the proposer. This creates some sense of entitlement to the proposer, and indeed leads to offers that are on average about 10 percent smaller. However, the offers remain substantially above zero, indicating that proposers have an element of altruism in their thinking. Remember that they do not know the identity of the responders, so this must be a generalized sense of altruism, not concern for the well-being of a particular person.

A third variation of individual preferences is also possible: contributions may be driven by a sense of shame. Jason Dana of the

University of Illinois, Daylian Cain of Yale School of Management, and Robyn Dawes of Carnegie-Mellon University performed an experiment with the following variation of the dictator game.[7] The dictator is asked to allocate $10. After the allocation is made, but before it is delivered to the other party, the dictator is given the following offer: You can have $9, the other party will get nothing, and they will never know that they were part of this experiment. Most dictators accept this offer. Thus they would rather give up a dollar to ensure that the other person never knows how greedy they were. (An altruistic person would prefer keeping $9 and giving away $1 to keeping $9 while the other person gets nothing.) Even when a dictator had offered $3, he would rather take that away to keep the other person in the dark. This is much like incurring a large cost to cross the street to avoid making a small donation to a beggar.

Observe two things about these experiments. First, they follow the standard methodology of science: hypotheses are tested by designing appropriate variations of controls in the experiment. We mention a few prominent variations of this kind here. (Many more are discussed in Colin Camerer's book cited in chapter 2, note 6.) Second, in the social sciences, multiple causes often coexist, each contributing part of the explanation for the same phenomenon. Hypotheses don't have to be either fully correct or totally wrong; accepting one need not mean rejecting all others.

Now consider the behavior of the responders. Why do they reject an offer when they know that the alternative is to get even less? The reason cannot be to establish a reputation for being a tough negotiator that may bear fruit in future plays of this game or other games of division. The same pair does not play repeatedly, and no track record of one player's past behavior is made available to future partners. Even if a reputational motive is implicitly present, it must take a deeper form: a general rule for action that the responder follows without doing any explicit thinking or calculation in each instance. It must be an instinctive action or an emotion-driven response. And that is indeed the case. In a new emerging line of experimental research called neuroeconomics, the subjects' brain activity is scanned using functional magnetic resonance imaging (fMRI) or positron emission tomography (PET) while they make various economic decisions. When ultimatum games are played under such conditions, it is found that the responders' anterior insula shows more activity as the proposers' offers become more unequal. Since the anterior insula is active for emotions, such as anger and disgust, this result helps explain why second movers reject unequal offers. Conversely, the left-side prefrontal cortex is more active when an unequal offer is accepted, indicating that conscious control is being exercised to balance between acting on one's disgust and getting more money.[8]

Many people (especially economists) argue that while responders may reject small shares of the small sums that are typically on offer in laboratory experiments, in the real world, where stakes are often much larger, rejection must be very unlikely. To test this, ultimatum game experiments have been conducted in poorer countries where the amounts were worth several months' income for the participants. Rejection does become somewhat less likely, but offers do not become significantly less generous. The consequences of rejection become more serious for the proposers just as they do for the responders, so proposers fearing rejection are likely to behave more cautiously.

Although behavior can be explained in part by instincts, hormones, or emotions hardwired into the brain, part of it varies from one culture to another. In experiments conducted across many countries, it was found that the perception of what constitutes a reasonable offer varied by up to 10 percent across cultures, but properties like aggressiveness or toughness varied less. Only one group differed substantially from the rest: among the Machiguenga of the Peruvian Amazon, the offers were much smaller (average 26 percent) and only one offer was rejected. Anthropologists explain that the Machiguenga live in small family units, are socially disconnected, and have no norms of sharing. Conversely, in two cultures the offers exceeded 50 percent; these have the custom of lavish giving when one has a stroke of good luck, which places an obligation on the recipients to return the favor even more generously in the future. This norm or habit seems to carry over to the experiment even though the players do not know whom they are giving to or receiving from.[9]

Evolution of Altruism and Fairness

What should we learn from the findings of these experiments on the ultimatum game, and others like them? Many of the outcomes do differ significantly from what we would expect based on the theory of backward reasoning with the assumption that each player cares only about his or her own reward. Which of the two—correct backward calculation or selfishness—is the wrong assumption, or is it a combination? And what are the implications?

Consider backward reasoning first. We saw the players on *Survivor* fail to do this correctly or fully in 21-flags. But they were playing the game for the first time, and even then, their discussion revealed glimpses of the correct reasoning. Our classroom experience shows that students learn the full strategy after playing the game, or watching it played, just three or four times. Many experiments inevitably and almost deliberately work with novice subjects, whose actions in the game are often steps in the process of learning the game. In the real world of business, politics, and professional sports, where people are experienced at playing the games they are involved in, we should expect that the players have accumulated much more learning and that they play generally good strategies either by calculation or by trained instinct. For somewhat more complex games, strategically aware players can use computers or consultants to do the calculations; this practice is still somewhat rare but is sure to spread. Therefore, we believe that backward reasoning should remain our starting point for analysis of such games and for predicting their outcomes. This first pass at the analysis can then be modified as necessary in a particular context, to recognize that beginners may make mistakes and that some games may become too complex to be solved unaided.

We believe that the more important lesson from the experimental research is that people bring many considerations and preferences into their choices besides their own rewards. This takes us beyond the scope of conventional economic theory. Game theorists should include in their analysis of games the players' concerns for fairness or altruism. "Behavioral game theory *extends* rationality rather than abandoning it."[10]

This is all to the good; a better understanding of people's motives enriches our understanding of economic decision making and strategic interactions alike. And that is already happening; frontier research in game theory increasingly includes in the players' objectives their concerns for equity, altruism, and similar concerns (and even a "second-round" concern to reward or punish others whose behavior reflects or violates these precepts).[11]

But we should not stop there; we should go one step further and think about why concerns for altruism and fairness, and anger or disgust when someone else violates these precepts, have such a strong hold on people. This takes us into the realm of speculation, but one plausible explanation can be found in evolutionary psychology. Groups that instill norms of fairness and altruism into their members will have less internal conflict than groups consisting of purely selfish individuals. Therefore they will be more successful in taking collective action, such as provision of goods that benefit the whole group and conservation of common resources, and they will spend less effort and resources in internal conflict. As a result, they will do better, both in absolute terms and in competition with groups that do not have similar norms. In other words, some measure of fairness and altruism may have evolutionary survival value.

Some biological evidence for rejecting unfair offers comes from an experiment run by Terry Burnham.[12] In his version of the ultimatum game, the amount at stake was $40 and the subjects were male Harvard graduate students. The divider was given only two choices: offer $25 and keep $15 or offer $5 and keep $35. Among those offered only $5, twenty students accepted and six rejected, giving themselves and the divider both zero. Now for the punch line. It turns out that the six who rejected the offer had testosterone levels 50 percent higher than those who accepted the offer.

To the extent that testosterone is connected with status and aggression, this could provide a genetic link that might explain an evolutionary advantage of what evolutionary biologist Robert Trivers has called "moralistic aggression."

In addition to a potential genetic link, societies have nongenetic ways of passing on norms, namely the processes of education and socialization of infants and children in families and schools. We see parents and teachers telling impressionable children the importance of caring for others, sharing, and being nice; some of this undoubtedly remains imprinted in their minds and influences their behavior throughout their lives.

Finally, we should point out that fairness and altruism have their limit. Long-run progress and success of a society need innovation and change. These in turn require individualism and a willingness to defy social norms and conventional wisdom; selfishness often accompanies these characteristics. We need the right balance between self-regarding and other-regarding behaviors.

VERY COMPLEX TREES

When you have acquired a little experience with backward reasoning, you will find that many strategic situations in everyday life or work lend themselves to "tree logic" without the need to draw and analyze trees explicitly. Many other games at an intermediate level of complexity can be solved using computer software packages that are increasingly available for this purpose. But for complex games such as chess, a complete solution by backward reasoning is simply not feasible.

In principle, chess is the ideal game of sequential moves amenable to solution by backward reasoning.[13] The players alternate moves; all previous moves are observable and irrevocable; there is no uncertainty about the position or the players' motives. The rule that the game is a draw if the same position is repeated ensures that the game ends within a finite total number of moves. We can start with the terminal nodes (or endpoints) and work backward. However, practice and principle are two different things. It has been estimated that the total number of nodes in chess is about 10^{120}, that is, 1 with 120 zeroes after it. A supercomputer 1,000 times as fast as the typical PC would need 10^{103} years to examine them all. Waiting for that is futile; foreseeable progress in computers is not likely to improve matters significantly. In the meantime, what have chess players and programmers of chess-playing computers done?

Chess experts have been successful at characterizing optimal strategies near the end of the game. Once the chessboard has only a small number of pieces on it, experts are able to look ahead to the end of the game and determine by backward reasoning whether one side has a guaranteed win or whether the other side can obtain a draw. But the middle of the game, when several pieces remain on the board, is far harder. Looking ahead five pairs of moves, which is about as much as can be done by experts in a reasonable amount of time, is not going to simplify the situation to a point where the endgame can be solved completely from there on.

The pragmatic solution is a combination of forward-looking analysis and value judgment. The former is the science of game theory—looking ahead and reasoning backward. The latter is the art of the practitioner—being able to judge the value of a position from the number and interconnections of the pieces without finding an explicit solution of the game from that point onward. Chess players often speak of this as "knowledge," but you can call it experience or instinct or art. The best chess players are usually distinguished by the depth and subtlety of their knowledge.

Knowledge can be distilled from the observation of many games and many players and then codified into rules. This has been done most extensively with regard to openings, that is, the first ten or even fifteen moves of a game. There are hundreds and hundreds of books that analyze different openings and discuss their relative merits and drawbacks.

How do computers fit into this picture? At one time, the project of programming computers to play chess was seen as an integral part of the emerging science of artificial intelligence; the aim was to design computers that would think as humans do. This did not succeed for many years. Then the attention shifted to using computers to do what they do best, crunch numbers. Computers can look ahead to more moves and do this more quickly than humans can.* Using pure number crunching, by the late 1990s dedicated chess computers like Fritz and Deep Blue could compete with the top human players. More recently, computers have been programmed with some knowledge of midgame positions, imparted by some of the best human players.

Human players have ratings determined by their performances; the best-ranked computers are already achieving ratings comparable to the 2800 enjoyed by the world's strongest human player, Garry Kasparov. In November 2003, Kasparov played a four-game match against the latest version of the Fritz computer, X3D. The result was one victory each and two draws. In July 2005, the Hydra chess computer demolished Michael Adams, ranked number 13 in the world, winning five games and drawing one in a six-game match. It may not be long before the rival computers rank at the top and play each other for world championships.

What should you take away from this account of chess? It shows the method for thinking about any highly complex games you may face. You should combine the rule of look ahead and reason back with your experience, which guides you in evaluating the intermediate positions reached at the end of your span of forward calculation. Success will come from such synthesis of the science of game theory and the art of playing a specific game, not from either alone.

BEING OF TWO MINDS

Chess strategy illustrates another important practical feature of looking forward and reasoning backward: you have to play the

---

* But good chess players can use their knowledge to disregard immediately those moves that are likely to be bad without pursuing their consequences four or five moves ahead, thereby saving their calculation time and effort for the moves that are more likely to be good ones.

game from the perspective of both players. While it is hard to calculate your best move in a complicated tree, it is even harder to predict what the other side will do.

If you really could analyze all possible moves and countermoves, and the other player could as well, then the two of you would agree up front as to how the entire game would play out. But once the analysis is limited to looking down only some branches of the tree, the other player may see something you didn't or miss something you've seen. Either way, the other side may then make a move you didn't anticipate.

To really look forward and reason backward, you have to predict what the other players will actually do, not what you would have done in their shoes. The problem is that when you try to put yourself in the other players' shoes, it is hard if not impossible to leave your own shoes behind. You know too much about what you are planning to do in your next move and it is hard to erase that knowledge when you are looking at the game from the other player's perspective. Indeed, that explains why people don't play chess (or poker) against themselves. You certainly can't bluff against yourself or make a surprise attack.

There is no perfect solution to this problem. When you try to put yourself in the other players' shoes, you have to know what they know and not know what they don't know. Your objectives have to be their objectives, not what you wish they had as an objective. In practice, firms trying to simulate the moves and countermoves of a potential business scenario will hire outsiders to play the role of the other players. That way, they can ensure that their game partners don't know too much. Often the biggest learning comes from seeing the moves that were not anticipated and then understanding what led to that outcome, so that it can be either avoided or promoted.

To end this chapter, we return to Charlie Brown's problem of whether or not to kick the football. This question became a real issue for football coach Tom Osborne in the final minutes of his championship game. We think he too got it wrong. Backward reasoning will reveal the mistake.

CASE STUDY: THE TALE OF TOM OSBORNE
AND THE '84 ORANGE BOWL

In the 1984 Orange Bowl the undefeated Nebraska Cornhuskers and the once-beaten Miami Hurricanes faced off. Because Nebraska came into the Bowl with the better record, it needed only a tie in order to finish the season with the number-one ranking.

Coming into the fourth quarter, Nebraska was behind 31–17. Then the Cornhuskers began a comeback. They scored a touchdown to make the score 31–23. Nebraska coach Tom Osborne had an important strategic decision to make.

In college football, a team that scores a touchdown then runs one play from a hash mark 2 1/2 yards from the goal line. The team has a choice between trying to run (or pass) the ball into the end zone, which scores two additional points, or trying the less risky strategy of kicking the ball through the goalposts, which scores one extra point.

Coach Osborne chose to play it safe, and Nebraska successfully kicked for the one extra point. Now the score was 31–24. The Cornhuskers continued their comeback. In the waning minutes of the game they scored a final touchdown, bringing the score to 31–30. A one-point conversion would have tied the game and landed them the title. But that would have been an unsatisfying victory. To win the championship with style, Osborne recognized that he had to go for the win.

The Cornhuskers went for the win with a two-point conversion attempt. Irving Fryar got the ball but failed to score. Miami and Nebraska ended the year with equal records. Since Miami beat Nebraska, it was Miami that was awarded the top place in the standings.

Put yourself in the cleats of Coach Osborne. Could you have done better?

Case Discussion

Many Monday morning quarterbacks fault Osborne for going for the win rather than the tie. But that is not our bone of contention. Given that Osborne was willing to take the additional risk for the win, he did it the wrong way. He would have done better to first try the two-point conversion. If it succeeded, then go for the one-point; if it failed, attempt a second two-pointer.

Let us look at this more carefully. When down by 14 points, he knew that he needed two touchdowns plus three extra points. He chose to go for the one-point and then the two. If both attempts succeeded, the order in which they were made becomes irrelevant. If the one-point conversion was missed but the two-point was successful, here too the order is irrelevant and the game ends up tied with Nebraska getting the championship. The only difference occurs if Nebraska misses the two-point conversion. Under Osborne's plan, that results in the loss of the game and the championship. If, instead, they had tried the two-point conversion first, then if it failed they would not necessarily have lost the game. They would have been behind 31–23. When they scored their next touchdown this would have brought them to 31–29. A successful two-point attempt would tie the game and win the number-one ranking!*

We have heard the counterargument that if Osborne went for the two-pointer first and missed, his team would have been playing for the tie. This would have provided less inspiration and perhaps they might not have scored the second touchdown. Moreover, by waiting until the end and going for the desperation win-lose two-pointer his team would rise to the occasion knowing everything was on the line. This argument is wrong for several reasons. Remember that if Nebraska waits until the second touchdown and then misses the two-point attempt, they lose. If they miss the two-point attempt on their first try, there is still a chance for a tie. Even though the chance may be diminished, something is better than nothing. The momentum argument is also flawed. While Nebraska's offense may rise to the occasion in a single play for the championship, we expect the Hurricanes' defense to rise as well.

---

* Furthermore, this would be a tie that resulted from the failed attempt to win, so no one would criticize Osborne for playing to tie.

The play is important for both sides. To the extent that there is a momentum effect, if Osborne makes the two-point conversion on the first touchdown, this should increase the chance of scoring another touchdown. It also allows him to tie the game with two field goals.

One of the general morals of this story is that if you have to take some risks, it is often better to do so as quickly as possible. This is obvious to those who play tennis: everyone knows to take more risk on the first serve and hit the second serve more cautiously. That way, if you fail on your first attempt, the game won't be over. You may still have time to take some other options that can bring you back to or even ahead of where you were. The wisdom of taking risks early applies to most aspects of life, whether it be career choices, investments, or dating.

For more practice using the principle of look forward, reason backward, have a look at the following case studies in chapter 14: "Here's Mud in Your Eye"; "Red I Win, Black You Lose"; "The Shark Repellent That Backfired"; "Tough Guy, Tender Offer"; "The Three-Way Duel"; and "Winning without Knowing How."

CHAPTER 3

Prisoners' Dilemmas
and How to
Resolve Them

MANY CONTEXTS, ONE CONCEPT

What do the following situations have in common?

- Two gas stations at the same corner, or two supermarkets in the same neighborhood, sometimes get into fierce price wars with each other.

- In general election campaigns, both the Democratic and the Republican parties in the United States often adopt centrist policies to attract the swing voters in the middle of the political spectrum, ignoring their core supporters who hold more extreme views to the left and the right, respectively.

- "The diversity and productivity of New England fisheries was once unequalled. A continuing trend over the past century has been the overexploitation and eventual collapse of species after species. Atlantic halibut, ocean perch, Haddock and Yellowtail Flounder . . . [have joined] the ranks of species written-off as commercially extinct."[1]

- Near the end of Joseph Heller's celebrated novel *Catch-22*, the Second World War is almost won. Yossarian does not want to be among the last to die; it won't make any difference to the outcome. He explains this to Major Danby, his superior officer.

When Danby asks, "But, Yossarian, suppose everyone felt that way?" Yossarian replies, "Then I'd certainly be a damned fool to feel any other way, wouldn't I?"[2]

Answer: They are all instances of the prisoners' dilemma.* As in the interrogation of Dick Hickock and Perry Smith from *In Cold Blood* recounted in chapter 1, each has a personal incentive to do something that ultimately leads to a result that is bad for everyone when everyone similarly does what his or her personal interest dictates. If one confesses, the other had better confess to avoid the really harsh sentence reserved for recalcitrants; if one holds out, the other can cut himself a much better deal by confessing. Indeed, the force is so strong that each prisoner's temptation to confess exists regardless of whether the two are guilty (as was the case in *In Cold Blood*) or innocent and being framed by the police (as in the movie *L.A. Confidential*).

Price wars are no different. If the Nexon gas station charges a low price, the Lunaco station had better set its own price low to avoid losing too many customers; if Nexon prices its gas high, Lunaco can divert many customers its way by pricing low. But when both stations price low, neither makes money (though customers are better off).

If the Democrats adopt a platform that appeals to the middle, the Republicans may stand to lose all these voters and therefore the election if they cater only to their core supporters in the economic and social right wings; if the Democrats cater to their core supporters in the minorities and the unions, then the Republicans can capture the middle and therefore win a large majority by being more centrist. If all others fish conservatively, one fisherman going

---

* No prizes for correct answers—after all, the prisoners' dilemma is the subject of this chapter. But we take this opportunity to point out, as we did in chapter 2, that the common conceptual framework of game theory can help us understand a great variety of diverse and seemingly unrelated phenomena. We should also point out that neighboring stores do not *constantly* engage in price wars, and political parties do not *always* gravitate to the center. In fact, analyses and illustrations of how the participants in such games can avoid or resolve the dilemma is an important part of this chapter.

for a bigger catch is not going to deplete the fishery to any significant extent; if all others are fishing aggressively, then any single
fisherman would be a fool to try single-handed conservation.[3] The
result is overfishing and extinction. Yossarian's logic is what makes
it so difficult to continue to support a failed war.

A LITTLE HISTORY

How did theorists devise and name this game that captures so
many economic, political, and social interactions? It happened
very early in the history of the subject. Harold Kuhn, himself one
of the pioneers of game theory, recounted the story in a symposium
held in conjunction with the 1994 Nobel Prize award ceremonies:

> Al Tucker was on leave at Stanford in the Spring of 1950 and,
> because of the shortage of offices, he was housed in the Psychol-
> ogy Department. One day a psychologist knocked on his door
> and asked what he was doing. Tucker replied: "I'm working on
> game theory," and the psychologist asked if he would give a sem-
> inar on his work. For that seminar, Al Tucker invented pris-
> oner's dilemma as an example of game theory, Nash equilibria,
> and the attendant paradoxes of non-socially-desirable equilibria.
> A truly seminal example, it inspired dozens of research papers
> and several entire books.[4]

Others tell a slightly different story. According to them, the
mathematical structure of the game predates Tucker and can be
attributed to two mathematicians, Merrill Flood and Melvin
Dresher, at the Rand Corporation (a cold war think tank).[5]
Tucker's genius was to invent the story illustrating the mathematics. And genius it was, because presentation can make or break an
idea; a memorable presentation spreads and is assimilated in the
community of thinkers far better and faster, whereas a dull and dry
presentation may be overlooked or forgotten.

A Visual Representation

We will develop the method for displaying and solving the game
using a business example. Rainbow's End and B. B. Lean are rival

mail-order firms that sell clothes. Every fall they print and mail their winter catalogs. Each firm must honor the prices printed in its catalog for the whole winter season. The preparation time for the catalogs is much longer than the mailing window, so the two firms must make their pricing decisions simultaneously and without knowing the other firm's choices. They know that the catalogs go to a common pool of potential customers, who are smart shoppers and are looking for low prices.

Both catalogs usually feature an almost identical item, say a chambray deluxe shirt. The cost of each shirt to each firm is $20.* The firms have estimated that if they each charge $80 for this item, each will sell 1,200 shirts, so each will make a profit of $(80 - 20) \times 1,200 = 72,000$ dollars. Moreover, it turns out that this price serves their joint interests best: if the firms can collude and charge a common price, $80 is the price that will maximize their combined profits.

The firms have estimated that if one of them cuts its price by $1 while the other holds its price unchanged, then the price cutter gains 100 customers, 80 of whom shift to it from the other firm, and 20 who are new—for example, they might decide to buy the shirt when they would not have at the higher price or might switch from a store in their local mall. Therefore each firm has the temptation to undercut the other to gain more customers; the whole purpose of this story is to figure out how these temptations play out.

We begin by supposing that each firm chooses between just two prices, $80 and $70.† If one firm cuts its price to $70 while the

* This includes not only the cost of buying the shirt from the supplier in China but also the cost of transporting it to the United States, any import duties, and the costs of stocking it and of order fulfillment. In other words, it includes all costs specifically attributable to this item. The intention is to have a comprehensive measure of what economists would call marginal cost.

† This specification, and in particular the assumption that there are just two possible choices for the price, is just to build the analytical method for such games in the simplest possible way. In the following chapter we will allow the firms much greater freedom in choosing their prices.

other is still charging $80, the price cutter gains 1,000 customers and the other loses 800. So the price cutter sells 2,200 shirts while the other's sales drop to 400; the profits are (70 − 20) × 2,200 = $110,000 for the price cutter, and (80 − 20) × 400 = $24,000 for the other firm.

What happens if both firms cut their price to $70 at the same time? If both firms reduce their price by $1, existing customers stay put, but each gains the 20 new customers. So when both reduce their price by $10, each gains 10 × 20 = 200 net sales above the previous 1,200. Each sells 1,400 and makes a profit of (70 − 20) × 1,400 = $70,000.

We want to display the profit consequences (the firms' payoffs in their game) visually. However, we cannot do this using a game tree like the ones in chapter 2. Here the two players act simultaneously. Neither can make his move knowing what the other has done or anticipating how the other will respond. Instead, each must think about what the other is thinking at the same time. A starting point for this thinking about thinking is to lay out all the consequences of all the combinations of the simultaneous choices the two could make. Since each has two alternatives, $80 or $70, there are four such combinations. We can display them most easily in a spreadsheet-like format of rows and columns, which we will generally refer to as a game table or payoff table. The choices of Rainbow's End (RE for short) are arrayed along the rows, and those of B. B. Lean (BB) along the columns. In each of the four cells corresponding to each choice of a row by RE and of a column by BB, we show two numbers—the profit, in thousands of dollars, from selling this shirt. In each cell, the number in the southwest corner belongs to the row player, and the number in the northeast corner belongs to the column player.* In the jargon of game theory, these

---

* Thomas Schelling invented this way of representing both players' payoffs in the same table while making clear which payoff belongs to which player. With excessive modesty, he writes: "If I am ever asked whether I ever made a contribution to game theory, I shall answer yes . . . the invention of staggered payoffs in a matrix." Actually Schelling developed many of the most important concepts of game theory—focal points, credibility, commitment, threats and promises, tip-

numbers are called payoffs.* To make it abundantly clear which payoffs belong to which player, we have also put the numbers in two different shades of gray for this example.

		B. B. Lean (BB)			
			80		70
			72,000		110,000
Rainbow's End (RE)	80	72,000		24,000	
			24,000		70,000
	70	110,000		70,000	

Before we "solve" the game, let us observe and emphasize one feature of it. Compare the payoff pairs across the four cells. A better outcome for RE does not always imply a worse outcome for BB, or vice versa. Specifically, both of them are better off in the top left cell than in the bottom right cell. This game need not end with a winner and a loser; it is not zero-sum. We similarly pointed out in chapter 2 that the Charlie Brown investment game was not zero-sum, and neither are most games we meet in reality. In many games, as in the prisoners' dilemma, the issue will be how to avoid a lose-lose outcome or to achieve a win-win outcome.

The Dilemma

Now consider the reasoning of RE's manager. "If BB chooses $80, I can get $110,000 instead of $72,000 by cutting my price to $70. If BB chooses $70, then my payoff is $70,000 if I also charge

---

ping, and much more. We will cite him and his work frequently in the chapters to come.

* Generally, higher payoff numbers are better for each player. Sometimes, as with prisoners under interrogation, the payoff numbers are years in jail, so each player prefers a smaller number for himself. The same can happen if the payoff numbers are rankings where 1 is best. When looking at a game table, you should check the interpretation of the payoff numbers for that game.

$70, but only $24,000 if I charge $80. So, in both cases, choosing $70 is better than choosing $80. My better choice (in fact my best choice, since I have only two alternatives) is the same no matter what BB chooses. I don't need to think through their thinking at all; I should just go ahead and set my price at $70."

When a simultaneous-move game has this special feature, namely that for a player the best choice is the same regardless of what the other player or players choose, it greatly simplifies the players' thinking and the game theorists' analysis. Therefore it is worth making a big deal of it, and looking for it to simplify the solution of the game. The name given by game theorists for this property is *dominant strategy*. A player is said to have a dominant strategy if that same strategy is better for him than all of his other available strategies no matter what strategy or strategy combination the other player or players choose. And we have a simple rule for behavior in simultaneous-move games:*

RULE 2: If you have a dominant strategy, use it.

The prisoners' dilemma is an even more special game—not just one player but both (or all) players have dominant strategies. The reasoning of BB's manager is exactly analogous to that of RE's manager, and you should fix the idea by going through it on your own. You will see that $70 is also the dominant strategy for BB.

The result is the outcome shown in the bottom right cell of the game table; both charge $70 and make a profit of $70,000 each. And here is the feature that makes the prisoners' dilemma such an important game. When both players use their dominant strategies, both do worse than they would have if somehow they could have jointly and credibly agreed that each would choose the other, dom-

---

* In chapter 2, we could offer a single, unifying principle to devise the best strategies for games with sequential moves. This was our Rule 1: Look forward and reason backward. It won't be so simple for simultaneous-move games. But the thinking about thinking required for simultaneous moves can be summarized in three simple rules for action. These rules in turn rest on two simple ideas—dominant strategies and equilibrium. Rule 2 is given here; Rules 3 and 4 will follow in the next chapter.

inated strategy. In this game, that would have meant charging $80 each to obtain the outcome in the top left cell of the game table, namely $72,000 each.*

It would not be enough for just one of them to price at $80; then that firm would do very badly. Somehow they must both be induced to price high, and this is hard to achieve given the temptation each of them has to try to undercut the other. Each firm pursuing its own self-interest does not lead to an outcome that is best for them all, in stark contrast to what conventional theories of economics from Adam Smith onward have taught us.†

This opens up a host of questions, some of which pertain to more general aspects of game theory. What happens if only one player has a dominant strategy? What if none of the players has a dominant strategy? When the best choice for each varies depending on what the other is choosing simultaneously, can they see through each other's choices and arrive at a solution to the game? We will take up these questions in the next chapter, where we develop a more general concept of solution for simultaneous-move games, namely Nash equilibrium. In this chapter we focus on questions about the prisoners' dilemma game per se.

In the general context, the two strategies available to each player are labeled "Cooperate" and "Defect" (or sometimes "Cheat"), and we will follow this usage. Defect is the dominant strategy for each, and the combination where both choose Defect yields a worse outcome for both than if both choose Cooperate.

---

* Actually, $80 is the common price that yields the two the highest possible joint profit; that is the price they would choose if they could get together and cartelize the industry. Rigorous proof of this statement requires some math, so just take our word for it. For readers who want to follow the calculation, it is on the book's web site.

† The beneficiaries from this price cutting by the firms are of course the consumers, who are not active players in this game. Therefore it is often in the larger society's interest to prevent the two firms from resolving their pricing dilemma. That is the role of antitrust policies in the United States and other countries.

Some Preliminary Ideas for Resolving the Dilemma

The players caught on the horns of this dilemma have strong incentives to make joint agreements to avoid it. For example, the fishermen in New England might agree to limit their catch to preserve the fish stocks for the future. The difficulty is to make such agreements stick, when each faces the temptation to cheat, for example, to take more than one's allotted quota of fish. What does game theory have to say on this issue? And what happens in the actual play of such games?

In the fifty years since the prisoners' dilemma game was invented, its theory has advanced a great deal, and much evidence has accumulated, both from observations about the real world and from controlled experiments in laboratory settings. Let us look at all this material and see what we can learn from it.

The flip side of achieving cooperation is avoiding defection. A player can be given the incentive to choose cooperation rather than the originally dominant strategy of defection by giving him a suitable reward, or deterred from defecting by creating the prospect of a suitable punishment.

The reward approach is problematic for several reasons. Rewards can be internal—one player pays the other for taking the cooperative action. Sometimes they can be external; some third party that also benefits from the two players' cooperation pays them for cooperating. In either case, the reward cannot be given before the choice is made; otherwise the player will simply pocket the reward and then defect. If the reward is merely promised, the promise may not be credible: after the promisee has chosen cooperation, the promisor may renege.

These difficulties notwithstanding, rewards are sometimes feasible and useful. At an extreme of creativity and imagination, the players could make simultaneous and mutual promises and make these credible by depositing the promised rewards in an escrow account controlled by a third party.[6] More realistically, sometimes the players interact in several dimensions, and cooperation in one can be rewarded with reciprocation in another. For example, among groups of female chimpanzees, help with grooming is recip-rocated by sharing food or help with child minding. Sometimes third parties may have sufficiently strong interests in bringing about cooperation in a game. For example, in the interest of bringing to an end various conflicts around the world, the United States and the European Union have from time to time promised economic assistance to combatants as a reward for peaceful resolutions of their disputes. The United States rewarded Israel and Egypt in this way for cooperating to strike the Camp David Accords in 1978.

Punishment is the more usual method of resolving prisoners' dilemmas. This could be immediate. In a scene from the movie *L.A. Confidential*, Sergeant Ed Exley promises Leroy Fontaine, one of the suspects he is interrogating, that if he turns state's wit-ness, he will get a shorter sentence than the other two, Sugar Ray Coates and Tyrone Jones. But Leroy knows that, when he emerges from jail, he may find friends of the other two waiting for him!

But the punishment that comes to mind most naturally in this context arises from the fact that most such games are parts of an ongoing relationship. Cheating may gain one player a short-term advantage, but this can harm the relationship and create a longer-run cost. If this cost is sufficiently large, that can act as a deterrent against cheating in the first place.*

A striking example comes from baseball. Batters in the American League are hit by pitches 11 to 17 percent more often than their colleagues in the National League. According to Sewanee profes-sors Doug Drinen and John-Charles Bradbury, most of this differ-ence is explained by the designated hitter rule.[7] In the American League, the pitchers don't bat. Thus an American League pitcher who plunks a batter doesn't have to fear direct retaliation from the opposing team's pitcher. Although pitchers are unlikely to get hit, the chance goes up by a factor of four if they have just plunked someone in the previous half inning. The fear of retaliation is clear.

---

* Robert Aumann was awarded the 2005 Nobel Prize in Economics for his instrumental work in the development of the general theory of tacit cooperation in repeated games.

As ace pitcher Curt Schilling explained: "Are you seriously going to throw at someone when you are facing Randy Johnson?"[8]

When most people think about one player punishing the other for past cheating, they think of some version of tit for tat. And that was indeed the finding of what is perhaps the most famous experiment on the prisoners' dilemma. Let us recount what happened and what it teaches.

TIT FOR TAT

In the early 1980s, University of Michigan political scientist Robert Axelrod invited game theorists from around the world to submit their strategies for playing the prisoners' dilemma in the form of computer programs. The programs were matched against each other in pairs to play a prisoners' dilemma game repeated 150 times. Contestants were then ranked by the sum of their scores.

The winner was Anatol Rapoport, a mathematics professor at the University of Toronto. His winning strategy was among the simplest: tit for tat. Axelrod was surprised by this. He repeated the tournament with an enlarged set of contestants. Once again Rapoport submitted tit for tat and beat the competition.

Tit for tat is a variation of the eye for an eye rule of behavior: Do unto others as they have done onto you.* More precisely, the strat-

---

* In Exodus (21:22–25), we are told: "If men who are fighting hit a pregnant woman and she gives birth prematurely but there is no serious injury, the offender must be fined whatever the woman's husband demands and the court allows. But if there is serious injury, you are to take life for life, eye for eye, tooth for tooth, hand for hand, foot for foot, burn for burn, wound for wound, bruise for bruise." The New Testament suggests more cooperative behavior. In Matthew (5:38–39) we find: "You have heard that it was said, 'Eye for eye, and tooth for tooth.' But I tell you, Do not resist an evil person. If someone strikes you on the right cheek, turn to him the other also." We move from "Do unto others as they have done onto you" to the golden rule: "Do to others as you would have them do to you" (Luke 6:31). If people were to follow the golden rule, there would be no prisoners' dilemma. And if we think in the larger perspective, although cooperation might lower your payoffs in any particular game, the potential reward in an afterlife may make this a rational strategy even for a selfish

egy cooperates in the first period and from then on mimics the rival's action from the previous period.

Axelrod argues that tit for tat embodies four principles that should be present in any effective strategy for the repeated prisoners' dilemma: clarity, niceness, provocability, and forgivingness. Tit for tat is as *clear* and simple as you can get; the opponent does not have to do much thinking or calculation about what you are up to. It is *nice* in that it never initiates cheating. It is *provocable*—that is, it never lets cheating go unpunished. And it is *forgiving,* because it does not hold a grudge for too long and is willing to restore cooperation.

One of the impressive features about tit for tat is that it did so well overall even though it did not (nor could it) beat any one of its rivals in a head-on competition. At best, tit for tat ties its rival. Hence if Axelrod had scored each competition as a winner-take-all contest, tit for tat would have only losses and ties and therefore could not have had the best track record.*

But Axelrod did not score the pairwise plays as winner-take-all: close counted. The big advantage of tit for tat is that it always comes close. At worst, tit for tat ends up getting beaten by one defection—that is, it gets taken advantage of once and then ties from then on.

The reason tit for tat won the tournament is that it usually managed to encourage cooperation whenever possible while avoiding exploitation. The other entries either were too trusting and open to exploitation or were too aggressive and knocked one another out.

In spite of all this, we believe that tit for tat is a flawed strategy. The slightest possibility of a mistake or a misperception results in a complete breakdown in the success of tit for tat. This flaw was not apparent in the artificial setting of a computer tournament,

---

individual. You don't think there is an afterlife? Pascal's Wager says that the consequences of acting on that assumption can be quite drastic, so why take the chance.

* Since every loser must be paired with a winner, it must be the case that some contestant will have more wins than losses, else there will be more losses than wins overall. (The only exception is when every single match is a tie.)

because mistakes and misperceptions did not arise. But when tit for tat is applied to real-world problems, mistakes and misperceptions cannot be avoided, and the result can be disastrous.

The problem with tit for tat is that any mistake "echoes" back and forth. One side punishes the other for a defection, and this sets off a chain reaction. The rival responds to the punishment by hitting back. This response calls for a second punishment. At no point does the strategy accept a punishment without hitting back.

Suppose, for example, that both Flood and Dresher start out playing tit for tat. No one initiates a defection, and all goes well for a while. Then, in round 11, say, suppose Flood chooses Defect by mistake, or Flood chooses Cooperate but Dresher mistakenly thinks Flood chose Defect. In either case, Dresher will play Defect in round 12, but Flood will play Cooperate because Dresher played Cooperate in round 11. In round 13 the roles will be switched. The pattern of one playing Cooperate and the other playing Defect will continue back and forth, until another mistake or misperception restores cooperation or leads both to defect.

Such cycles or reprisals are often observed in real-life feuds between Israelis and Arabs in the Middle East, or Catholics and Protestants in Northern Ireland, or Hindus and Muslims in India. Along the West Virginia–Kentucky border, we had the memorable feud between the Hatfields and the McCoys. And in fiction, Mark Twain's Grangerfords and Shepherdsons offer another vivid example of how tit for tat behavior can end in a cycle of reprisals. When Huck Finn tries to understand the origins of the Grangerford-Shepherdson feud, he runs into the chicken-or-egg problem:

> "What was the trouble about, Buck?—land?"
>
> "I reckon maybe—I don't know."
>
> "Well, who done the shooting? Was it a Grangerford or a Shepherdson?"
>
> "Laws, how do I know? It was so long ago."
>
> "Don't anybody know?"
>
> "Oh, yes, pa knows, I reckon, and some of the other old people; but they don't know now what the row was about in the first place."

What tit for tat lacks is a way of saying "Enough is enough." It is too provocable, and not forgiving enough. And indeed, subsequent versions of Axelrod's tournament, which allowed possibilities of mistakes and misperceptions, showed other, more generous strategies to be superior to tit for tat.*

Here we might even learn something from monkeys. Cotton-top tamarin monkeys were placed in a game where each had the opportunity to pull a lever that would give the other food. But pulling the lever required effort. The ideal for each monkey would be to shirk while his partner pulled the lever. But the monkeys learned to cooperate in order to avoid retaliation. Tamarin cooperation remained stable as long as there were no more than two consecutive defections by one player, a strategy that resembles tit for two tats.[9]

MORE RECENT EXPERIMENTS

Thousands of experiments on the prisoners' dilemma have been performed in classrooms and laboratories, involving different numbers of players, repetitions, and other treatments. Here are some important findings.[10]

First and foremost is that cooperation occurs significantly often, even when each pair of players meets only once. On average, almost half of the players choose the cooperative action. Indeed, the most striking demonstration of this was on the Game Show Network's production of *Friend or Foe*. In this show, two-person

---

* In 2004, Graham Kendall at Nottingham ran a contest to celebrate the twentieth anniversary of Axelrod's original tournament. It was "won" by a group from England's Southampton University. The Southampton group submitted multiple entries, sixty in all. There were 59 drones and 1 queen. All their entries started with an unusual pattern so they would recognize each other. Then the drone programs sacrificed themselves so that the queen would do well. The drone programs also refused to cooperate with any rival program so as to knock down the opponents' scores. While having an army of drones prepared to sacrifice themselves on your behalf is one way to increase your payoff, it doesn't tell us much about how to play a prisoners' dilemma.

teams were asked trivia questions. The money earned from correct answers went into a "trust fund," which over the 105 episodes ranged from $200 to $16,400. To divide the trust fund, the two contestants played a one-shot dilemma.

Each privately wrote down "friend" or "foe." When both wrote down friend, the pot was split evenly. If one wrote down foe while the other wrote friend, the person writing foe would get the whole pot. But if both wrote foe, then neither would get anything. Whatever the other side does, you get at least as much, and possibly more, by writing down foe than if you wrote friend. Yet almost half the contestants wrote down friend. Even as the pot grew larger there was no change in the likelihood of cooperation. People were as likely to cooperate when the fund was below $3,000 as they were when it was above $5,000. These were some of the findings in a pair of studies by Professors Felix Oberholzer-Gee, Joel Waldfogel, Matthew White, and John List.[11]

If you are wondering how watching television counts as academic research, it turns out that more than $700,000 was paid out to contestants. This was the best-funded experiment on the prisoners' dilemma, ever. There was much to learn. It turns out that women were much more likely to cooperate than men, 53.7 percent versus 47.5 percent in season 1. The contestants in season 1 didn't have the advantage of seeing the results from the other matches before making their decision. But in season 2, the results of the first 40 episodes had been aired and this pattern became apparent. The contestants had learned from the experience of others. When the team consisted of two women, the cooperation rate rose to 55 percent. But when a woman was paired with a guy, her cooperation rate fell to 34.2 percent. And the guy's rate fell, too, down to 42.3 percent. Overall, cooperation dropped by ten points.

When a group of subjects is assembled and matched pairwise a number of times, with different pairings at different times, the proportion choosing cooperation generally declines over time. However, it does not go to zero, settling instead on a small set of persistent cooperators.

If the same pair plays the basic dilemma game repeatedly, they often build up to a significant sequence of mutual cooperation, until one player defects near the end of the sequence of repetitions. This happened in the very first experiment conducted on the dilemma. Almost immediately after they had thought up the game, Flood and Dresher recruited two of their colleagues to play the dilemma game a hundred times.[12] On 60 of these rounds, both players chose Cooperate. A long stretch of mutual cooperation lasted from round 83 to round 98, until one player sneaked in a defection in round 99.

Actually, according to the strict logic of game theory, this should not have happened. When the game is repeated exactly 100 times, it is a sequence of simultaneous-move games, and we can apply the logic of backward reasoning to it. Look ahead to what will happen on the 100th play. There are no more games to come, so defection cannot be punished in any future rounds. Dominant strategy calculations dictate that both players should choose Defect on the last round. But once that is a given, the 99th round becomes effectively the last round. Although there is one more round to come, defection on the 99th round is not going to be selectively punished by the other player in the 100th round because his choice in that round is foreordained. Therefore the logic of dominant strategies applies to the 99th round. And one can work back this sequential logic all the way to round 1. But in actual play, both in the laboratory and the real world, players seem to ignore this logic and achieve the benefits of mutual cooperation. What may seem at first sight to be irrational behavior—departing from one's dominant strategy—turns out to be a good choice, so long as everyone else is similarly "irrational."

Game theorists suggest an explanation for this phenomenon. The world contains some "reciprocators," people who will cooperate so long as the other does likewise. Suppose you are not one of these relatively nice people. If you behaved true to your type in a finitely repeated game of prisoners' dilemma, you would start cheating right away. That would reveal your nature to the other player. To hide the truth (at least for a while), you have to behave nicely. Why would you want to do that? Suppose you started by acting nicely. Then the other player, even if he is not a reciprocator, would think it possible that you are one of the few nice people around. There are real gains to be had by cooperating for a while, and the other player would plan to reciprocate your niceness to achieve these gains. That helps you, too. Of course you are planning to sneak in a defection near the end of the game, just as the other player is. But you two can still have an initial phase of mutually beneficial cooperation. While each side is waiting to take advantage of the other, both are benefiting from this mutual deception.

In some experiments, instead of pairing each subject in the group with another person and playing several two-person dilemmas, the whole group is engaged in one large multiperson dilemma. We mention a particularly entertaining and instructive instance from the classroom. Professor Raymond Battalio of Texas A&M University had his class of 27 students play the following game.[13] Each student owned a hypothetical firm and had to decide (simultaneously and independently, by writing on a slip of paper) whether to produce 1 and help keep the total supply low and the price high or produce 2 and gain at the expense of others. Depending on the total number of students producing 1, money would be paid to students according to the following table:

Number of students who write 1	Payoff to each student who writes 1	Payoff to each student who writes 2
0		$0.50
1	$0.04	$0.54
2	$0.08	$0.58
3	$0.12	$0.62
. . .	. . .	. . .
25	$1.00	$1.50
26	$1.04	$1.54
27	$1.08	

This is easier to see and more striking in a chart:

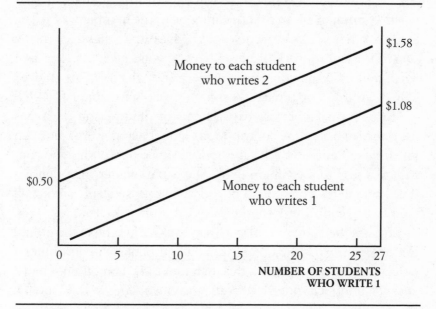

The game is "rigged" so that students who write 2 (Defect) always get 50 cents more than those who write 1 (Cooperate), but the more of them that write 2, the less their collective gain. Suppose all 27 start planning to write 1, so each would get $1.08. Now one thinks of sneaking a switch to 2. There would be 26 1s, and each would get $1.04 (4 cents less than in the original plan), but the switcher would get $1.54 (46 cents more). The same is true irrespective of the initial number of students thinking of writing 1 versus 2. Writing 2 is a dominant strategy. Each student who switches from writing 1 to writing 2 increases his own payout by 46 cents but decreases that of each of his 26 colleagues by 4 cents—the group as a whole loses 58 cents. By the time everyone acts selfishly, each maximizing his own payoff, they each get 50 cents. If they could have successfully conspired and acted so as to minimize their individual payoff, they would each receive $1.08. How would you play?

In some practice plays of this game, first without classroom discussion and then with some discussion to achieve a "conspiracy," the number of cooperative students writing 1 ranged from 3 to a maximum of 14. In a final binding play, the number was 4. The total payout was $15.82, which is $13.34 less than that from totally successful collusion. "I'll never trust anyone again as long as I live," muttered the conspiracy leader. And what was his choice? "Oh, I wrote 2," he replied. Yossarian would have understood.

More recent laboratory experiments of multiperson dilemmas use a format called the contribution game. Each player is given an initial stake, say $10. Each can choose to keep part of this and contribute a part to a common pool. The experimenter then doubles the accumulated common pool and divides this equally among all the players, contributors and noncontributors alike.

Suppose there are four players, say A, B, C, and D, in the group. Regardless of what the others are doing, if person A contributes a dollar to the common pool, this increases the common pool by $2 after the doubling. But $1.50 of the increment goes to B, C, and D; A gets only 50 cents. Therefore A loses out by raising his contribution; conversely he would gain by lowering it. And that is true no matter how much, if anything, the others are contributing. In other words, contributing nothing is the dominant strategy for A. The same goes for B, C, and D. This logic says that each should hope to become a "free rider" on the efforts of the others. If all four play their dominant strategy, the common pool is empty and each simply keeps the initial stake of $10. When everyone tries to be a free rider, the bus stays in the garage. If everyone had put all of their initial stakes in the common pool, the pool after doubling would be $80 and the share of each would be $20. But each has the personal incentive to cheat on such an arrangement. This is their dilemma.

The contribution game is not a mere curiosity of the laboratory or theory; it is played in the real world in social interactions where some communal benefit can be achieved by voluntary contributions from members of the group, but the benefit cannot be withheld from those who did not contribute. Flood control in a village, or conservation of natural resources, are cases in point: it is not possible to build levees or dams so that flood waters will selectively go to the fields of those who did not help in the construction, and it is not practicable to withhold gas or fish in the future from someone who consumed too much in the past. This creates a multiperson dilemma: each player has the temptation to shirk or withhold his contribution, hoping to enjoy the benefits of the others' contributions. When they all think this way, the total of contributions is meager or even zero, and they all suffer. These situations are ubiquitous, and of such magnitude that all of social theory and policy needs a good understanding of how the dilemmas might be resolved.

In what is perhaps the most interesting variant of the game, players are given an opportunity to punish those who cheat on an implicit social contract of cooperation. However, they must bear a personal cost to do so. After the contribution game is played, the players are informed about the individual contributions of other players. Then a second phase is played, where each player can take an action to lower the payoffs of other players at a cost to himself of so many cents (typically 33) per dollar reduction chosen. In other words, if player A chooses to reduce B's payoff by three dollars, then A's payoff is reduced by one dollar. These reductions are not reallocated to anyone else; they simply return to the general funds of the experimenter.

The results of the experiment show that people engage in a significant amount of punishment of "social cheaters," and that the prospect of the punishment increases the contributions in the first phase of the game dramatically. Such punishments seem to be an effective mechanism for achieving cooperation that benefits the whole group. But the fact that individuals carry them out is surprising at first. The act of punishing others at a personal cost is itself a contribution for the general benefit, and it is a dominated strategy; if it succeeds in eliciting better behavior from the cheater in the future, its benefits will be for the group as a whole, and the punisher will get only his small share of this benefit. Therefore the punishment has to be the result of something other than a selfish calculation. That is indeed the case. Experiments on this game have been conducted while the players' brains were being imaged by PET scan.[14] These revealed that the act of imposing the penalty activated a brain region called the dorsal striatum, which is involved in experiencing pleasure or satisfaction. In other words, people actually derive a psychological benefit or pleasure from punishing social cheaters. Such an instinct must have deep biological roots and may have been selected for an evolutionary advantage.[15]

HOW TO ACHIEVE COOPERATION

These examples and experiments have suggested several preconditions and strategies for successful cooperation. Let us develop the concepts more systematically and apply them to some more examples from the real world.

Successful punishment regimes must satisfy several requirements. Let us examine these one by one.

Detection of cheating: Before cheating can be punished, it must be detected. If detection is fast and accurate, the punishment can be immediate and accurate. That reduces the gain from cheating while increasing its cost, and thus increases the prospects for successful cooperation. For example, airlines constantly monitor each other's fares; if American were to lower its fare from New York to Chicago, United can respond in under five minutes. But in other contexts, firms that want to cut their prices can do so in secret deals with the customers, or hide their price cuts in a complicated deal involving many dimensions of delivery time, quality, warranties, and so on. In extreme situations, each firm can only observe its own sales and profits, which can depend on some chance elements as well as on other firms' actions. For example, how much one firm sells can depend on the vagaries of demand, not just on other firms' secret price cuts. Then detection and punishment become not only slow but also inaccurate, raising the temptation to cheat.

Finally, when three or more firms are simultaneously in the market, they must find out not only whether cheating has occurred but who has cheated. Otherwise any punishments cannot be targeted to hurt the miscreant but have to be blunt, perhaps unleashing a price war that hurts all.

Nature of punishment: Next, there is the choice of punishment. Sometimes the players have available to them actions that hurt others, and these can be invoked after an instance of cheating even in a one-time interaction. As we pointed out in the dilemma in *L.A. Confidential*, the friends of Sugar and Tyrone will punish Leroy when he emerges from jail after his light sentence for turning state's witness. In the Texas A&M classroom experiment, if the students could detect who had reneged on the conspiracy for all of them to write 1, they could inflict social sanctions such as ostracism on the cheaters. Few students would risk that for the sake of an extra 50 cents.

Other kinds of punishments arise within the structure of the game. Usually this happens because the game is repeated, and the gain from cheating in one play leads to a loss in future plays. Whether this is enough to deter a player who is contemplating cheating depends on the sizes of the gains and losses and on the importance of the future relative to the present. We will return to this aspect soon.

Clarity: The boundaries of acceptable behavior, and the consequences of cheating, should be clear to a prospective cheater. If these things are complex or confusing, the player may cheat by mistake or fail to make a rational calculation and play by some hunch. For example, suppose Rainbow's End and B. B. Lean are playing their price-setting game repeatedly, and RE decides that it will infer that BB has cheated if RE's discounted mean of profits from the last seventeen months is 10 percent less than the average real rate of return to industrial capital over the same period. BB does not know this rule directly; it must infer what rule RE is using by observing RE's actions. But the rule stated here is too complicated for BB to figure out. Therefore it is not a good deterrent against BB's cheating. Something like tit for tat is abundantly clear: if BB cheats, it will see RE cutting its price the very next time.

Certainty: Players should have confidence that defection will be punished and cooperation rewarded. This is a major problem in some international agreements like trade liberalization in the

World Trade Organization (WTO). When one country complains that another has cheated on the trade agreement, the WTO initiates an administrative process that drags on for months or years. The facts of the case have little bearing on the judgment, which usually depends more on dictates of international politics and diplomacy. Such enforcement procedures are unlikely to be effective.

Size: How harsh should such punishments be? It might seem that there is no limit. If the punishment is strong enough to deter cheating, it need never actually be inflicted. Therefore it may as well be set at a sufficiently high level to ensure deterrence. For example, the WTO could have a provision to nuke any nation that breaks its undertakings to keep its protective tariffs at the agreed low levels. Of course you recoil in horror at the suggestion, but that is at least in part because you think it possible that some error may cause this to happen. When errors are possible, as they always are in practice, the size of the punishment should be kept as low as is compatible with successful deterrence in most circumstances. It may even be optimal to forgive occasional defection in extreme situations—for example, a firm that is evidently fighting for its survival may be allowed some price cuts without triggering reactions from rivals.

Repetition: Look at the pricing game between Rainbow's End and B. B. Lean. Suppose they are going merrily along from one year to the next, holding prices at their joint best, $80. One year the management of RE considers the possibility of cutting the price to $70. They reckon that this will yield them an extra profit of $110,000 − $72,000 = $38,000. But that can lead to a collapse of trust. RE should expect that in future years BB will also choose $70, and each will make only $70,000 each year. If RE had kept to the original arrangement, each would have kept on making $72,000. Thus RE's price cutting will cost it $72,000 − $70,000 = $2,000 every year in the future. Is a one-time gain of $38,000 worth the loss of $2,000 every year thereafter?

One key variable that determines the balance of present and

future considerations is the interest rate. Suppose the interest rate is 10% per year. Then RE can stash away its extra $38,000 and earn $3,800 every year. That comfortably exceeds the loss of $2,000 in each of those years. Therefore cheating is in RE's interest. But if the interest rate is only 5% per year, then the $38,000 earns only $1,900 in each subsequent year, less than the loss of $2,000 due to the collapse of the arrangement; so RE does not cheat. The interest rate at which the two magnitudes just balance is 2/38 = 0.0526, or 5.26% per year.

The key idea here is that when interest rates are low, the future is relatively more valuable. For example, if the interest rate is 100%, then the future has low value relative to the present—a dollar in a year's time is worth only 50 cents right now because you can turn the 50 cents into a dollar in a year by earning another 50 cents as interest during the year. But if the interest rate is zero, then a dollar in a year's time is worth the same as a dollar right away.*

In our example, for realistic interest rates a little above 5%, the temptation for each firm to cut the price by $10 below their joint best price of $80 is quite finely balanced, and collusion in a repeated game may or may not be possible. In chapter 4 we will see how low the price can fall if there is no shadow of the future and the temptation to cheat is irresistible.

Another relevant consideration is the likelihood of continuation of the relationship. If the shirt is a transient fashion item that may not sell at all next year, then the temptation to cheat this year is not offset by any prospect of future losses.

But Rainbow's End and B. B. Lean sell many items besides this shirt. Won't cheating on the shirt price bring about retaliation on all those other items in the future? And isn't the prospect of this huge retaliation enough to deter the defection? Alas, the usefulness of multiproduct interactions for sustaining cooperation is not

* If you read the financial press, you have often seen the statement: "Interest rates and bond prices move in opposite directions." The lower the interest rate, the higher the prices of bonds. And bonds, being promises of future income, reflect the importance of the future. This is another way to remember the role of interest rates.

so simple. The prospect of multiproduct retaliation goes hand in hand with that of immediate gains from simultaneous cheating in all of those dimensions, not just one. If all the products had identical payoff tables, then the gains and losses would both increase by a factor equal to the number of products, and so whether the balance is positive or negative would not change. Therefore successful punishments in multiproduct dilemmas must depend in a more subtle way on differences among the products.

A third relevant consideration is the expected variation in the size of the business over time. This has two aspects—steady growth or decline, and fluctuations. If the business is expected to grow, then a firm considering defection now will recognize that it stands to lose more in the future due to the collapse of the cooperation and will be more hesitant to defect. Conversely, if the business is on a path of decline, then firms will be more tempted to defect and take what they can now, knowing that there is less at stake in the future. As for fluctuations, firms will be more tempted to cheat when a temporary boom arrives; cheating will bring them larger immediate profits, whereas the downside from the collapse of the cooperation will hit them in the future, when the volume of business will be only the average, by definition of the average. Therefore we should expect that price wars will break out during times of high demand. But this is not always the case. If a period of low demand is caused by a general economic downturn, then the customers will have lower incomes and may become sharper shoppers as a result—their loyalties to one firm or the other may break down, and they may respond more quickly to price differences. In that case, a firm cutting its price can expect to attract more customers away from its rival, and thereby reap a larger immediate gain from defection.

Finally, the composition of the group of players is important. If this is stable and expected to remain so, that is conducive to the maintenance of cooperation. New players who do not have a stake or a history of participation in the cooperative arrangement are less likely to abide by it. And if the current group of players expects new ones to enter and shake up the tacit cooperation in the future, that increases their own incentive to cheat and take some extra benefit right now.

SOLUTION BY KANTIAN CATEGORICAL IMPERATIVE?

It is sometimes said that the reason some people cooperate in the prisoners' dilemma is that they are making the decision not only for themselves but for the other player. That is wrong in point of fact, but the person is acting as if this is the case.

The person truly wants the other side to cooperate and reasons to himself that the other side is going through the same logical decision process that he is. Thus the other side must come to the same logical conclusion that he has. Hence if the player cooperates, he reasons that the other side will do so as well, and if he defects, he reasons that it will cause the other side to defect. This is similar to the categorical imperative of the German philosopher Immanuel Kant: "Take only such actions as you would like to see become a universal law."

Of course, nothing could be further from the truth. The actions of one player have no effect whatsoever on the other player in the game. Still people think that somehow their actions can influence the choice of others, even when their actions are invisible.

The power of this thinking was revealed in an experiment done with Princeton undergraduates by Eldar Shafir and Amos Tversky.[16] In their experiment, they put students in a prisoners' dilemma game. But unlike the usual dilemma, in some treatments they told one side what the other had done. When students were told that the other side had defected on them, only 3 percent responded with cooperation. When told that the other side had cooperated, this increased cooperation levels up to 16 percent. It was still the case that the large majority of students were willing to act selfishly. But many were willing to reciprocate the cooperative behavior exhibited by the other side, even at their own expense.

What do you think would happen when the students were not told anything about the other player's choice at all? Would the percentage of cooperators be between 3 and 16 percent? No; it rose to

37 percent. At one level, this makes no sense. If you wouldn't cooperate when you learned that the other side had defected and you wouldn't cooperate when you learned that the other side had cooperated, why would you then cooperate when you don't know what the other side had done?

Shafir and Tversky call this "quasi-magical" thinking—the idea that by taking some action, you can influence what the other side will do. People realize they can't change what the other side has done once they've been told what the other side has done. But if it remains open or undisclosed, then they imagine that their actions might have some influence—or that the other side will somehow be employing the same reasoning chain and reach the same outcome they do. Since Cooperate, Cooperate is preferred to Defect, Defect, the person chooses Cooperate.

We want to be clear that such logic is completely illogical. What you do and how you get there has no impact at all on what the other side thinks and acts. They have to make up their mind without reading your mind or seeing your move. However, the fact remains that if the people in a society engage in such quasi-magical thinking, they will not fall victim to many prisoners' dilemmas and all will reap higher payoffs from their mutual interactions. Could it be that human societies deliberately instill such thinking into their members for just such an ultimate purpose?

DILEMMAS IN BUSINESS

Armed with the tool kit of experimental findings and theoretical ideas in the previous sections, let us step outside the laboratory and look at some instances of prisoners' dilemmas in the real world and attempts at resolving them.

Let us begin with the dilemma of rival firms in an industry. Their joint interests are best served by monopolizing or cartelizing the industry and keeping prices high. But each firm can do better for itself by cheating on such an agreement and sneaking in price cuts to steal business from its rivals. What can the firms do? Some factors conducive to successful collusion, such as growing demand or lack of disruptive entry, may be at least partially outside their control. But they can try to facilitate the detection of cheating and devise effective punishment strategies.

Collusion is easier to achieve if the firms meet regularly and communicate. Then they can negotiate and compromise on what are acceptable practices and what constitutes cheating. The process of negotiation and its memory contributes to clarity. If something occurs that looks prima facie like cheating, another meeting can help clarify whether it is something extraneous, an innocent error by a participant, or deliberate cheating. Therefore unnecessary punishments can be avoided. And the meeting can also help the group implement the appropriate punishment actions.

The problem is that the group's success in resolving their dilemma harms the general public's interest. Consumers must pay higher prices, and the firms withhold some supply from the market to keep the price high. As Adam Smith said, "People of the same trade seldom meet together, even for merriment and diversion, but the conversation ends in a conspiracy against the public, or in some contrivance to raise prices."[17] Governments that want to protect the general public interest get into the game and enact antitrust laws that make it illegal for firms to collude in this way.* In the United States, the Sherman Antitrust Act prohibits conspiracies "in restraint of trade or commerce," of which price fixing or market-share fixing conspiracies are the prime instance and the ones most frequently attempted. In fact the Supreme Court has ruled that not only are explicit agreements of this kind forbidden, but also any explicit or tacit arrangement among firms that has the effect of price fixing is a violation of the Sherman Act, regardless of its primary intent. Violation of these laws can lead to jail terms for the firms' executives, not just fines for the corporations that are impersonal entities.

Not that firms don't try to get away with the illegal practices. In

---

* Not all governments care enough about the general interest. Some are beholden to the producers' special interests and ignore or even facilitate cartels. We won't name any, lest they ban our book in their countries!

1996 Archer Daniels Midland (ADM), a leading American proces-
sor of agricultural products, and their Japanese counterpart, Aji-
nomoto were caught in just such a conspiracy. They had arranged
market sharing and pricing agreements for various products such as
lysine (which is produced from corn and used for fattening up
chickens and pigs). The aim was to keep the prices high at the
expense of their customers. Their philosophy was: "The competitors are our friends, and the customers are our enemies." The companies' misdeeds came to light because one of the ADM negotiators
became an informant for the FBI and arranged for many of the
meetings to be recorded for audio and sometimes also video.[18]

An instance famous in antitrust history and business school case
studies concerns the large turbines that generate electricity. In the
1950s, the U.S. market for these turbines consisted of three firms:
GE was the largest, with a market share of around 60 percent,
Westinghouse was the next, with approximately 30 percent, and
Allied-Chalmers had about 10 percent. They kept these shares,
and obtained high prices, using a clever coordination device.
Here's how it worked. Electric utilities invited bids for the turbines they intended to buy. If the bid was issued during days 1–17
of a lunar month, Westinghouse and Allied-Chalmers had to put in
very high bids that would be sure losers, and GE was the conspiracy's chosen winner by making the lowest bid (but still at a monop-
olist's price allowing big profits). Similarly, Westinghouse was the
designated winner in the conspiracy if the bid was issued during
days 18–25, and Allied-Chalmers for days 26–28. Since the utilities
did not issue their solicitations for bids according to the lunar cal-
endar, over time each of the three producers got the agreed market
share. Any cheating on the agreement would have been immediately visible to the rivals. But, so long as the Department of Justice
did not think of linking the winners to the lunar cycles, it was safe
from detection by the law. Eventually the authorities did figure it
out, some executives of the three firms went to jail, and the profitable conspiracy collapsed. Different schemes were tried later.[19]

A variant of the turbine scheme later appeared in the bidding at
the airwave spectrum auctions in 1996–1997. A firm that wanted

the right for the licenses in a particular location would signal to the other firms its determination to fight for that right by using the telephone area code for that location as the last three digits of its bid. Then the other firms would let it win. So long as the same set of firms interacts in a large number of such auctions over time and so long as the antitrust authorities do not figure it out, the scheme may be sustainable.[20]

More commonly, the firms in an industry try to attain and sustain implicit or tacit agreements without explicit communication. This eliminates the risk of criminal antitrust action, although the antitrust authorities can take other measures to break up even implicit collusion. The downside is that the arrangement is less clear and cheating is harder to detect, but firms can devise methods to improve both.

Instead of agreeing on the prices to be charged, the firms can agree on a division of the market, by geography, product line, or some similar measure. Cheating is then more visible—your salespeople will quickly come to know if another company has stolen some of your assigned market.

Detection of price cuts, especially in the case of retail sales, can be simplified, and retaliation made quick and automatic, by the use of devices like "matching or beating competition" policies and most-favored-customer clauses. Many companies selling household and electronic goods loudly proclaim that they will beat any competitor's price. Some even guarantee that if you find a better price for the same product within a month after your purchase, they will refund the difference, or in some cases even double the difference. At first sight, these strategies seem to promote competition by guaranteeing low prices. But a little game-theoretic thinking shows that in reality they can have exactly the opposite effect. Suppose Rainbow's End and B. B. Lean had such policies, and their tacit agreement was to price the shirt at $80. Now each firm knows that if it sneaks a cut to $70, the rival will find out about it quickly; in fact the strategy is especially clever in that it puts the customers, who have the best natural incentive to locate low prices, in charge of detecting cheating. And the prospective defector also knows that the rival can retaliate instantaneously by cutting its own price; it does not have to wait until next year's catalog is printed. Therefore the cheater is more effectively deterred.

Promises to meet or beat the competition can be clever and indirect. In the competition between Pratt & Whitney (P&W) and Rolls-Royce (RR) for jet aircraft engines to power Boeing 757 and 767 planes, P&W promised all prospective purchasers that its engines would be 8 percent more fuel-efficient than those of RR, otherwise P&W would pay the difference in fuel costs.[21]

A most-favored-customer clause says that the seller will offer to all customers the best price they offer to the most favored ones. Taken at face value, it seems that the manufacturers are guaranteeing low prices. But let's look deeper. The clause means that the manufacturer cannot compete by offering selective discounts to attract new customers away from its rival, while charging the old higher price to its established clientele. They must make general price cuts, which are more costly, because they reduce the profit margin on all sales. You can see the advantage of this clause to a cartel: the gain from cheating is less, and the cartel is more likely to hold.

A branch of the U.S. antitrust enforcement system, the Federal Trade Commission, considered such a clause that was being used by DuPont, Ethyl, and other manufacturers of antiknock additive compounds in gasoline. The commission ruled that there was an anticompetitive effect and forbade the companies from using such clauses in their contracts with customers.*

TRAGEDIES OF THE COMMONS

Among the examples at the start of this chapter, we mentioned problems like overfishing that arise because each person stands to benefit by taking more, while the costs of his action are visited upon

* This ruling was not without some controversy. The commission's chairman, James Miller, dissented. He wrote that the clauses "arguably reduce buyers' search costs and facilitate their ability to find the best price-value among buyers." For more information, see "In the matter of Ethyl Corporation et al.," FTC Docket 9128, *FTC Decisions* 101 (January–June 1983): 425–686.

numerous others or on future generations. University of California biologist Garrett Harding called this the "tragedy of the commons," using among his examples the overgrazing of commonly owned land in fifteenth- and sixteenth-century England.[22] The problem has become well known under this name. Today the problem of global warming is an even more serious example; no one gets enough private benefit from reducing carbon emissions, but all stand to suffer serious consequences when each follows his self-interest.

This is just a multiperson prisoners' dilemma, like the one Yossarian faced in *Catch-22* about risking his life in wartime. Of course societies recognize the costs of letting such dilemmas go unresolved and make attempts to achieve better outcomes. What determines whether these attempts succeed?

Indiana University political scientist Elinor Ostrom and her collaborators and students have conducted an impressive array of case studies of attempts to resolve dilemmas of the tragedy of the commons—that is, to use and conserve common property resources in their general interest and avoid overexploitation and rapid depletion. They studied some successful and some unsuccessful attempts of this kind and derived some prerequisites for cooperation.[23]

First, there must be clear rules that identify who is a member of the group of players in the game—those who have the right to use the resource. The criterion is often geography or residence but can also be based on ethnicity or skills, or membership may be sold by auction or for an entry fee.*

---

* The establishment of property rights is what actually happened in England. In two waves of "enclosures," first by local aristocrats during the Tudor period and later by acts of Parliament in the eighteenth and nineteenth centuries, previously common land was given to private owners. When land is private property, the invisible hand will shut the gate to just the right extent. The owner will charge grazing fees to maximize his rental income, and this will cut back on use. This will enhance overall economic efficiency but alter the distribution of income; the grazing fees will make the owner richer and the herdsmen poorer. Even absent concern for the distributional consequences, this approach is not always feasible. Property rights over the high seas or SO_2 and CO_2 emissions are hard to define and enforce in the absence of an international government: fish

Second, there must be clear rules defining permissible and forbidden actions. These include restrictions on time of use (open or closed seasons for hunting or fishing, or what kinds of crops can be planted and any requirements to keep the land fallow in certain years), location (a fixed position or a specified rotation for inshore fishing), the technology (size of fishing nets), and, finally, the quantity or fraction of the resource (amount of wood from a forest that each person is allowed to gather and take away).

Third, a system of penalties for violation of the above rules must be clear and understood by all parties. This need not be an elaborate written code; shared norms in stable communities can be just as clear and effective. The sanctions used against rule breakers range from verbal chastisement or social ostracism to fines, the loss of future rights, and, in some extreme cases, incarceration. The severity of each type of sanction can also be adjusted. An important principle is graduation. The first instance of suspected cheating is most commonly met simply by a direct approach to the violator and a request to resolve the problem. The fines for a first or second offense are low and are ratcheted up only if the infractions persist or get more blatant and serious.

Fourth, a good system to detect cheating must be in place. The best method is to make detection automatic in the course of the players' normal routine. For example, a fishery that has good and bad areas may arrange a rotation of the rights to the good areas. Anyone assigned to a good spot will automatically notice if a violator is using it and has the best incentive to report the violator to others and get the group to invoke the appropriate sanctions. Another example is the requirement that harvesting from forests or similar common areas must be done in teams; this facilitates mutual monitoring and eliminates the need to hire guards.

Sometimes the rules on what is permissible must be designed in

---

and pollutants move from one ocean to another, SO_2 is carried by the wind across borders, and CO_2 from any country rises to the same atmosphere. For this reason, whaling, acid rain, or global warming must be handled by more direct controls, but securing the necessary international agreements is no easy matter.

the light of feasible methods of detection. For example, the size of a fisherman's catch is often difficult to monitor exactly and difficult even for a well-intentioned fisherman to control exactly. Therefore rules based on fish quantity quotas are rarely used. Quantity quotas perform better when quantities are more easily and accurately observable, as in the case of water supplied from storage and harvesting of forest products.

Fifth, when the above categories of rules and enforcement systems are being designed, information that is easily available to the prospective users proves particularly valuable. Although each may have the temptation after the fact to cheat, they all have a common prior interest to design a good system. They can make the best use of their knowledge of the resource and of the technologies for exploiting it, the feasibility of detecting various infractions, and the credibility of various kinds of sanctions in their group. Centralized or top-down management has been demonstrated to get many of these things wrong and therefore perform poorly.

While Ostrom and her collaborators are generally optimistic about finding good solutions to many problems of collective action using local information and systems of norms, she gives a salutary warning against perfection: "The dilemma never fully disappears, even in the best operating systems. . . . No amount of monitoring or sanctioning reduces the temptation to zero. Instead of thinking of overcoming or conquering tragedies of the commons, effective governance systems cope better than others."

NATURE RED IN TOOTH AND CLAW

As you might expect, prisoners' dilemmas arise in species other than humans. In matters like building shelter, gathering food, and avoiding predators, an animal can act either selfishly in the interest of itself or its immediate kin, or in the interest of a larger group. What circumstances favor good collective outcomes? Evolutionary biologists have studied this question and found some fascinating examples and ideas. Here is a brief sample.[24]

The British biologist J. B. S. Haldane was once asked whether he would risk his life to save a fellow human being and replied: "For more than two brothers, or more than eight cousins, yes." You share half of your genes with a brother (other than an identical twin), and one-eighth of your genes with a cousin; therefore such action increases the expected number of copies of your genes that propagate to the next generation. Such behavior makes excellent biological sense; the process of evolution would favor it. This purely genetic basis for cooperative behavior among close kin explains the amazing and complex cooperative behavior observed in ant colonies and beehives.

Among animals, altruism without such genetic ties is rare. But reciprocal altruism can arise and persist among members of a group of animals with much less genetic identity, if their interaction is sufficiently stable and long-lasting. Hunting packs of wolves and other animals are examples of this. Here is an instance that is a bit gruesome but fascinating: Vampire bats in Costa Rica live in colonies of a dozen or so but hunt individually. On any day, some may be lucky and others unlucky. The lucky ones return to the hollow trees where the whole group lives and can share their luck by regurgitating the blood they have brought from their hunt. A bat that does not get a blood meal for three days is at risk of death. The colonies develop effective practices of mutual "insurance" against this risk by such sharing.[25]

University of Maryland biologist Gerald Wilkinson explored the basis of this behavior by collecting bats from different locations and putting them together. Then he systematically withheld blood from some of them and saw whether others shared with them. He found that sharing occurred only when the bat was on the verge of death, and not earlier. Bats seem to be able to distinguish real need from mere temporary bad luck. More interesting, he found that sharing occurred only among bats that already knew each other from their previous group, and that a bat was much more likely to share with another bat that had come to its aid in the past. In other words, the bats are able to recognize other individual bats and keep score of their past behavior in order to develop an effective system of reciprocal altruism.

CASE STUDY: THE EARLY BIRD KILLS THE GOLDEN GOOSE

The Galápagos Islands are the home of Darwin's finches. Life on these volcanic islands is difficult and so evolutionary pressures are high. Even a millimeter change in the beak of a finch can make all the difference in the competition for survival.*

Each island differs in its food sources, and finches' beaks reflect those differences. On Daphne Major, the primary food source is a cactus. Here the aptly named cactus finch has evolved so that its beak is ideally suited to gather the pollen and nectar of the cactus blossom.

The birds are not consciously playing a game against each other. Yet each adaptation of a bird's beak can be seen as its strategy in life. Strategies that provide an advantage in gathering food will lead to survival, a choice of mating partners, and more offspring. The beak of the finch is a result of this combination of natural and sexual selection.

Even when things seem to be working, genetics throws a few curveballs into the mix. There is the old saying that the early bird gets the worm. On Daphne Major, it was the early finch that got the nectar. Rather than wait until nine in the morning when the cactus blossoms naturally open for business, a dozen finches were trying something new. They were prying open the cactus blossom to get a head start.

At first glance, this would seem to give these birds an edge over their late-coming rivals. The only problem is that in the process of prying open the blossom, the birds would often snip the stigma. As Weiner explains:

> [The stigma] is the top of the hollow tube that pokes out like a tall straight straw from the center of each blossom. When the stigma is cut, the flower is sterilized. The male sex cells in the

---

* This example is motivated by Jonathan Weiner's wonderful book, *The Beak of the Finch: A Story of Evolution in Our Time* (New York: Knopf, 1994). See especially chapter 20: "The Metaphysical Crossbeak."

pollen cannot reach the female sex cells in the flower. The cactus flower withers without bearing fruit.[26]

When the cactus flowers wither, the main source of food disappears for the cactus finch. You can predict the end result of this strategy: no nectar, no pollen, no seeds, no fruit, and then no more cactus finch. Does that mean that evolution has led the finches into a prisoners' dilemma where the eventual outcome is extinction?

Case Discussion

Not quite, on two counts. Finches are territorial and so the finches (and their offspring) whose local cactus shut down may end up as losers. Killing next year's neighborhood food supply is not worth today's extra sip of pollen. Therefore these deviant finches would not appear to have a fitness advantage over the others. But that conclusion changes if this strategy ever becomes pervasive. The deviant finches will expand their search for food and even those finches that wait will not save their cactus's stigma. Given the famine that is sure to follow, the birds most likely to survive are those who started in the strongest position. The extra sip of nectar could make the difference.

What we have here is a cancerous adaptation. If it stays small, it can die out. But if it ever grows too large, it will become the fittest strategy on a sinking ship. Once it ever becomes advantageous even on a relative scale, the only way to get rid of it is to eliminate the entire population and start again. With no finches left on Daphne Major, there will be no one left to snip the stigmas and the cacti will bloom again. When two lucky finches alight on this island, they will have an opportunity to start the process from scratch.

The game we have here is a cousin to the prisoners' dilemma, a life and death case of the "stag hunt" game analyzed by the philosopher Jean-Jacques Rousseau.* In the stag hunt, if everyone works together to capture the stag, they succeed and all eat well. A

* There are other interpretations of Rousseau's stag hunt, to which we return in the history section of the next chapter.

problem arises if some hunters come across a hare along the way. If too many hunters are sidetracked chasing after hares, there won't be enough hunters left to capture the stag. In that case, everyone will do better chasing after rabbits. The best strategy is to go after the stag *if and only if* you can be confident that most everyone is doing the same thing. You have no reason not to chase after the stag, except if you lack confidence in what others will do.

The result is a confidence game. There are two ways it can be played. Everyone works together and life is good. Or everyone looks out for themselves and life is nasty, brutish, and short. This is not the classic prisoners' dilemma in which each person has an incentive to cheat no matter what others do. Here, there is no incentive to cheat, so long as you can trust others to do the same. But can you trust them? And even if you do, can you trust them to trust you? Or can you trust them to trust you to trust them? As FDR famously said (in a different context), we have nothing to fear but fear itself.

For more practice with prisoners' dilemmas, have a look at the following case studies in chapter 14: "What Price a Dollar?" and "The King Lear Problem."

CHAPTER 4

A Beautiful
Equilibrium

BIG GAME OF COORDINATION

Fred and Barney are Stone Age rabbit hunters. One evening, while carousing, they happen to engage in some shop talk. As they exchange information and ideas, they realize that by cooperating they could hunt much bigger game, such as stag or bison. One person on his own cannot expect any success hunting either stag or bison. But done jointly, each day's stag or bison hunting is expected to yield six times as much meat as a day's rabbit hunting by one person. Cooperation promises great advantage: each hunter's share of meat from a big-game hunt is three times what he can get hunting rabbits on his own.

The two agree to go big-game hunting together the following day and return to their respective caves. Unfortunately, they caroused too well, and both have forgotten whether they decided to go after stag or bison. The hunting grounds for the two species are in opposite directions. There were no cell phones in those days, and this was before the two became neighbors, so one could not quickly visit the other's cave to ascertain where to go. Each would have to make the decision the next morning in isolation.

Therefore the two end up playing a simultaneous-move game of deciding where to go. If we call each hunter's quantity of meat from a day's rabbit hunting 1, then the share of each from successful coordination in hunting either stag or bison is 3. So the payoff table of the game is as shown here:

		Barney's choice		
		Stag	Bison	Rabbit
Fred's choice	Stag	3 / 3	0 / 0	1 / 0
	Bison	0 / 0	3 / 3	1 / 0
	Rabbit	0 / 1	0 / 1	1 / 1

This game differs from the prisoners' dilemma of the previous chapter in many ways. Let us focus on one crucial difference. Fred's best choice depends on what Barney does, and vice versa. For neither player is there a strategy that is best regardless of what the other does; unlike in the prisoners' dilemma, this game has no dominant strategies. So each player has to think about the other's choice and figure out his own best choice in light of that.

Fred's thinking goes as follows: "If Barney goes to the grounds where the stags are, then I will get my share of the large catch if I go there too, but nothing if I go to the bison grounds. If Barney goes to the bison grounds, things are the other way around. Rather than take the risk of going to one of these areas and finding that Barney has gone to the other, should I go by myself after rabbits and make sure of my usual, albeit small, quantity of meat? In other words, should I take 1 for sure instead of risking either 3 or nothing? It depends on what I think Barney is likely to do, so let me put myself in his shoes (bare feet?) and think what he is thinking. Oh, he is wondering what I am likely to do and is trying to put himself in my shoes! Is there any end to this circular thinking about thinking?"

SQUARING THE CIRCLE

John Nash's beautiful equilibrium was designed as a theoretical way to square just such circles of thinking about thinking about other people's choices in games of strategy.* The idea is to look for an outcome where each player in the game chooses the strategy that best serves his or her own interest, in response to the other's strategy. If such a configuration of strategies arises, neither player has any reason to change his choice unilaterally. Therefore, this is a potentially stable outcome of a game where the players make individual and simultaneous choices of strategies. We begin by illustrating the idea with some examples of it in action. Later in this chapter we discuss how well it predicts outcomes in various games; we find reasons for cautious optimism and for making Nash equilibrium a starting point of the analysis of almost all games.

Let us develop the concept by considering a more general version of the pricing game between Rainbow's End and B. B. Lean. In chapter 3 we allowed each company the choice of just two prices for the shirt, namely $80 and $70. We also recognized the strength of the temptation for each to cut the price. Let us therefore allow more choices in a lower range, going in $1 steps from $42 to $38.† In the earlier example, when both charge $80, each sells 1,200 shirts. If one of them cuts its price by $1 while the other holds its price unchanged, then the price cutter gains 100 cus-

---

* For those readers who have not seen the movie *A Beautiful Mind*, starring Russell Crowe as Nash, or read Sylvia Nasar's best-selling book of the same name, we should add that John Nash developed his fundamental concept of equilibrium in games around 1950 and went on to make contributions of equal or greater importance in mathematics. After several decades of severe mental illness, he recovered and was awarded the 1994 Nobel Prize in Economics. This was the first Nobel Prize for game theory.

† The $1 increment and the restricted range of prices are chosen merely to simplify our entrée into this game by keeping the number of strategies available to each player finite. Later in the chapter we will consider briefly the case where each firm can choose its price from a continuous range of values.

tomers, 80 of whom shift from the other firm and 20 of whom shift from some other firm that is not a part of this game or decide to buy a shirt when they would otherwise not have done so. If both firms reduce their price by $1, existing customers stay put, but each gains 20 new ones. So when both firms charge $42 instead of $80, each gains 38 × 20 = 760 customers above the original 1,200. Then each sells 1,960 shirts and makes a profit of (42 × 20) × 1,960 = 43,120 dollars. Doing similar calculations for the other price combinations, we have the game table below.

B. B. Lean's price

		42	41	40	39	38
Rainbow's End's price	42	43,120 / 43,120	*43,260* / 41,360	43,200 / 39,600	42,940 / 37,840	42,480 / 36,080
	41	41,360 / *43,260*	41,580 / 41,580	*41,600* / 39,900	41,420 / 38,220	41,040 / 36,540
	40	39,600 / 43,200	39,900 / *41,600*	*40,000* / *40,000*	39,900 / *38,400*	39,600 / 36,800
	39	37,840 / 42,940	38,220 / 41,420	*38,400* / 39,900	38,380 / 38,380	38,160 / *36,860*
	38	36,080 / 42,480	36,540 / 41,040	36,800 / 39,600	*36,860* / 38,160	36,700 / 36,700

The table may seem daunting but is in fact easy to construct using Microsoft Excel or any other spreadsheet program.

TRIP TO THE GYM NO. 2

Try your hand at constructing this table in Excel.

Best Responses

Consider the thinking of RE's executives in charge of setting prices. (From now on, we will simply say "RE's thinking," and similarly for BB.) If RE believes that BB is choosing $42, then RE's profits from choosing various possible prices are given by the numbers in the southwest corners of the first column of profits in the above table. Of those five numbers, the highest is $43,260, corresponding to RE's price $41. Therefore this is RE's "best response" to BB's choice of $42. Similarly, RE's best response is $40 if it believes that BB is choosing $41, $40, or $39, and $39 if it believes BB is choosing $38. We show these best-response profit numbers in bold italics for clarity. We also show BB's best responses to the various possible prices of RE, using bold, italicized numbers in the northeast corners of the appropriate cells.

Before proceeding, we must make two remarks about best responses. First, the term itself requires clarification. The two firms' choices are simultaneous. Therefore, unlike the situation in chapter 2, each firm is not observing the other's choice and then "responding" with its own best choice given the other firm's actual choice. Rather, each firm is formulating a belief (which may be based on thinking or experience or educated guesswork) about what the other firm is choosing, and responding to this belief.

Second, note that it is not always best for one firm to undercut the other's price. If RE believes that BB is choosing $42, RE should choose a lower price, namely $41; but if RE believes that BB is choosing $39, RE's best response is higher, namely $40. In choosing its best price, RE has to balance two opposing considerations: undercutting will increase the quantity it sells, but will leave it a lower profit margin per unit sold. If RE believes that BB is setting a very low price, then the reduction in RE's profit margin from undercutting BB may be too big, and RE's best choice may be to accept a lower sales volume to get a higher profit margin on each shirt. In the extreme case where RE thinks BB is pricing at cost, namely $20, matching this price will yield RE zero profit. RE does better to choose a higher price, keeping some loyal customers and extracting some profit from them.

Nash Equilibrium

Now return to the table and inspect the best responses. One fact immediately stands out: one cell, namely the one where each firm charges $40, has both of its numbers in bold italics, yielding a profit of $40,000 to each firm. If RE believes that BB is choosing the price of $40, then its own best price is $40, and vice versa. If the two firms choose to price their shirts at $40 each, the beliefs of each about the other's price are confirmed by the actual outcome. Then there would be no reason for one firm to change its price if the truth about the other firm's choice were somehow revealed. Therefore these choices constitute a stable configuration in the game.

Such an outcome in a game, where the action of each player is best for him given his beliefs about the other's action, and the action of each is consistent with the other's beliefs about it, neatly squares the circle of thinking about thinking. Therefore it has a good claim to be called a resting point of the players' thought processes, or an equilibrium of the game. Indeed, this is just a definition of Nash equilibrium.

To highlight the Nash equilibrium, we shade its cell in gray and will do the same in all the game tables that follow.

The price-setting game in chapter 3, with just two price choices of $80 and $70, was a prisoners' dilemma. The more general game with several price choices shares this feature. If both firms could make a credible, enforceable agreement to collude, they could both charge prices considerably higher than the Nash equilibrium price of $40, and this would yield larger profits to both. As we saw in chapter 3, a common price of $80 gives each of them $72,000, as opposed to only $40,000 in the Nash equilibrium. The result should impress upon you how consumers can suffer if an industry is a monopoly or a producers' cartel.

In the above example, the two firms were symmetrically situated in all relevant matters of costs and in the quantity sold for each combination of own and rival prices. In general this need not be so, and in the resulting Nash equilibrium the two firms' prices can be differ-

TRIP TO THE GYM NO. 3

Suppose Rainbow's End locates a cheaper source for its shirts, so its cost per shirt goes down from $20 to $11.60, while B. B. Lean's cost remains at $20. Recalculate the payoff table and find the new Nash equilibrium.

ent. For those of you who want to acquire a better grasp of the methods and the concepts, we offer this as an "exercise"; casual readers should feel free to peek at the answer in the workouts.

The pricing game has many other features, but they are more

complex than the material so far. Therefore we postpone them to a position later in this chapter. To conclude this section, we make a few general remarks about Nash equilibria.

Does every game have a Nash equilibrium? The answer is essentially yes, provided we generalize the concept of actions or strategies to allow mixing of moves. This was Nash's famous theorem. We will develop the idea of mixing moves in the next chapter. Games that have no Nash equilibrium, even when mixing is allowed, are so complex or esoteric that we can safely leave them to very advanced treatments of game theory.

Is Nash equilibrium a good solution for simultaneous-move games? We will discuss some arguments and evidence bearing on this issue later in this chapter, and our answer will be a guarded yes.

Does every game have a unique Nash equilibrium? No. In the rest of this chapter we will look at some important examples of games with multiple Nash equilibria and discuss the new issues they raise.

Which Equilibrium?

Let us try Nash's theory on the hunting game. Finding best responses in the hunting game is easy. Fred should simply make the same choice that he believes Barney is choosing. Here is the result.

		Barney's choice		
		Stag	Bison	Rabbit
Fred's choice	Stag	3 **3**	0 0	1 0
	Bison	0 0	**3** 3	1 0
	Rabbit	0 1	0 1	***1*** *1*

So the game has three Nash equilibria.* Which of these will emerge as the outcome? Or will the two fail to reach any of the equilibria at all? The idea of Nash equilibrium does not by itself give the answers. Some additional and different consideration is needed.

If Fred and Barney had met at the stag party of a mutual friend, that might make the choice of Stag more prominent in their minds. If the ritual in their society is that as the head of the family sets out for the day's hunting he calls out in farewell, "Bye, son," the choice of Bison might be prominent. But if the ritual is for the family to call out in farewell "Be safe," the prominence might attach to the safer choice that guarantees some meat regardless of what the other chooses, namely rabbit hunting.

But what, precisely, constitutes "prominence"? One strategy, say Stag, may be prominent in Fred's mind, but that is not enough for him to make that choice. He must ask himself whether the same strategy is also prominent for Barney. And that in turn involves asking whether Barney will think it prominent to Fred. Selecting among multiple Nash equilibria requires resolution of a similar problem of thinking about thinking as does the concept of Nash equilibrium itself.

To square the circle, the "prominence" must be a multilevel back-and-forth concept. For the equilibrium to be selected successfully when the two are thinking and acting in isolation, it must be obvious to Fred that it is obvious to Barney that it is obvious to Fred . . . that is the right choice. If an equilibrium is obvious ad infinitum in this way, that is, if the players' expectations converge upon it, we call it a *focal point*. The development of this concept was just one of Thomas Schelling's many pioneering contributions to game theory.

Whether a game has a focal point can depend on many circumstances, including most notably the players' common experiences, which may be historical, cultural, linguistic, or purely accidental. Here are some examples.

---

* If mixing moves is allowed, there are other Nash equilibria as well. But they are somewhat strange and mostly of academic interest. We discuss them briefly in chapter 5.

We begin with one of Schelling's classics. Suppose you are told to meet someone in New York City on a specific day but without being told where or at what time. You don't even know who the other person is, so you cannot contact him/her in advance (but you are told how you would identify each other if and when you do meet). You are also told that the other person has been given identical instructions.

Your chances of success might seem slim; New York City is huge, and the day is long. But in fact people in this situation succeed surprisingly often. The time is simple: noon is the obvious focal point; expectations converge on it almost instinctively. The location is harder, but there are just a few landmark locations on which expectations can converge. This at least narrows down the choices considerably and improves the chances of a successful meeting.

Schelling conducted experiments in which the subjects were from the Boston or New Haven areas. In those days they traveled to New York by train and arrived at Grand Central Station; for them the clock in that station was focal. Nowadays, many people would think the Empire State Building is a focal point because of the movie *Sleepless in Seattle* (or *An Affair to Remember*); others would think Times Square the obvious "crossroads of the world."

One of us (Nalebuff) performed this experiment in an ABC *Primetime* program titled *Life: The Game*.[1] Six pairs of mutual strangers were taken to different parts of New York and told to find others about whom they had no information except that the other pair would be looking for them under similar conditions. The discussions within each pair followed Schelling's reasoning remarkably well. Each thought about what they believed would be the obvious places to meet and about what others would think they were thinking: each team, say team A, in its thinking recognized the fact that another team, say B, was simultaneously thinking about what was obvious to A. Eventually, three of the pairs went to the Empire State Building and the other three to Times Square. All chose noon for the time. There remained some further issues to be sorted out: the Empire State Building has observation decks on two different levels, and Times Square is a big place. But with a little ingenuity, including a display of signs, all six pairs were successful in meeting.*

What is essential for success is not that the place is obvious to you, or obvious to the other team, but that it is obvious to each that it is obvious to the others that . . . And, if the Empire State Building has this property, then each team has to go there even though it may be inconvenient for them to get there, because it is the only place each can expect the other team to be. If there were just two teams, one of them might think the Empire State Building the obvious focal point and the other might think Times Square equally obvious; then the two would fail to meet.

Professor David Kreps of Stanford Business School conducted the following experiment in his class. Two students were chosen to play the game, and each had to make his/her choice without any possibility of communication with the other. Their job was to divide up a list of cities between them. One student was assigned Boston, and the other was assigned San Francisco (and these assignments were public so that each knew the other's city). Each was then given a list of nine other U.S. cities—Atlanta, Chicago, Dallas, Denver, Houston, Los Angeles, New York, Philadelphia, and Seattle—and asked to choose a subset of these cities. If their choices resulted in a complete and nonoverlapping division, both got a prize. But if their combined list missed a city or had any duplicates, then they both got nothing.

How many Nash equilibria does this game have? If the student assigned Boston chooses, say, Atlanta and Chicago, while the student assigned San Francisco chooses the rest (Dallas, Denver, Houston, Los Angeles, New York, Philadelphia, and Seattle), that

* One of the pairs sat outside the Empire State Building for almost an hour, waiting for noon. If they had decided to wait inside, they would have done much better. It was also instructive that the teams of men went running from one site to another (Port Authority, Penn Station, Times Square, Grand Central, Empire State Building) without any sign that would help them be found by another team. As might be expected, the male teams even crossed paths without recognizing each other. In contrast, the all-women teams made signs and hats. They picked a single spot and waited to be found.

is a Nash equilibrium: given the choice of one, any change in the choice of the other will create either an omission or an overlap and would lower the payoff to the deviator. The same argument applies if, say, one chooses Dallas, Los Angeles, and Seattle while the other chooses the other six. In other words, there are as many Nash equilibria as there are ways of dividing up the list of nine numbers into two distinct subsets. There are 2^9, or 512, such ways; therefore the game has a huge number of Nash equilibria.

Can the players' expectations converge to create a focal point? When both players were Americans or long-time U.S. residents, over 80 percent of the time they chose the division geographically; the student assigned Boston chose all the cities east of the Mississippi and the student assigned San Francisco chose those west of the Mississippi.* Such coordination was much less likely when one or both students were non-U.S. residents. Thus nationality or culture can help create a focal point. When Kreps's pairs lacked such common experience, choices were sometimes made alphabetically, but even then there was no clear dividing point. If the total number of cities was even, an equal split might be focal, but with nine cities, that is not possible. Thus one should not assume that players will always find a way to select one of multiple Nash equilibria by a convergence of expectations; failure to find a focal point is a distinct possibility.†

Next, suppose each of two players is asked to choose a positive integer. If both choose the same number, both get a prize. If the two choose different numbers, neither gets anything. The overwhelmingly frequent choice is 1: it is the first among the whole

---

* Perhaps in a few years' time this will no longer work, if the news stories about the deterioration of geographic knowledge among American schoolchildren are true.

† The game of dividing cities might seem uninteresting or irrelevant, but think of two firms that are trying to divide up the U.S. market between them to allow each to enjoy an uncontested monopoly in its assigned territory. U.S. antitrust laws forbid explicit collusion. To arrive at a tacit understanding requires a convergence of expectations. Kreps's experiment suggests that two American firms may achieve this better than could an American firm and a foreign firm.

numbers (positive integers), it is the smallest, and so on; therefore, it is focal. Here the reason for its salience is basically mathematical.

Schelling gives the example of two or more people who have gone to a crowded place together and get separated. Where should each go in the expectation of finding the other? If the place, say a department store or a railway station, has a Lost and Found window, it has a good claim to be focal. Here the reason for its salience is linguistic. Sometimes meeting places are deliberately created to guarantee a convergence of expectations; for example, many railway stations in Germany and Switzerland have a well-signposted *Treffpunkt* (meeting point).

What is neat about the game of meeting is not just that the two players find each other but that the focal point ends up being relevant to so many strategic interactions. Probably the most important is the stock market. John Maynard Keynes, arguably the twentieth century's most famous economist, explained its behavior by analogy with a newspaper contest that was common in his time, where a number of photographs of faces were presented, and readers had to guess which face the majority of other voters would judge the most beautiful.[2] When everyone thinks along these lines, the question becomes which face most people think that most others will think that most others will think . . . is the most beautiful. If one contestant was significantly more beautiful than all the others, this could provide the necessary focal point. But the reader's job was rarely that easy. Imagine instead that the hundred finalists were practically indistinguishable except for the color of their hair. Of the hundred, only one is a redhead. Would you pick the redhead?

The aim becomes not to make any absolute judgment of beauty but to find a focal point of this process of thinking. How do we agree on that? The reader must figure out the realized convention without the benefit of communication. "Pick the most beautiful" might be the stated rule, but that could be significantly more difficult than picking the redhead, or the one with an interesting gap between her two front teeth (Lauren Hutton) or the mole (Cindy Crawford). Anything that distinguishes becomes a focal point and allows people's expectations to converge. For this reason, we should not be surprised that many of the world's top models do not have perfect features; rather, they are almost perfect but have some interesting flaw that gives their look a personality and a focal point.

Keynes used the beauty contest as a metaphor for the stock market, where each investor wants to buy the stocks that will rise in price, which means the stocks that investors, in general, think will appreciate. The hot stock is the one that everyone thinks that everyone else thinks . . . is the hot stock. There can be different reasons why different sectors or stocks become hot at different times—a well-publicized initial public offering, a famous analyst's recommendation, and so on. The focal point concept also explains the attention paid to round numbers: 10,000 for the Dow, or 2,500 for the Nasdaq. These indexes are just values of a specified portfolio of stocks. A number like 10,000 does not have any intrinsic meaning; it serves as a focal point only because expectations can converge more easily on round numbers.

The point of all this is that equilibrium can easily be determined by whim or fad. There is nothing fundamental that guarantees the most beautiful contestant will be chosen or the best stock will appreciate the fastest. There are some forces that work in the right direction. High forecast earnings are similar to the beauty contestant's complexion—one of the many necessary but by no means sufficient conditions needed to anchor otherwise arbitrary whims and fads.

Many mathematical game theorists dislike the dependence of an outcome on historical, cultural, or linguistic aspects of the game or on purely arbitrary devices like round numbers; they would prefer the solution be determined purely by the abstract mathematical facts about the game—the number of players, the strategies available to each, and the payoffs to each in relation to the strategy choices of all. We disagree. We think it entirely appropriate that the outcome of a game played by humans interacting in a society should depend on the social and psychological aspects of the game.

Think of the example of bargaining. Here the players' interests seem to be totally conflicting; a larger share for one means a smaller share for the other. But in many negotiations, if the two parties fail to agree, neither will get anything and both may suffer serious damage, as happens when wage bargaining breaks down and a strike or a lockout ensues. The two parties' interests are aligned to the extent that both want to avoid such disagreement. They can do so if they can find a focal point, with the common expectation that neither will concede anything beyond that point. That is why a 50:50 split is so often observed. It is simple and clear, it has the advantage of appearing fair, and, once such considerations get a foothold, it serves for the convergence of expectations.

Consider the problem of excessive compensation of CEOs. Often a CEO really cares about prestige. Whether the person gets paid $5 million or $10 million won't really have a big impact on the person's life. (That's easy for us to say from where we sit, where both numbers are quite abstract.) What's the meeting place that the CEOs care about? It is being better than average. Everyone wants to be in the top half. They all want to meet there. The problem is that this meeting spot only allows in half of the folks. But the way they get around this is via escalating pay. Every firm pays its CEO above last year's average, so everyone can think they have an above-average CEO. The end result is wildly escalating CEO salaries. To solve the problem, we need to find some other focal meeting point. For example, historically CEOs got prestige in their community via public service. Competing in that dimension was good all around. The current focal point on pay was created by *Business Week* surveys and compensation consultants. Changing it won't be easy.

The issue of fairness is also one of choosing a focal point. The Millennium Development Goals and Jeff Sachs's book *The End of Poverty* emphasize that contributing 1 percent of gross domestic product (GDP) to development will end poverty by 2025. The key point here is that the focal point of contributions is based on a percentage of income, not an absolute amount. Thus rich countries have a bigger obligation to contribute than the less rich. The apparent fairness of this can contribute to the convergence of expectations. Whether the promised funds will actually materialize remains to be seen.

BATTLES AND CHICKENS

In the hunting game, the two players' interests are perfectly aligned; both prefer one of the big-game equilibria, and the only question is how they can coordinate their beliefs on a focal point. We now turn to two other games, which also have non-unique Nash equilibria, but have an element of conflicting interests. Each leads to different ideas about strategy.

Both of these games date from the 1950s and have stories that fit those times. We will illustrate them using variants of the game between our Stone Age hunters, Fred and Barney. But we will relate the original sexist stories too, partly because they explain the names that have come to be attached to these games and partly for the amusement value of looking back on the quaint thoughts and norms of old times.

The first game is generically called battle of the sexes. The idea is that a husband and wife have different preferences in movies, and the two available choices are very different. The husband likes lots of action and fighting; he wants to see *300*. The wife likes three-handkerchief weepies; her choice is *Pride & Prejudice* (or *A Beautiful Mind*). But both prefer watching either movie in the other's company to watching any movie on their own.

In the hunting version, remove the Rabbit choice and keep only Stag and Bison. But suppose Fred prefers stag meat and rates the outcome of a jointly conducted stag hunt 4 instead of 3, while Barney has the opposite preference. The revised game payoff table is as shown below.

		Barney's choice	
		Stag	Bison
Fred's choice	Stag	4 *3*	0 0
	Bison	0 0	*3* 4

As usual, best responses are shown in bold italics. We see at once that the game has two Nash equilibria, one where both choose Stag, and the other where both choose Bison. Both players prefer to have either equilibrium outcome than to hunt alone in one of the two nonequilibrium outcomes. But they have conflicting preferences over the two equilibria: Fred would rather be in the Stag equilibrium and Barney in the Bison equilibrium.

How might one or the other outcome be sustained? If Fred can somehow convey to Barney that he, Fred, is credibly and unyieldingly determined to choose Stag, then Barney must make the best of the situation by complying. However, Fred faces two problems in using such a strategy.

First, it requires some method of communication before the actual choices are made. Of course, communication is usually a two-way process, so Barney might try the same strategy. Fred would ideally like to have a device that will let him send messages but not receive them. But that is not without its own problems; how can Fred be sure that Barney has received and understood the message?

Second, and more important, is the problem of credibly conveying an unyielding determination. This can be faked, and Barney might put it to the test by defying Fred and choosing Bison, which would leave Fred with a pair of bad choices: give in and choose Bison, which leads to humiliation and destruction of reputation, or go ahead with the original choice of Stag, which means missing the opportunity of the joint hunt, getting zero meat, and ending up with a hungry family.

In chapter 7 we will examine some ways that Fred could make his determination credible and achieve his preferred outcome. But we will also examine some ways that Barney could undermine Fred's commitment.

If they have two-way communication before the game is played, this is essentially a game of negotiation. The two prefer different outcomes, but both prefer some agreement to complete disagreement. If the game is repeated, they may be able to agree to a compromise—for example, alternate between the two grounds on alternate days. Even in a single play, they may be able to achieve a compromise in the sense of a statistical average by tossing a coin and choosing one equilibrium if it comes up heads and the other equilibrium if it comes up tails. We will devote an entire chapter to the important subject of negotiation.

The second classic game is called chicken. In the standard telling of this story, two teenagers drive toward each other on a straight road, and the first one to swerve to avoid a collision is the loser, or chicken. If both keep straight, however, they crash, and that is the worst outcome for both. To create a game of chicken out of the hunting situation, remove the Stag and Bison choices, but suppose there are two areas for rabbit hunting. One, located to the south, is large but sparse; both can go there and each will get 1 of meat. The other, located to the north, is plentiful but small. If just one hunter goes there, he can get 2 of meat. If both go there, they will merely interfere and start fighting with each other and get nothing. If one goes north and the other goes south, the one who goes north will enjoy his 2 of meat. The one going south will get his 1. But his and his family's feeling of envy for the other who comes back at the end of the day with 2 will reduce his enjoyment, so we will give him a payoff of only 1/2 instead of 1. This yields the game payoff table shown below.

		Barney's choice	
		North	South
Fred's choice	North	0 0	*1/2* *2*
	South	*2* *1/2*	1 1

As usual, best responses are shown in bold italics. We see at once that the game has two Nash equilibria, with one player going north and the other going south. The latter is then the chicken; he has made the best of a bad situation in responding to the other's choice of North.

Both games, the battle of the sexes and chicken, have a mixture of common and conflicting interests: in both, the two players agree in preferring an equilibrium outcome to a nonequilibrium outcome, but they disagree as to which equilibrium is better. This conflict is sharper in chicken, in the sense that if each player tries to achieve his preferred equilibrium, both end up in their worst outcome.

Methods for selecting one of the equilibria in chicken are similar to those in the battle of the sexes. One of the players, say Fred, may make a commitment to choosing his preferred strategy, namely going north. Once again, it is important to make this commitment credible and to ensure that the other player knows it. We will consider commitments and their credibility more fully in chapters 6 and 7.

There is also the possibility of compromise in chicken. In a repeated interaction Fred and Barney may agree to alternate between North and South; in a single play, they may use a coin toss or other randomizing method to decide who gets North.

Finally, chicken shows a general point about games: even though the players are perfectly symmetric as regards their strategies and payoffs, the Nash equilibria of the game can be asymmetric, that is, the players choose different actions.

A LITTLE HISTORY

In the course of developing examples in this chapter and the one before it, we have introduced several games that have become classics. The prisoners' dilemma, of course, everyone knows. But the game of the two Stone Age hunters trying to meet is almost equally well known. Jean-Jacques Rousseau introduced it in an almost identical setting—of course he did not have *Flintstones* characters to add color to the story.

The hunters' meeting game differs from the prisoners' dilemma because Fred's best response is to take the same action as Barney does (and vice versa), whereas in a prisoners' dilemma game Fred would have a dominant strategy (just one action—for example, Rabbit—would be his best choice regardless of what Barney does) and so would Barney. Another way to express the difference is to say that in the meeting game, Fred would go stag hunting if he had the assurance, whether by direct communication or because of the existence of a focal point, that Barney would also go stag hunting, and vice versa. For this reason, the game is often called the *assurance game*.

Rousseau did not put his idea in precise game-theoretic language, and his phrasing leaves his meaning open to different interpretations. In Maurice Cranston's translation, the large animal is a deer, and the statement of the problem is as follows: "If it was a matter of hunting a deer, everyone well realized that he must remain faithfully at his post; but if a hare happened to pass within the reach of one of them, we cannot doubt that he would have gone off in pursuit of it without scruple and, having caught his own prey, he would have cared very little about having caused his companions to lose theirs."[3] Of course if the others were going for the hare, then there would be no point in any one hunter's attempting the deer. So the statement seems to imply that each hunter's dominant strategy is to go after a hare, which makes the game a prisoners' dilemma. However, the game is more commonly interpreted as an assurance game, where each hunter prefers to join the stag hunt if all the others are doing likewise.

In the version of chicken made famous by the movie *Rebel Without a Cause*, two teenagers drive their cars in parallel toward a cliff; the one who first jumps out of his car is the chicken. The metaphor of this game was used for nuclear brinkmanship by Bertrand Russell and others. The game was discussed in detail by Thomas Schelling in his pioneering game-theoretic analysis of strategic moves, and we will pick this back up in chapter 6.

To the best of our knowledge, the battle of the sexes game does not have such roots in philosophy or popular culture. It appears in the book *Games and Decisions* by R. Duncan Luce and Howard Raiffa, an early classic on formal game theory.[4]

FINDING NASH EQUILIBRIA

How can we find Nash equilibrium for a game? In a table, the worst-case method is cell-by-cell inspection. If both of the payoff entries in a cell are best responses, the strategies and payoffs for that cell constitute a Nash equilibrium. If the table is large, this procedure can get tedious. But God made computers precisely to rescue humans from the tedium of inspection and calculation. Software packages to find Nash equilibria are readily available.[5]

Sometimes there are shortcuts; we now describe one that is often useful.

Successive Elimination

Return to the pricing game between Rainbow's End and B. B. Lean. Here again is the table of payoffs:

		B. B. Lean's price								
			42		41		40		39		38
	42	43,120	43,120	41,360	**43,260**	39,600	43,200	37,840	42,940	36,080	42,480
	41	41,360	**43,260**	41,580	41,580	39,900	**41,600**	38,220	41,420	36,540	41,040
	40	39,600	43,200	39,900	**41,600**	**40,000**	**40,000**	**38,400**	39,900	36,800	39,600
	39	37,840	42,940	38,220	41,420	**38,400**	39,900	38,380	38,380	**36,860**	38,160
	38	36,080	42,480	36,540	41,040	36,800	39,600	**36,860**	38,160	36,700	36,700

(Rainbow's End's price labels the rows: 42, 41, 40, 39, 38)

RE does not know what price BB is choosing. But it can figure out what price or prices BB is not choosing: BB will never set its price at \$42 or \$38. There are two reasons (both of which apply in our example, but in other situations only one may apply).[6]

First, each of these strategies is uniformly worse for BB than another available strategy. No matter what it thinks RE is choosing, \$41 is better for BB than \$42, and \$39 is better than \$38. To see this, consider the \$41 versus \$42 comparison; the other is similar. Look at the five numbers for BB's profits from choosing \$41 (shaded in dark gray) versus those from choosing \$42 (shaded in light gray).

For each of RE's five possible choices, BB's profit from choosing $42 is smaller than that from choosing $41:

> 43,120 < 43,260,
> 41,360 < 41,580,
> 39,600 < 39,900,
> 37,840 < 38,220,
> 36,080 < 36,540.

So no matter what BB expects RE to do, BB will never choose $42, and RE can confidently expect BB to rule out the $42 strategy, and, likewise, $38.

When one strategy, say A, is uniformly worse for a player than another, say B, we say that A is *dominated* by B. If such is the case, that player will never use A, although whether he uses B remains to be seen. The other player can confidently proceed in thinking on this basis; in particular, he need not consider playing a strategy that is the best response only to A. When solving the game, we can remove dominated strategies from consideration. This reduces the size of the game table and simplifies the analysis.*

The second avenue for elimination and simplification is to look for strategies that are *never best responses* to anything the other player might be choosing. In this example, $42 is never BB's best response to anything RE might be choosing within the range we are considering. So, RE can confidently think, "No matter what BB is thinking about my choice, it will never choose $42."

Of course, anything that is dominated is a never best response. It is more instructive to look at BB's option to price at $39. This can *almost* be eliminated for being a never best response. A price of $39 is only a best response to an RE price of $38. Once we know that $38 is dominated, then we can conclude that a BB price of $39 will never be a best response to anything RE will ever play. The

---

* If A is dominated by B, then conversely, B dominates A. So if A and B were the only two strategies available to that player, B would be a dominant strategy. With more than two strategies available, it is possible that A is dominated by B, but B is not dominant, because it does not dominate some third strategy C. In general, elimination of dominated strategies may be possible even in games that do not have any dominant strategies.

advantage, then, of looking for never best responses is that you are able to eliminate strategies that are not dominated but would still never be chosen.

We can perform a similar analysis for the other player. RE's $42 and $38 strategies are eliminated, leaving us with a 3-by-3 game table:

		B. B. Lean's price		
		41	**40**	**39**
Rainbow's End's price	**41**	41,580 / 41,580	*41,600* / 39,900	41,420 / 38,220
	40	39,900 / *41,600*	*40,000* / 40,000	39,900 / *38,400*
	39	38,220 / 41,420	*38,400* / 39,900	38,380 / 38,380

In this simplified game, each firm has a dominant strategy, namely $40. Therefore our Rule 2 (from chapter 3) indicates that as a solution for the game.

The $40 strategy was not dominant in the original larger game; for example, if RE thought that BB would charge $42, then its profits from setting its own price at $41, namely $43,260, would be more than its profits from choosing $40, namely $43,200. The elimination of some strategies can open up the way to eliminate more in a second round. Here just two rounds sufficed to pin down the outcome. In other examples it may take more rounds, and even then the range of outcomes may be narrowed somewhat but not all the way to uniqueness.

If successive elimination of dominated strategies (or never-best-response strategies) and choice of dominant strategies does lead to a unique outcome, that is a Nash equilibrium. When this works, it is an easy way to find Nash equilibria. Therefore we summarize our discussion of finding Nash equilibria into two rules:

RULE 3: Eliminate from consideration any dominated strategies and strategies that are never best responses, and go on doing so successively.

RULE 4: Having exhausted the simple avenues of looking for dominant strategies or ruling out dominated ones, next search all the cells of the game table for a pair of mutual best responses in the same cell, which is a Nash equilibrium of the game.

GAMES WITH INFINITELY MANY STRATEGIES

In each of the versions of the pricing game we discussed so far, we allowed each firm only a small number of price points: only $80 and $70 in chapter 3, and only between $42 and $38 in $1 steps in this chapter. Our purpose was only to convey the concepts of the prisoners' dilemma and Nash equilibrium in the simplest possible context. In reality, prices can be any number of dollars and cents, and for all intents and purposes it is as if they can be chosen over a continuous range of numbers.

Our theory can cope with this further extension quite easily, using nothing more than basic high-school algebra and geometry. We can show the prices of the two firms in a two-dimensional graph, measuring RE's price along the horizontal or X axis and BB's price along the vertical or Y axis. We can show the best

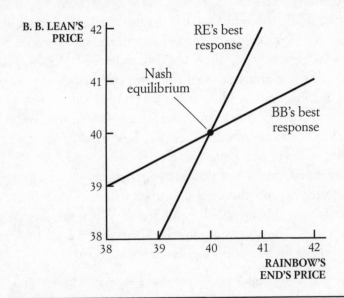

responses in this graph instead of showing bold italic profit numbers in a game table of discrete price points.

We do this for the original example where the cost of each shirt to each store was $20. We omit the details of the mathematics and merely tell you the result.[7] The formula for BB's best response in terms of RE's price (or BB's belief about the price RE is setting) is

> BB's best response price = 24 + 0.4 × RE's price (or BB's belief about it).

This is shown as the flatter of the two lines in the above graph. We see that for each $1 cut in RE's price, BB's best response should be to cut its own price but by less, namely 40 cents. This is the result of BB's calculation, striking the best balance between losing customers to RE and accepting a lower profit margin.

The steeper of the two curves in the figure is RE's best response to its belief about BB's price. Where the two curves intersect, the best response of each is consistent with the other's beliefs; we have a Nash equilibrium. The figure shows that this occurs when each firm charges $40. Moreover, it shows that this particular game has exactly one Nash equilibrium. Our finding a unique Nash equilibrium in the table where prices had to be multiples of $1 was not an artificial consequence of that restriction.

Such graphs or tables that allow much more detail than we could in the simple examples are a standard method for computing Nash equilibria. The calculation or graphing can quickly get too complicated for paper-and-pencil methods, and too boring besides, but that's what computers are for. The simple examples give us a basic understanding of the concept, and we should reserve our human thinking skills for the higher-level activity of assessing its usefulness. Indeed, that is our very next topic.

A BEAUTIFUL EQUILIBRIUM?

John Nash's equilibrium has a lot of conceptual claim to be the solution of a game where each player has the freedom of choice. Perhaps the strongest argument in its favor takes the form of a counterargument to any other proposed solution. A Nash equilibrium is a configuration of strategies where each player's choice is his best response to the other player's choice (or the other players' choices when there are more than two players in the game). If some outcome is not a Nash equilibrium, at least one player must be choosing an action that is not his best response. Such a player has a clear incentive to deviate from that action, which would destroy the proposed solution.

If there are multiple Nash equilibria, we do need some additional method for figuring out which one will emerge as the outcome. But that just says we need Nash plus something else; it does not contradict Nash.

So we have a beautiful theory. But does it work in practice? One answers this question by looking for instances where such games are played in the real world, or by creating them in a laboratory setting and then comparing the actual outcomes against the predictions of the theory. If the agreement is sufficiently good, that supports the theory; if not, the theory should be rejected. Simple, right? In fact the process turns complicated very quickly, both in implementation and in interpretation. The results are mixed, with some reasons for optimism for the theory but also some ways in which the theory must be augmented or altered.

The two methods—observation and experiment—have different merits and flaws. Laboratory experiments allow proper scientific "control." The experimenters can specify the rules of the game and the objectives of the participants quite precisely. For example, in pricing games where the subjects play the roles of the managers of the firms, we can specify the costs of the two firms and the equations for the quantities each would sell in relation to the prices both charge, and give the players the appropriate motivation by paying them in proportion to the profits they achieve for their firm in the game. We can study the effects of a particular factor, keeping all other things constant. By contrast, games that occur in real life have too many other things going on that we cannot control and too many things about the players—their true motivations, the firms' costs of production, and so on—that we do not know. That makes it hard to make inferences about the underlying conditions and causes by observing the outcomes.

On the other hand, real-world observations do have some advantages. They lack the artificiality of laboratory experiments, in which the subjects are usually students, who have no previous experience in business or the similar applications that motivate the games. Many are novices even to the setting of the laboratory where the games are staged. They have to understand the rules of the game and then play it, all in a matter of an hour or two. Think how long it took you to figure out how to play even simple board games or computer games; that will tell you how naïve the play in such settings can be. We already discussed some examples of this problem in chapter 2. A second issue concerns incentives. Although the experimenter can give the students the correct incentives by designing the structure of their monetary payments to fit their performance in the game, the sizes of the payments are usually small, and even college students may not take them sufficiently seriously. By contrast, business games and even professional sports in the real world are played by experienced players for large stakes.

For these reasons, one should not rely solely on any one form of evidence, whether it supports or rejects a theory, but should use both kinds and learn from each. With these cautions in mind, let us see how the two types of empirical approaches do.

The field of industrial organization in economics provides the largest body of empirical testing of game-theoretic competition among firms. Industries like auto manufacturing have been studied in depth. These empirical investigators start with several handicaps. They do not know the firms' costs and demands from any independent source, and must estimate these things from the same data that they want to use for testing the pricing equilibrium. They do not know precisely how the quantities sold by each firm depend on the prices charged by all. In the examples in this chapter, we simply assumed a linear relationship, but the real-world counterparts (demand functions, in the jargon of economics) can be nonlinear in quite complicated ways. The investigator must assume some specific form of the nonlinearity. Real-life competition among firms is not just about prices; it has many other dimensions—advertising, investment, research and development. Real-life managers may not have the pure and simple aims of profit (or shareholder value) maximization that economic theory usually assumes. And competition among firms in real life extends over several years, so an appropriate combination of backward reasoning and Nash equilibrium concepts must be specified. And many other conditions, such as income and costs, change from one year to the next, and firms enter or exit the industry. The investigator must think about what all these other things might be and make proper allowance for (control for, in statistical jargon) their effects on quantities and prices. Real-world outcomes are also affected by many random factors and so, uncertainty must be allowed for.

A researcher must make a choice in each of these matters and then derive equations that capture and quantify all the relevant effects. These equations are then fitted to the data, and statistical tests performed to see how well they do. Then comes an equally difficult problem: What does one conclude from the findings? For example, suppose the data do not fit your equations very well. Something in your specification that led to the equations was not correct, but what was it? It could be the nonlinear form of the equations you chose; it could be the exclusion of some relevant variable, like income, or of some relevant dimension of competition, like advertising; or it could be that the Nash equilibrium concept used in your derivations is invalid. Or, it could be a combination of all these things. You cannot conclude that Nash equilibrium is incorrect when something else might be wrong. (But you would be right to raise your level of doubt about the equilibrium concept.)

Different researchers have made different choices in all these matters and, predictably, have found different results. After a thorough survey of this research, Peter Reiss and Frank Wolak of Stanford University give a mixed verdict: "The bad news is that the underlying economics can make the empirical models extremely complex. The good news is that the attempts so far have begun to define the issues that need to be addressed."[8] In other words, more research is needed.

Another active area for empirical estimation concerns auctions where a small number of strategically aware firms interact in bidding for things like bandwidths in the airwave spectrum. In these auctions, asymmetry of information is a key issue for the bidders and also for the auctioneer. Therefore we postpone the discussion of auctions to chapter 10, after we have examined the general issues of information in games in chapter 8. Here we merely mention that empirical estimation of auction games is already having considerable success.[9]

What do laboratory experiments have to say about the predictive power of game theory? Here the record is also mixed. Among the earliest experiments were the markets set up by Vernon Smith. He found surprisingly good results for game theory as well as for economic theory: small numbers of traders, each with no direct knowledge of the others' costs or values, could achieve equilibrium exchanges very quickly.

Other experiments with different kinds of games yielded outcomes that seemed contradictory to theoretical predictions. For example, in the ultimatum game, where one player makes a take-it-or-leave-it offer to the other for dividing a given sum between the two, the offers were surprisingly generous. And in prisoners' dilemmas, good behavior occurred far more frequently than theory might lead people to believe. We discussed some of these findings in chapters 2 and 3. Our general conclusion was that the participants in these games had different preferences or valuations than the purely selfish ones that used to be the natural assumption in economics. This is an interesting and important finding on its own; however, once the realistic "social" or "other-regarding" preferences are allowed for, the theoretical concepts of equilibrium— backward reasoning in sequential-move games and Nash in simultaneous-move games—yield generally good explanations of the observed outcomes.

When a game does not have a unique Nash equilibrium, the

players have the additional problem of locating a focal point or some other method of selection among the possible equilibria. How well they succeed depends on the context, in just the way that theory suggests. If the players have sufficiently common understanding for their expectations to converge, they will succeed in settling on a good outcome; otherwise disequilibrium may persist.

Most experiments work with subjects who have no prior experience playing the particular game. The behavior of these novices does not initially conform to equilibrium theory, but it often converges to equilibrium as they gain experience. But some uncertainty about what the other player will do persists, and a good concept of equilibrium should allow players to recognize such uncertainty and respond to it. One such extension of the Nash equilibrium concept has become increasingly popular; this is the *quantal response equilibrium,* developed by professors Richard McKelvey and Thomas Palfrey of Caltech. This is too technical for a book like ours, but some readers may be inspired to read and study it.[10]

After a detailed review of the relevant work, two of the top researchers in the field of experimental economics, Charles Holt of the University of Virginia and Alvin Roth of Harvard University, offer a guardedly optimistic prognosis: "In the last 20 years, the notion of Nash equilibrium has become a required part of the tool kit for economists and other social and behavioral scientists. . . . There have been modifications, generalizations, and refinements, but the basic equilibrium analysis is the place to begin (and sometimes end) the analysis of strategic interactions."[11] We think that to be exactly the right attitude and recommend this approach to our readers. When studying or playing a game, begin with the Nash equilibrium, and then think of reasons why, and the manner in which, the outcome may differ from the Nash predictions. This dual approach is more likely to give you a good understanding or success in actual play than either a totally nihilistic—anything goes—attitude or a slavishly naïve adherence to the Nash equilibrium with additional assumptions, such as selfishness.

CASE STUDY: HALF WAY

A Nash equilibrium is a combination of two conditions:

i. Each player is choosing a best response to what he believes the other players will do in the game.

ii. Each player's beliefs are correct. The other players are doing just what everyone else thinks they are doing.

It is easier to describe this outcome in a two-player game. Our two players, Abe and Bea, each have beliefs about what the other will do. Based on those beliefs, Abe and Bea each choose to take an action that maximizes their payoffs. The beliefs prove right: Abe's best response to what he thinks Bea is doing is just what Bea thought Abe would do, and Bea's best response to what she thought Abe would do is indeed just what Abe expected her to do.

Let's look at these two conditions separately. The first condition is quite natural. If otherwise, then you'd have to argue that someone is not taking the best action given what he or she believes. If he or she had something better, why not do it?

Mostly, the rub comes in the second condition—that everyone is correct in what they believe. For Sherlock Holmes and Professor Moriarty this was not a problem:

"'All that I have to say has already crossed your mind,' said he.
'Then possibly my answer has crossed yours,' I replied.
'You stand fast?'
'Absolutely.'"

For the rest of us, correctly anticipating what the other side will do is often a challenge.

The following simple game will help illustrate the interplay between these two conditions and why you might or might not want to accept them.

Abe and Bea are playing a game with the following rules: Each player is to pick a number between 0 and 100, inclusive. There is a $100 prize to the player whose number is closest to half the other person's number.

We'll be Abe and you can play Bea. Any questions?

What if there's a tie?

Okay, in that case we split the prize. Any other questions?

No.

Great, then let's play. We've picked our number. Time for you to pick yours. What is your number? To help keep yourself honest, write it down.

Case Discussion

We picked 50. No, we didn't. To see what we actually picked, you'll have to read on.

Let's start by taking a step back and use the two-step approach to finding a Nash equilibrium. In step 1, we believe that your strategy had to be an optimal response to something we might have done. Since our number has to be something between 0 and 100, we figure that you couldn't have picked any number bigger than 50. For example, the number 60 is only an optimal response if you thought we would pick 120, something we couldn't do under the rules.

What that tells us is that if your choice was truly a best response to something we might have done, you had to pick a number between 0 and 50. By the same token, if we picked a number based on something that you might have done, we would have picked something between 0 and 50.

Believe it or not, many folks stop right there. When this game is played among people who haven't read this book, the most common response is 50. Frankly, we think that is a pretty lame answer (with apologies if that's what you picked). Remember that 50 is only the best choice if you think that the other side was going to pick 100. But, in order for the other side to pick 100, they would have to have misunderstood the game. They would have had to pick a number that had (almost) no chance of winning. Any number less than 100 will beat 100.

We will assume that your strategy was a best response to some-thing we might have done and so it is between 0 and 50. That means our best choice should be something between 0 and 25.

Note that at this juncture, we have taken a critical step. It may seem so natural that you didn't even notice. We are no longer relying on our first condition that our strategy is a best response. We have taken the next step and proposed that our strategy should be a best response to something that is a best response from you.

If you are going to do something that is a best response, we should be doing something that is a best response to a best response.

At this point, we are beginning to form some beliefs about your actions. Instead of imagining that you can do anything allowed by the rules, we are going to assume that you will actually have picked a move that is a best response. Given the quite sensible belief that you are not going to do something that doesn't make sense, it then follows that we should only pick a number between 0 and 25.

Of course, by the same token, you should be realizing that we won't be picking a number bigger than 50. If you think that way, then you won't pick a number bigger than 25.

As you might have guessed, the experimental evidence shows that after 50, 25 is the most common guess in this game. Frankly, 25 is a much better guess than 50. At least it has a chance of win-ning if the other player was foolish enough to pick 50.

If we take the view that you are only going to pick a number between 0 and 25, then our best response is now limited to num-bers between 0 and 12.5. In fact, 12.5 is our guess. We'll win if our guess is closer to half your number than your number is to half ours. That means we win if you picked anything higher than 12.5.

Did we win?

Why did we pick 12.5? We thought you would pick a number between 0 and 25, and that's because we thought you'd think we'd pick a number between 0 and 50. We could of course go on with our reasoning and conclude that you'd figure we'd pick a number between 0 and 25, leading you to choose something between 0 and 12.5. If you had thought that, then you'd be one step ahead of us and would have won. Our experience suggests that most people don't think more than two or three levels, at least on their first go-around.

Now that you've had some practice and better understand the game, you might want a rematch. That's fair. So write down your number again—we promise not to peek.

We are pretty confident that you expect us to pick something less than 12.5. That means you'll pick something less than 6.25. And if we think you'll pick something less than 6.25, we should pick a number less than 3.125.

Now if this were the first go-around, we might stop there. But we just explained that most folks stop after two levels of reasoning, and this time we expect that you are determined to beat us, so you'll engage in at least one more level of thinking ahead. If you expect us to pick 3.125, then you'll pick 1.5625, which leads us to think of 0.78125.

At this point, we are guessing that you can see where this is all heading. If you think we are going to pick a number between 0 and X, then you should pick something between 0 and X/2. And if we think you are going to pick something between 0 and X/2, then we should pick something between 0 and X/4.

The only way that we can both be right is if we both pick 0. That's what we've done. This is the Nash equilibrium. If you pick 0, we want to pick 0; if we pick 0, you want to pick 0. Thus if we both correctly anticipate what the other will do, we both do best picking 0, just what we expected the other to do.

We should have picked 0 the first time around as well. If you pick X and we pick 0, then we win. That is because 0 is closer to X/2 than X is to 0/2 = 0. We knew this all along but didn't want to give it away the first time we played.

As it turned out, we didn't actually need to know anything about what you might be doing to pick 0. But that is a highly unusual case and an artifact of having only two players in the game.

Let's modify the game to add more players. Now the person

whose number is closest to half the average number wins. Under these rules, it is no longer the case that 0 always wins.* But it is still the case that the best responses converge to zero. In the first round of reasoning, all players will pick something between 0 and 50. (The average number picked can't be above 100, so half the average is bounded by 50.) In the second iteration of logic, if everyone thinks others will play a best response, then in response everyone should pick something between 0 and 25. In the third iteration of logic, they'll all pick something between 0 and 12.5.

How far people are able to go in this reasoning is a judgment call. Again, our experience suggests that most people stop at two or three levels of reasoning. The case of a Nash equilibrium requires that the players follow the logic all the way. Each player picks a best response to what he or she believes that the other players are doing. The logic of Nash equilibrium leads us to the conclusion that all players will pick 0. Everyone picking 0 is the only strategy where each of the players is choosing a best response to what they believe other players are doing and each is right about what they believe the others will be doing.

When people play this game, they rarely pick zero on the first go-around. This is convincing evidence against the predictive power of Nash equilibrium. On the other hand, when they play the game even two or three times, they get very close to the Nash result. That is convincing evidence in favor of Nash.

Our view is that both perspectives are correct. To get to a Nash equilibrium, all players have to choose best responses—which is relatively straightforward. They also all have to have correct beliefs about what the other players will be doing in the game. This is much harder. It is theoretically possible to develop a set of internally consistent beliefs without playing the game, but it is often easier to play the game. To the extent that players learn that their beliefs were wrong by playing the game and then learn how to do a

---

* If there are three players and the other two have picked 1 and 5, then the average of the three numbers (0, 1, and 5) is 2, half the average is 1, and the person picking 1 will win.

better job predicting what others will do, they will converge to a Nash equilibrium.

While experience is helpful, it is no guarantee of success. One problem arises when there are multiple Nash equilibria. Consider the annoying problem of what to do when a mobile phone call gets dropped. Should you wait for the other person to call you, or should you call? Waiting is a best response if you think the other person will call, and calling is a best response if you think the other person will wait. The problem here is that there are two equally attractive Nash equilibria: You call and the other person waits; or you wait and the other person calls.

Experience doesn't always help get you there. If you both wait, then you might decide to call, but if you both happen to call at the same time, then you get busy signals (or at least you did in the days before call waiting). To resolve this dilemma, we often turn to social conventions, such as having the person who first made the call do the callback. At least that way you know the person has the number.

EPILOGUE TO PART I

In the previous four chapters, we introduced several concepts and methods, using examples from business, sports, politics, and so forth as vehicles. In the chapters to follow, we will put the ideas and techniques to work. Here we recapitulate and summarize them for ready reference.

A *game* is a situation of strategic interdependence: the outcome of your choices (strategies) depends upon the choices of one or more other persons acting purposely. The decision makers involved in a game are called players, and their choices are called *moves*. The interests of the players in a game may be in strict conflict; one person's gain is always another's loss. Such games are called *zero-sum*. More typically, there are zones of commonality of interests as well as of conflict and so, there can be combinations of mutually gainful or mutually harmful strategies. Nevertheless, we usually refer to the other players in a game as one's rivals.

The moves in a game may be *sequential* or *simultaneous*. In a game of sequential moves, there is a linear chain of thinking: If I do this, my rival can do that, and in turn I can respond in the following way. Such a game is studied by drawing a *game tree*. The best choices of moves can be found by applying *Rule 1: Look forward and reason backward.*

In a game with simultaneous moves, there is a logical circle of reasoning: I think that he thinks that I think that . . . and so on. This circle must be squared; one must see through the rival's action even though one cannot see it when making one's own move. To tackle such a game, construct a *table* that shows the outcomes corresponding to all conceivable combinations of choices. Then proceed in the following steps.

Begin by seeing if either side has a *dominant strategy*—one that outperforms all of that side's other strategies, irrespective of the rival's choice. This leads to *Rule 2: If you have a dominant strategy, use it.* If you don't have a dominant strategy, but your rival does, then count on his using it, and choose your best response accordingly.

Next, if neither side has a dominant strategy, see if either has a *dominated strategy*—one that is uniformly worse for the side playing it than all the rest of its strategies. If so, apply *Rule 3: Eliminate dominated strategies from consideration*. Go on doing so successively. If during the process any dominant strategies emerge in the smaller games, they should be chosen. If this procedure ends in a unique solution, you have found the prescriptions of action for the players and the outcome of the game. Even if the procedure does not lead to a unique outcome, it can reduce the size of the game to a more manageable level. Finally, if there are neither dominant nor dominated strategies, or after the game has been simplified as far as possible using the second step, apply *Rule 4: Look for an equilibrium, a pair of strategies in which each player's action is the best response to the other's*. If there is a unique equilibrium of this kind, there are good arguments why all players should choose it. If there are many such equilibria, one needs a commonly understood rule or convention for choosing one over the others. If there is no such equilibrium, that usually means that any systematic behavior can be exploited by one's rivals, which indicates the need for *mixing one's plays*, the subject of the next chapter.

In practice, games can have some sequential moves and some simultaneous moves; in that case a combination of these techniques must be employed to think about and determine one's best choice of actions.

Part II

CHAPTER 5

Choice
and Chance

WIT'S END

The Princess Bride is a brilliant whimsical comedy; among its many memorable scenes, the battle of wits between the hero (Westley) and a villain (the Sicilian Vizzini) ranks high. Westley challenges Vizzini to the following game. Westley will poison one of two glasses of wine out of Vizzini's sight. Then Vizzini will choose to drink from one, and Westley must drink from the other. Vizzini claims to be far smarter than Westley: "Have you ever heard of Plato, Aristotle, Socrates? . . . Morons." He therefore believes he can win by reasoning.

> All I have to do is divine from what I know of you: are you the sort of man who would put the poison into his own goblet or his enemy's? Now, a clever man would put the poison into his own goblet, because he would know that only a great fool would reach for what he was given. I am not a great fool, so I can clearly not choose the wine in front of you. But you must have known I was not a great fool, you would have counted on it, so I can clearly not choose the wine in front of me.

He goes on to other considerations, all of which go in similar logical circles. Finally he distracts Westley, switches the goblets, and laughs confidently as both drink from their respective glasses. He says to Westley: "You fell victim to one of the classic blunders. The most famous is 'Never get involved in a land war in Asia,' but only slightly less well known is this: 'Never go in against a Sicilian when death is on the line.'" Vizzini is still laughing at his expected victory when he suddenly falls over dead.

Why did Vizzini's reasoning fail? Each of his arguments was innately self-contradictory. If Vizzini reasons that Westley would poison goblet A, his deduction is that he should choose goblet B. But Westley can also make the same logical deduction, in which case he should poison goblet B. But Vizzini should foresee this, and therefore should choose goblet A. But . . . There is no end to this circle of logic.*

Vizzini's dilemma arises in many games. Imagine you are about to shoot a penalty kick in soccer. Do you shoot to the goalie's left or right? Suppose some consideration—your being left-footed versus right-footed, the goalie being left-handed versus right-handed, or which side you chose the last time you took a penalty kick—suggests that you should choose left. If the goalie is able to think through this thinking, then he will mentally and even physically prepare to cover that side, so you will do better by choosing his right instead. But what if the goalie raises his level of thinking one notch? Then you would have done better by sticking to the initial idea of kicking to his left. And so on. Where does it end?

---

* Those of you who have seen the movie or read the book know that Vizzini's reasoning had a more basic flaw. Westley had over the years built up immunity to the Iocane powder and had poisoned both glasses. Thus Vizzini was doomed no matter which one he chose, and Westley was safe. Vizzini did not know this and was playing the game under an impossible informational handicap. More generally, when someone else proposes a game or a deal to you, you should always think: "Do they know something I don't?" Recall the advice from Sky Masterson's father: Do not bet the man who would make the jack of spades jump out of the deck and squirt cider in your ear (tale 9 in chapter 1). We will return later in the book to a fuller consideration of such issues of information asymmetries in games. Here we will stay with the flaw of the circular logic, since that has independent interest and numerous applications of its own.

The only logically valid deduction in such situations is that if you follow any system or pattern in your choices, it will be exploited by the other player to his advantage and to your disadvantage; therefore you should not follow any such system or pattern. If you are known to be a left-side kicker, goalies will cover that side better and save your kicks more often. You have to keep them guessing by being unsystematic, or random, on any single occasion. Deliberately choosing your actions at random may seem irrational in something that purports to be rational strategic thinking, but there is method in this apparent madness. The value of randomization can be quantified, not merely understood in a vague general sense. In this chapter we will explicate this method.

MIXING IT UP ON THE SOCCER FIELD

The penalty kick in soccer is indeed the simplest and the best-known example of the general situation requiring random moves or, in game-theoretic jargon, *mixed strategies*. It has been much studied in theoretical and empirical research on games and discussed in the media.[1]

A penalty is awarded for a specified set of prohibited actions or fouls by the defense in a marked rectangular area in front of its goal. Penalty kicks are also used as a tiebreaker of last resort at the end of a soccer match. The goal is 8 yards wide and 8 feet high. The ball is put on a spot 12 yards from the goal line directly in front of the midpoint of the goal. The kicker has to shoot the ball directly from this spot. The goalie has to stand on the goal line at the midpoint of the goal and is not allowed to leave the goal line until the kicker strikes the ball.

A well-kicked ball takes only two-tenths of a second to go from the spot to the goal line. A goalie who waits to see which way the ball has been kicked cannot hope to stop it unless it happens to be aimed directly at him. The goal area is wide; therefore the goalie must decide in advance whether to jump to cover one side and, if so, whether to jump left or right. The kicker in his run up to the spot must also decide which way to kick before he sees which way the goalie is leaning. Of course each will do his best to disguise his choice from the other. Therefore the game is best regarded as one with simultaneous moves. In fact, it is rare for the goalie to stand in the center without jumping left or right, and also relatively rare for the kicker to kick to the center of the goal, and such behavior can also be explained theoretically. Therefore we will simplify the exposition by limiting each player to just two choices. Since kickers usually kick using the inside of their foot, the natural direction of kicking for a right-footed kicker is to the goalie's right, and for a left-footed kicker it is to the goalie's left. For simplicity of writing we will refer to the natural side as "Right." So the choices are Left and Right for each player. When the goalie chooses Right, it means the kicker's natural side.

With two choices for each player and simultaneous moves, we can depict the outcomes in the usual 2-by-2 game payoff table. For each combination of choices of Left and Right by each of the two players, there is still some element of chance; for example, the kick may sail over the crossbar, or the goalie may touch the ball only to deflect it into the net. We measure the kicker's payoff by the percentage of times a goal is scored for that combination of choices, and the goalie's payoff by the percentage of times a goal is not scored.

Of course these numbers are specific to the particular kicker and the goalie, and detailed data are available from the top professional soccer leagues in many countries. For illustrative purposes, consider the average over a number of different kickers and goalies, collected by Ignacio Palacios-Huerta, from the top Italian, Spanish, and English leagues for the period 1995–2000. Remember that in each cell, the payoff shown in the southwest corner belongs to the row player (kicker), and that shown in the northeast corner belongs to the column player (goalie). The kicker's payoffs are higher when the two choose the opposite sides than when they choose the same side. When the two choose opposite sides, the kicker's success rate is almost the same whether the side is natural or not; the only reason for failure is a shot that goes too wide or too high. Within the pair of outcomes when the two choose the same side, the kicker's payoff is higher when he chooses his natural side than when he chooses his non-natural side. All of this is quite intuitive.

		Goalie	
		Left	Right
Kicker	Left	42 / 58	5 / 95
	Right	7 / 93	30 / 70

Let us look for a Nash equilibrium of this game. Both playing Left is not an equilibrium because when the goalie is playing Left, the kicker can improve his payoff from 58 to 93 by switching to Right. But that cannot be an equilibrium either, because then the goalie can improve his payoff from 7 to 30 by switching to Right also. But in that case the kicker does better by switching to Left, and then the goalie does better by also switching to Left. In other words, the game as depicted does not have a Nash equilibrium at all.

The cycles of switching neatly follow the cycles of Vizzini's circular logic as to which goblet would contain the poison. And the fact that the game has no Nash equilibrium in the stated pairs of strategies is exactly the game-theoretic statement of the importance of mixing one's moves. What we need to do is to introduce mixing as a new kind of strategy and then look for a Nash equilibrium in this expanded strategy set. To prepare for that, we will refer to the strategies originally specified—Left and Right for each player—as the *pure strategies.*

Before we proceed with the analysis, let us simplify the game table. This game has the special feature that the two players' interests are exactly opposed. In each cell, the goalie's payoff is always 100 minus the kicker's payoff. Therefore, comparing cells, whenever the kicker has a higher payoff the goalie has a lower payoff, and vice versa.

Many people's raw intuition about games, derived from their experience of sports just like this one, is that each game must have a winner and a loser. However, in the general world of games of strategy, such games of pure conflict are relatively rare. Games in economics, where the players engage in voluntary trade for mutual benefit, can have outcomes where everyone wins. Prisoners' dilemmas illustrate situations where everyone can lose. And bargaining and chicken games can have lopsided outcomes in which one side wins at the expense of the other. So most games involve a mixture of conflict and common interest. However, the case of pure conflict was the first to be studied theoretically and retains some special interest. As we have seen, such games are called zero-sum, the idea being that the payoff of one player is always exactly the negative of that of the other player, or, more generally, *constant-sum*, as in the present case, where the two players' payoffs always sum to 100.

The game table for such games can be simplified in appearance by showing only one player's payoff, since the other's payoff may be understood as the negative of that of the first player or as a constant (such as 100) minus the first player's payoff, as is the case in this example. Usually the row player's payoff is shown explicitly. In that case, the row player prefers outcomes with the larger numbers, and the column player prefers outcomes with the smaller numbers. With this convention, the payoff table for the penalty kick game looks like this:

		Goalie	
		Left	Right
Kicker	Left	58	95
	Right	93	70

If you are the kicker, which of the two pure strategies would you prefer? If you choose your Left, the goalie can keep your success percentage down to 58 by choosing his Left; if you choose your Right, the goalie can keep your success percentage down to 70 by choosing his Right also.* Of the two, you prefer the (Right, Right) combination.

Can you do better? Suppose you choose Left or Right at random in proportions of 50:50. For example, as you stand ready to run up and kick, you toss a coin in the palm of your hand out of the goalie's sight and choose Left if the coin shows tails and Right if it shows heads. If the goalie chooses his Left, your mixture will succeed $1/2 \times 58 + 1/2 \times 93 = 75.5$ percent of the time; if the goalie chooses his Right, your mixture will succeed $1/2 \times 95 + 1/2 \times 70 = 82.5$ percent of the time. If the goalie believes you are making your choice according to such a mixture, he will choose his Left to hold your success rate down to 75.5 percent. But that is still better than the 70 you would have achieved by using the better of your two pure strategies.

An easy way to check whether randomness is needed is to ask whether there is any harm in letting the other player find out your actual choice *before* he responds. When this would be disadvantageous to you, there is advantage in randomness that keeps the other guessing.

Is 50:50 the best mixture for you? No. Try a mixture where you choose your Left 40 percent of the time and your Right 60 percent of the time. To do so, you might take a small book in your pocket, and as you stand ready to run up for your kick, take it out and open it at a random page out of the goalie's sight. If the last digit of the page number is between 1 and 4, choose your Left; if it is between 5 and 0, choose your Right. Now the success rate of your mixture against the goalie's Left is $0.4 \times 58 + 0.6 \times 93 = 79$, and against the goalie's Right it is $0.4 \times 95 + 0.6 \times 70 = 80$. The goalie can hold you down to 79 by choosing his Left, but that is better than the 75.5 percent you could have achieved with a 50:50 mix.

* This might come about because you acquire a reputation for being "always a Left-chooser" or "always a Right-chooser." Of course you don't want to establish such a pattern and a reputation, but that is exactly the point of the merit of randomization that we are in the process of developing.

Observe how the successively better mixture proportions for the kicker are narrowing the difference between the success rates against the goalie's Left and Right choices: from the 93 to 70 difference for the better of the kicker's two pure strategies, to the 82.5 to 75.5 difference for the 50:50 mix, to the 80 to 79 difference for the 40:60 mix. It should be intuitively clear that your best mixture proportion achieves the same rate of success whether the goalie chooses his Left or his Right. That also fits with the intuition that mixing moves is good because it prevents the other player from exploiting any systematic choice or pattern of choices.

A little calculation, which we postpone to a later section of this chapter, reveals that the best mixture for the kicker is to choose his Left 38.3 percent of the time and his Right 61.7 percent of the time. This achieves a success rate of $0.383 \times 58 + 0.617 \times 93 = 79.6$ percent against the goalie's Left, and $0.383 \times 95 + 0.617 \times 70 = 79.6$ percent against the goalie's Right.

What about the goalie's strategy? If he chooses the pure strategy Left, the kicker can achieve 93 percent success by choosing his own Right; if the goalie chooses his pure strategy Right, the kicker can achieve 95 percent success by choosing his own Left. By mixing, the goalie can hold the kicker down to a much lower success rate. The best mixture for the goalie is one that gives the kicker the same success rate whether he chooses to kick to the Left or the Right. It turns out that the goalie should choose the proportions of his Left and Right at 41.7 and 58.3, respectively, and this gives the kicker a success rate of 79.6 percent.

Notice one seeming coincidence: the success percentage the kicker can ensure by choosing his best mixture, namely 79.6, is the same as the success percentage to which the goalie can hold down the kicker by choosing his own best mixture. Actually this is no coincidence; it is an important general property of mixed strategy equilibria in games of pure conflict (zero-sum games).

This result, called the minimax theorem, is due to Princeton mathematician and polymath John von Neumann. It was later elaborated by him, in coauthorship with Princeton economist Oscar Morgenstern, in their classic book *Theory of Games and*

Economic Behavior,[2] which can be said to have launched the whole subject of game theory.

The theorem states that in zero-sum games in which the players' interests are strictly opposed (one's gain is the other's loss), one player should attempt to minimize his opponent's maximum payoff while his opponent attempts to maximize his own minimum payoff. When they do so, the surprising conclusion is that the minimum of the maximum (minimax) payoffs equals the maximum of the minimum (maximin) payoffs. The general proof of the minimax theorem is quite complicated, but the result is useful and worth remembering. If all you want to know is the gain of one player or the loss of the other when both play their best mixes, you need only compute the best mix for one of them and determine its result.

Theory and Reality

How close is the performance of actual kickers and goalies to our theoretical calculations about the respective best mixtures? The following table is constructed from Palacios-Huerta's data and our calculations.[3]

Proportion of Left in mixture for			The percent of times a goal results when the other player chooses his	
			Left	Right
Kicker	Best	38.3%	79.6%	79.6%
	Actual	40.0%	79.0%	80.0%
Goalie	Best	41.7%	79.6%	79.6%
	Actual	42.3%	79.3%	79.7%

Pretty good, huh? In each case, the actual mixture proportions are quite close to the best. The actual mixtures yield almost equal success rates regardless of the other player's choice and therefore are close to being immune to exploitation by the other.

Similar evidence of agreement between actual play and theoretical predictions comes from top-level professional tennis matches.[4]

This is to be expected. The same people regularly play against one another and study their opponents' methods; any reasonably obvious pattern would be noticed and exploited. And the stakes are large, in terms of money, achievement, and fame; therefore the players have strong incentives not to make mistakes.

However, the success of game theory is not complete or universal. Later in this chapter we will examine how well or poorly the theory of mixed strategies succeeds in other games and why. First let us summarize the general principle expressed here in the form of a rule for action:

> **RULE 5: In a game of pure conflict (zero-sum game), if it would be disadvantageous for you to let the opponent see your actual choice in advance, then you benefit by choosing at random from your available pure strategies. The proportions in your mix should be such that the opponent cannot exploit your choice by pursuing any particular pure strategy from the ones available to him—that is, you get the same average payoff when he plays any of his pure strategies against your mixture.**

When one player follows this rule, the other cannot do any better by using one of his own pure strategies than another. Therefore he is indifferent between them and cannot do any better than to use the mixture prescribed for him by the same rule. When both follow the rule, then neither gets any better payoff by deviating from this behavior. This is just the definition of Nash equilibrium from chapter 4. In other words, what we have when both players use this rule is a Nash equilibrium in mixed strategies. So, the von Neumann-Morgenstern minimax theorem can be regarded as a special case of the more general theory of Nash. The minimax theorem works only for two-player zero-sum games, while the Nash equilibrium concept can be used with any number of players and any mixture of conflict and common interest in the game.

Equilibria of zero-sum games don't necessarily have to involve mixed strategies. As a simple example, suppose the kicker has very low success rates when kicking to the Left (his non-natural side) even when the goalie guesses wrong. This can happen because there is a significant probability that the kicker will miss the target anyway when kicking with the outside of his foot. Specifically, suppose the payoff table is:

		Goalie	
		Left	Right
Kicker	Left	38	65
	Right	93	70

Then the strategy Right is dominant for the kicker, and there is no reason to mix. More generally, there can be equilibria in pure strategies without dominance. But this is no cause for concern; the methods for finding mixed strategy equilibria will also yield such pure strategy equilibria as special cases of mixtures, where the proportion of the one strategy in the mixture is the entire 100 percent.

CHILD'S PLAY

On October 23, 2005, Andrew Bergel of Toronto was crowned that year's Rock Paper Scissors International World Champion and received the Gold Medal of the World RPS Society. Stan Long of Newark, California, won the silver medal, and Stewart Waldman of New York the bronze.

The World RPS Society maintains a web site, www.worldrps .com, where the official rules of play and various guides to strategy are posted. It also holds an annual world championship event. Did you know that the game you played as a child has become this big?

The rules of the game are the same as the ones you followed as a kid and as described in chapter 1. Two players simultaneously choose ("throw," in the technical jargon of the game) one of three hand signals: Rock is a fist, Paper is a horizontal flat palm, and Scissors is signified by holding out the index and middle fingers at an angle to each other and pointing toward the opponent. If the two players make the same choice, it is a tie. If the two make different choices, Rock wins against (breaks) Scissors, Scissors wins against (cuts) Paper, and Paper wins against (covers) Rock. Each pair plays many times in succession, and the winner of a majority of these plays is the winner of that match.

The elaborate rules set out on the web page of the World RPS Society ensure two things. First, they describe in precise terms the hand shapes that constitute each type of throw; this prevents any attempts at cheating, where one player makes some ambiguous gesture, which he later claims to be the one that beats what his opponent chose. Second, they describe a sequence of actions, called the prime, the approach, and the delivery, intended to ensure that the two players' moves are simultaneous; this prevents one player from seeing in advance what the other has done and making the winning response to it.

Thus we have a two-player simultaneous-move game with three pure strategies for each. If a win counts as 1 point, a loss as −1, and a tie as 0, the game table is as follows, with the players named Andrew and Stan in honor of their achievements in the 2005 World Championships:

		Stan's choice		
		Rock	Paper	Scissors
Andrew's choice	Rock	0 / 0	−1 / 1	1 / −1
	Paper	1 / −1	0 / 0	−1 / 1
	Scissors	−1 / 1	1 / −1	0 / 0

What would game theory recommend? This is a zero-sum game, and revealing your move in advance can be disadvantageous. If Andrew chooses just one pure move, Stan can always make a winning response and hold Andrew's payoff down to −1. If Andrew mixes the three moves in equal proportions of 1/3 each, it gives him the average payoff of $(1/3) \times 1 + (1/3) \times 0 + (1/3) \times (-1) = 0$ against any one of Stan's pure strategies. With the symmetric structure of the game, this is quite obviously the best that Andrew can do, and calculation confirms this intuition. The same argument goes for Stan. Therefore mixing all three strategies in equal proportions is best for both and yields a Nash equilibrium in mixed strategies.

However, this is not how most participants in the championships play. The web site labels this the Chaos Play and advises against it. "Critics of this strategy insist that there is no such thing as a random throw. Human beings will always use some impulse or inclination to choose a throw, and will therefore settle into unconscious but nonetheless predictable patterns. The Chaos School has been dwindling in recent years as tournament statistics show the greater effectiveness of other strategies."

The problem of "settling into unconscious but nonetheless predictable patterns" is indeed a serious one deserving further discussion, to which we will turn in a moment. But first let us see what kinds of strategies are favored by participants in the World RPS Championship.

The web site lists several "gambits," like the cleverly named strategy Bureaucrat, which consists of three successive throws of Paper, or Scissor Sandwich, which consists of Paper, Scissors, Paper. Another is the Exclusion Strategy, which leaves out one of the throws. The idea behind these is that opponents will focus their entire strategy on predicting when the pattern will change, or when the missing throw will appear, and you can exploit this weakness in their reasoning.

There are also physical skills of deception, and detection of the opponent's deception. The players watch each other's body language and hands for signals of what they are about to throw; they also try to deceive the opponent by acting in a way that suggests one throw and choosing a different one instead. Soccer penalty kickers and goalies similarly watch each other's legs and body movements to guess which way the other will go. Such skills matter; for example, in the penalty shoot-out that decided the 2006 World Cup quarter-final match between England and Portugal, the Portuguese goalie guessed correctly every time and saved three of the kicks, which clinched the victory for his team.

MIXING IT UP IN THE LABORATORY

By contrast with the remarkable agreement between theory and reality of mixed strategies on the soccer field and the tennis court, the evidence from laboratory experiments is mixed or even negative. The first book-length treatment of experimental economics declared flatly: "Subjects in experiments are rarely (if ever) observed flipping coins."[5] What explains this difference?

Some of the reasons are the same as those discussed in chapter 4 when we contrasted the two kinds of empirical evidence. The laboratory setting involves somewhat artificially structured games played by novice subjects for relatively small stakes, whereas the field setting has experienced players engaged in familiar games, for stakes that are huge in terms of fame and prestige and often also in terms of money.

Another limitation of the experimental setting may be at work. The experiments always begin with a session where the rules are carefully explained, and the experimenters go to great lengths to ensure that the subjects understand the rules. The rules make no explicit mention of the possibility of randomization and don't provide coins or dice or the instruction, "You are allowed, if you wish, to flip the coins or roll the dice to decide what you are going to do." Then it is hardly surprising that the subjects, instructed to follow the rules precisely as stated, don't flip coins. We have known ever since Stanley Milgram's renowned experiment that subjects treat experimenters as authority figures to be obeyed.[6] It is hardly surprising that they follow the rules literally and do not think of randomizing.

However, the fact remains that even when the laboratory games were structured to be similar to soccer penalty kicks, where the value of mixing moves is evident, the subjects do not seem to have used randomization either correctly or appropriately over time.[7]

Thus we have a mixed record of success and failure for the theory of mixed strategies. Let us develop some of these findings a little further, both to understand what we should expect in games we observe and to learn how to play better.

HOW TO ACT RANDOMLY

Randomization does not mean alternating between the pure strategies. If a pitcher is told to mix fastballs and forkballs in equal proportions, he should not throw a fastball, then a forkball, then a fastball again, and so on in strict rotation. The batters will quickly notice the pattern and exploit it. Similarly, if the proportion of fastballs to forkballs is to be 60:40, that does not mean throwing six fastballs followed by four forkballs and so on.

What should the pitcher do when mixing fastballs and forkballs randomly in equal proportions? One way is to pick a number at random between 1 and 10. If the number is 5 or less, throw a fastball; if the number is 6 or above, go for the forkball. Of course, this only reduces the problem one layer. How do you go about picking a random number between 1 and 10?

Let us start with the simpler problem of trying to write down what a random sequence of coin tosses will look like. If the sequence is truly random, then anyone who tries to guess what you write down will be correct no more than 50 percent on average. But writing down such a "random" sequence is more difficult than you might imagine.

Psychologists have found that people tend to forget that heads is just as likely to be followed by heads as by tails; therefore they have too many reversals, and too few strings of heads, in their successive guesses. If a fair coin toss comes up heads thirty times in a row, the next toss is still equally likely to be heads or tails. There is no such thing as "being due" for tails. Similarly, in the lottery, last week's number is just as likely to win again as any other number.

The knowledge that people fall into the error of too many reversals explains many of the stratagems and gambits used by participants in the World RPS Championships. Players attempt to exploit this weakness and, at the next higher level, attempt to exploit these attempts in turn. The player who throws Paper thrice in succession is looking for the opponent to think that a fourth Paper is unlikely, and the player who leaves out one of the throws and mixes among just the other two in many successive plays is trying to exploit the opponent's thinking that the missing throw is "due."

To avoid getting caught putting order into the randomness, you need a more objective or independent mechanism. One such trick is to choose some fixed rule, but one that is both secret and sufficiently complicated that it is difficult to discover. Look, for example, at the length of our sentences. If the sentence has an odd number of words, call it heads; if the sentence length is even, call it tails. That should be a good random number generator. Working backward over the previous ten sentences yields T, H, H, T, H, H, H, H, T, T. If our book isn't handy, don't worry; we carry random number sequences with us all the time. Take a succession of your friends' and relatives' birthdates. For even dates, guess heads; for odd, tails. Or look at the second hand on your watch. Provided your watch is not too accurate, no one else will know the current position of the second hand. Our advice to the pitcher who must mix in proportions of 50:50 or to the catcher who is calling the pitches is to glance at his wristwatch just before each pitch. If the second hand points toward an even number, then throw a fastball; an odd number, then throw a forkball. The second hand can be used to achieve any ratio. To throw fastballs 40 percent of the time and forkballs 60 percent, choose fastball if the second hand is between 1 and 24, and forkball if it is between 25 and 60.

Just how successful were the top professionals in tennis and soccer at correct randomization? The analysis of data in grand slam tennis finals revealed that there was indeed some tendency to reverse between serves to the forehand and the backhand more frequently than appropriate for true randomness; in the jargon of statistics, there was negative serial correlation. But it seems to have been too weak to be successfully picked up and exploited by the opponents, as seen from the statistically insignificant difference of success rates of the two kinds of serves. In the case of soccer penalty kicks, the randomization was close to being correct; the incidence of reversals (negative serial correlation) was statistically insignificant. This is understandable; successive penalty kicks taken by the same player come several weeks apart, so the tendency to reverse is likely to be less pronounced.

The championship-level players of Rock Paper Scissors seem to place a lot of importance on strategies that deliberately depart from randomization, and try to exploit the other player's attempts to interpret patterns. How successful are these attempts? One kind of evidence would come from consistency of success. If some players are better at deploying nonrandom strategies, they should do well in contest after contest, year after year. The World RPS Society does not "have the manpower to record how each competitor does at the Championships and the sport is not developed enough so that others track the info. In general, there have not been too many consistent players in a statistically significant way, but the Silver medalist from 2003 made it back to the final 8 the following year."[8] This suggests that the elaborate strategies do not give any persistent advantage.

Why not rely on the other player's randomization? If one player is using his best mix, then his success percentage is the same no matter what the other does. Suppose you are the kicker in the soccer example, and the goalie is using his best mix: Left 41.7 percent and Right 58.3 percent of the time. Then you will score a goal 79.6 percent of the time whether you kick to the Left, the Right, or any mixture of the two. Observing this, you might be tempted to spare yourself the calculation of your own best mix, just stick to any one action, and rely on the other player using his best mix. The problem is that unless you use your best mix, the other does not have the incentive to go on using his. If you stick to the Left, for example, the goalie will switch to covering the Left also. The reason why you *should* use your best mix is to *keep* the other player using his.

Unique Situations

All of this reasoning makes sense in games like football, baseball, or tennis, in which the same situation arises many times in one game, and the same players confront each other from one game to the next. Then there is time and opportunity to observe any systematic behavior and respond to it. Correspondingly, it is important to avoid patterns that can be exploited and stick to the best mix. But what about games that are played just once?

Consider the choices of points of attack and defense in a battle. Here the situation is usually unique, and the other side cannot infer any systematic pattern from your previous actions. But a case for random choice arises from the possibility of espionage. If you choose a definite course of action, and the enemy discovers what you are going to do, he will adapt his course of action to your maximum disadvantage. You want to surprise the enemy; the surest way to do so is to surprise yourself. You should keep your options open as long as possible, and at the last moment choose between them by an unpredictable and, therefore, espionage-proof device. The relative proportions of the device should also be such that if the enemy discovered them, he would not be able to turn the knowledge to his advantage. This is just the best mix calculated in the description above.

Finally, a warning. Even when you are using your best mix, there will be occasions when you have a poor outcome. Even if the kicker is unpredictable, sometimes the goalie will still guess right and save the shot. In football, on third down and a yard to go, a run up the middle is the percentage play; but it is important to throw an occasional bomb to keep the defense honest. When such a pass succeeds, fans and sportscasters will marvel at the cunning choice of play and say the coach is a genius. When it fails, the coach will come in for a lot of criticism: how could he gamble on a long pass instead of going for the percentage play?

The time to justify the coach's strategy is *before* using it on any particular occasion. The coach should publicize the fact that mixing is vital; that the run up the middle remains such a good percentage play precisely because some defensive resources must be diverted to guard against the occasional costly bomb. However, we suspect that even if the coach shouts this message in every newspaper and on every TV channel before the game, and then uses a bomb in such a situation and it fails, he will come in for just as much criticism as if he had not tried to educate the public in the elements of game theory.

MIXING STRATEGIES IN MIXED-MOTIVES GAMES

In this chapter thus far, we have considered only games where the players' motives are in pure conflict, that is, zero-sum or constant-sum games. But we have always emphasized that most games in reality have aspects of common interests as well as conflict. Does mixing have a role in these more general non-zero-sum games? Yes, but with qualifications.

To illustrate this, let us consider the hunting version of the battle of the sexes game from chapter 4. Remember our intrepid hunters Fred and Barney, who are deciding separately, each in his own cave, whether to go stag hunting or bison hunting that day. A successful hunt requires effort from both, so if the two make opposite choices, neither gets any meat. They have a common interest in avoiding such outcomes. But between the two successful possibilities where they are in the same hunting ground, Fred prefers stag meat and rates the outcome of a jointly conducted stag hunt 4 instead of 3, while Barney has the opposite preferences. Therefore the game table is as shown below.

		Barney's choice	
		Stag	Bison
Fred's choice	Stag	3 4 0	0 0
	Bison	0 0 3	4

We saw that the game has two Nash equilibria, shown shaded. We would now call these equilibria in pure strategies. Can there be equilibria with mixing?

Why would Fred choose a mixture? Perhaps he is uncertain about Barney's choice. If Fred's subjective uncertainty is such that he thinks the probabilities of Barney choosing Stag and Bison are y and $(1 - y)$, respectively, then he expects the payoff of $4y + 0(1 - y)$ $= 4y$ if he himself chooses Stag, and $0y + 3(1 - y)$ if he himself chooses Bison. If y is such that $4y = 3(1 - y)$, or $3 = 7y$, or $y = 3/7$, then Fred gets the same payoff whether he chooses Stag or Bison, and also if he chooses to mix between the two in any proportions at all. But suppose Fred's mixture of Stag and Bison is such that Barney is indifferent between his pure strategies. (This game is very symmetric, so you can guess, and also calculate, that this means Fred choosing Stag a fraction $x = 4/7$ of the time.) Then Barney could be mixing in just the right proportions to keep Fred indifferent, and therefore willing to choose just the right mixture of his own. The two mixtures $x = 4/7$ and $y = 3/7$ constitute a Nash equilibrium in mixed strategies.

Is such an equilibrium satisfactory in any way? No. The problem is that the two are making these choices independently. Therefore Fred will choose Stag when Barney is choosing Bison $(4/7) \times$ $(4/7) = 16/49$ of the time, and the other way around $(3/7) \times (3/7)$ $= 9/49$ of the time. Thus in 25/49 or just over half of the times the two will find themselves in separate places and get zero payoffs. Using the formulas in our calculation, we see that each gets the payoff $4 \times (3/7) + 0 \times (4/7) = 12/7 = 1.71$, which is less than the 3 of the unfavorable pure strategy equilibrium.

To avoid such errors, what they need is coordinated mixing. Can they do this while they are in their separate caves with no immediate means of communication? Perhaps they can make an agreement in advance based on something they know they are both going to observe as they set out. Suppose in their area there is a morning shower on half of the days. They can make an agreement that they both will go stag hunting if it is raining and bison hunting if it is dry. Then each will get an average payoff of $1/2 \times 3 + 1/2 \times 4 = 3.5$. Thus coordinated randomization provides them with a neat way to split the difference between the favored and unfavored pure strategy Nash equilibria, that is, as a negotiation device.

The uncoordinated Nash equilibrium in mixed strategies not

only has low payoff, but it is also fragile or unstable. If Fred's estimate of Barney's choosing Stag tips ever so slightly above $3/7 = 0.42857$, say to 0.43, then Fred's payoff from his own Stag, namely $4 \times 0.43 + 0 \times 0.57 = 1.72$, exceeds that from his own Bison, namely $0 \times 0.43 + 3 \times 0.57 = 1.71$. Therefore Fred no longer mixes but chooses pure Stag instead. Then Barney's best response is also pure Stag, and the mixed strategy equilibrium breaks down.

Finally, the mixed strategy equilibrium has a strange and unintuitive feature. Suppose we change Barney's payoffs to 6 and 7 instead of 3 and 4, respectively, leaving Fred's payoff numbers unchanged. What does that do to the mixture proportions? Again write y for the fraction of the time Barney is thought to choose Stag. Then Fred still gets 4y from his own choice of pure Stag and $3(1 - y)$ from his own choice of pure Bison, leading to $y = 3/7$ to keep Fred indifferent and therefore willing to mix. However, writing x for the proportion of Stag in Fred's mixture, Barney gets $6x + 0(1 - x) = 6x$ from his own pure Stag and $0x + 7(1 - x) = 7(1 - x)$ from his own pure Bison. Equating the two, we have $x = 7/13$. Thus the change in Barney's payoffs leaves his own equilibrium mixture unaffected, but changes Fred's equilibrium mixture proportions!

On further reflection, this is not so strange. Barney may be willing to mix only because he is unsure about what Fred is doing. So the calculation involves Barney's payoffs and Fred's choice probabilities. If we set the resulting expressions equal and solve, we see that Fred's mixture probabilities are "determined by" Barney's payoffs. And vice versa.

However, this reasoning is so subtle, and at first sight so strange, that most players in experimental situations fail to figure it out even when prompted to randomize. They change their mixture probabilities when their own payoffs change, not when the other player's payoffs change.

MIXING IN BUSINESS AND OTHER WARS

Our examples of the use of mixed strategies came from the sporting world. Why are there so few instances of randomized behavior out in the "real" worlds of business, politics, or war? First, most of those games are non-zero-sum, and we saw that the role of mixing in those situations is more limited and fragile, and not necessarily conducive to good outcomes. But other reasons also exist.

It may be difficult to build in the idea of leaving the outcome to chance in a corporate culture that wants to maintain control over the outcome. This is especially true when things go wrong, as they must occasionally when moves are chosen randomly. While (some) people understand that a football coach has to fake a punt once in a while in order to keep the defense honest, a similarly risky strategy in business can get you fired if it fails. But the point isn't that the risky strategy will always work, but rather that it avoids the danger of set patterns and predictability.

One application in which mixed strategies improve business performance is price discount coupons. Companies use these to build market share. The idea is to attract new customers without giving the same discount to your existing market. If competitors simultaneously offer coupons, then customers don't have any special incentive to switch brands. Instead, they stay with their current brand and take the discount. Only when one company offers coupons while the others don't are new customers attracted to try the product.

The price coupon game for competitors such as Coke and Pepsi is then analogous to the coordination problem of the hunters. Each company wants to be the only one to give coupons, just as Fred and Barney each want to choose his own favored hunting ground. But if they try to do this simultaneously, the effects cancel out and both are worse off. One solution would be to follow a predictable pattern of offering coupons every six months, and the competitors could learn to alternate. The problem with this approach is that when Coke predicts Pepsi is just about to offer coupons, Coke should step in first to preempt. The only way to avoid preemption is to keep the element of surprise that comes from using a randomized strategy.

Of course, independent randomization runs the risk of "mistakes" exactly as in our story of the Stone Age hunters Fred and

Barney. Competitors can do much better by cooperating instead, and there is strong statistical evidence that Coke and Pepsi reached just such a cooperative solution. There was a span of 52 weeks in which Coke and Pepsi each offered price promotions for 26 weeks, without any overlap. If each was choosing to run a promotion in any one week at random with a 50 percent chance, and choosing this independently of the other, the chance of there being zero overlaps is 1/495918532948104, or less than 1 in a quadrillion (a billion billion)! This was such a startling finding that it made its way to the media, including the CBS program *60 Minutes*.[9]

The purpose of the coupons is to expand market share. But each firm realizes that to be successful, it has to offer promotions when the other is not offering similar promotions. The strategy of randomly choosing weeks for promotion offers may have the intention of catching the other off-guard. But when both firms are following similar strategies, there are many weeks when both offer promotions. In those weeks their campaigns will merely cancel each other out; neither firm increases its share and both make a lower profit. The strategies thus create a prisoners' dilemma. The firms, being in an ongoing relationship, recognize that both can do better by resolving the dilemma. The way to do this is for each company to take a turn at having the lowest price, and then once the promotion ends, everyone goes back to their regular brands. That is just what they did.

There are other cases in which businesses must avoid set patterns and predictability. Some airlines offer discount tickets to travelers who are willing to buy tickets at the last minute. But they won't tell you how many seats are left in order to help you estimate the chances of success. If last-minute ticket availability were more predictable, then there would be a much greater possibility of exploiting the system, and the airlines would lose more of their otherwise regular paying customers.

The most widespread use of randomized strategies in business is to motivate compliance at a lower monitoring cost. This applies to everything from tax audits to drug testing to parking meters. It also explains why the punishment should not necessarily fit the crime.

The typical fine for illegal parking at a meter is many times the meter fee. If the meter rate is a dollar per hour, would a fine of $1.01 suffice to keep people honest? It would, provided the traffic police were *sure* to catch you each time you parked without putting money in the meter. Such enforcement would be very costly. The salaries of the traffic wardens would be the largest item, but the cost of administering the collection mechanism needed to keep the policy credible would be quite substantial, too.

Instead, the authorities use an equally effective and less costly strategy, namely to have larger fines and relax the enforcement efforts. When the fine is $25, a 1 in 25 risk of being caught is enough to keep you honest. A much smaller police force will do the job, and the fines collected will come closer to covering the administrative costs.

This is another instance of the usefulness of mixed strategies. It is similar to the soccer example in some ways, and different in other respects. Once again, the authorities choose a random strategy because it is better than any systematic action: no enforcement at all would mean misuse of scarce parking places, and a 100 percent enforcement would be too costly. However, the other side, the parking public, does not necessarily have a random strategy. In fact the authorities want to make the enforcement probability and the fine large enough to induce the public to comply with the parking regulations.

Random drug testing has many of the same features as parking meter enforcement. It is too time-consuming and costly to test every employee every day for evidence of drug use. It is also unnecessary. Random testing will uncover those who are unable to work drug free and discourage others from recreational use. Again, the probability of detection is small, but the fine when caught is high. This is one of the problems with the IRS audit strategy. The penalties are small given the chances of getting caught. When enforcement is random, it must be that the punishment is worse than the crime. The rule should be that the *expected* punishment should fit the crime, where expectation is in the statistical sense, taking into account the chance of being caught.

Those hoping to defeat enforcement can also use random strategies to their benefit. They can hide the true crime in the midst of many false alarms or decoys, and the enforcer's resources become spread too thin to be effective. For example, an air defense must be able to destroy nearly 100 percent of all incoming missiles. A cost-effective way of defeating the air defense is for the attacker to surround the real missile with a bodyguard of decoys. It is much cheaper to build a decoy missile than the real thing. Unless the defender can perfectly distinguish among them, he will be required to stop all incoming missiles, real and fake.

The practice of shooting dud shells began in World War II, not by the intentional design of building decoy missiles, but as a response to the problem of quality control. "The elimination of defective shells in production is expensive. Someone got the idea then of manufacturing duds and shooting them on a random basis. A military commander cannot afford to have a delayed time bomb buried under his position, and he never knew which was which. The bluff made him work at every unexploded shell that came over."10

When the cost of defense is proportional to the number of missiles that must be shot down, attackers can make this enforcement cost unbearably high. This problem is one of the major challenges in the design of the Star Wars defense; it may have no solution.

HOW TO FIND MIXED STRATEGY EQUILIBRIA

Many readers will be content to understand mixed strategies at a qualitative conceptual level and leave the calculation of the actual numbers to a computer program, which can handle mixed strategies when each player has any number of pure strategies, some of which may not even be used in the equilibrium.11 These readers can skip the rest of this chapter without any loss of continuity. But for those readers who know a little high-school algebra and geometry and want to know more about the method of calculation, we provide a few details.12

First consider the algebraic method. The proportion of Left in the kicker's mixture is the unknown we want to solve for; call it x. This is a fraction, so the proportion of Right is $(1 - x)$. The success rate of the mixture against the goalie's Left is $58x + 93(1 - x) = 93 - 35x$ percent, and that against the goalie's Right is $95x + 70(1 - x) = 70 + 25x$. For these two to be equal, $93 - 35x = 70 + 25x$, or $23 = 60x$, or $x = 23/60 = 0.383$.

We can also find the solution graphically by showing the consequences of various mixes in a chart. The fraction of times Left figures in the kicker's mixture, which we have labeled x, goes horizontally from 0 to 1. For each of these mixtures, one of the two lines shows the kicker's success rate when the goalie chooses his pure strategy Left (L), and the other shows the kicker's success rate when the goalie chooses his pure strategy Right (R). The former line starts at the height 93, namely the value of the expression $93 - 35x$ when x equals zero, and descends to 58, the value of the same expression when x equals 1. The latter line starts at the vertical position of 70, namely the value of the expression $70 + 25x$

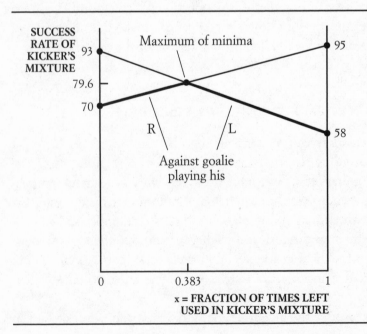

when x equals zero, and rises to 95, the value of the same expression when x equals 1.

The goalie wants to keep the kicker's success rate as low as possible. Therefore if the composition of the kicker's mixture were revealed to the goalie, he would choose L or R, whichever gives the lower of the two lines. These portions of the two lines are shown thicker, forming an inverted V of the minimum success rates the kicker can expect when the goalie exploits the kicker's choice optimally for his own purpose. The kicker wants to choose the highest success rate among these minima. He does this at the apex of the inverted V, where the two lines intersect. Close inspection, or algebraic solution, shows this to be where x = 0.383, and the success rate is 79.6 percent.

We can similarly analyze the goalie's mixing. Let y denote the fraction of times Left figures in the goalie's mixture. Then $(1 - y)$ is the fraction of times the goalie uses his Right. If the kicker plays his L against this mixture, his average success rate is $58y + 95(1 - y) = 95 - 37y$. If the kicker plays his R against this mixture, his average success rate is $93y + 70(1 - y) = 70 + 23y$. For the two expressions to be equal, $95 - 37y = 70 + 23y$, or $25 = 60y$, or y = $25/60 = 0.417$.

The graphical analysis from the goalie's perspective is a simple modification of that for the kicker. We show the consequences of various mixtures chosen by the goalie graphically. The fraction y of times the goalie's Left is included in his mixture goes horizontally from 0 to 1. The two lines show the kicker's success rate against these mixtures, one corresponding to the kicker's choice of his L and the other corresponding to the kicker's choice of his R. For any mixture chosen by the goalie, the kicker does best by choosing L or R, whichever gives him the higher success rate. The thicker portions of the lines show these maxima as a V shape. The goalie wants to keep the kicker's success rate as low as possible. He does so by setting y at the bottom of the V—that is, by choosing the minimum of the maxima. This occurs at y = 0.417, and the kicker's success rate is 79.6 percent.

The equality of the kicker's maximum of minima (maximin) and

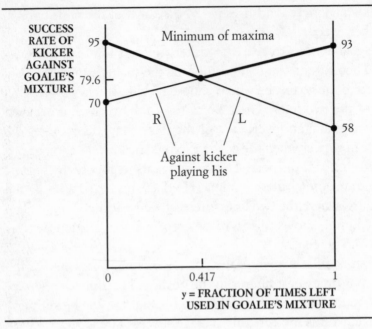

SUCCESS RATE OF KICKER AGAINST GOALIE'S MIXTURE

95

79.6

70

Minimum of maxima

93

R   L

58

Against kicker
playing his

0   0.417   1

y = FRACTION OF TIMES LEFT
USED IN GOALIE'S MIXTURE

the goalie's minimum of maxima (minimax) is just von Neumann and Morgenstern's minimax theorem in action. Perhaps more accurately it should be called the "maximin-equals-minimax theorem," but the common name is shorter and easier to remember.

Surprising Changes in Mixtures

Even within the domain of zero-sum games, mixed strategy equilibria have some seemingly strange properties. Return to the soccer penalty kick and suppose the goalie improves his skill at saving penalties struck to the natural (Right) side, so the kicker's success rate there goes down from 70 percent to 60 percent. What does this do to the goalie's mixture probabilities? We get the answer by shifting the relevant line in the graph. We see that the goalie's use of Left in his equilibrium mix goes up from 41.7 percent to 50 percent. When the goalie improves his skill at saving penalties struck to the right, he uses that side less frequently!

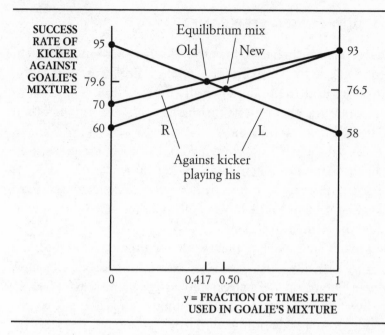

Although this seems strange at first sight, the reason is easy to understand. When the goalie gets better at saving penalties struck to the right, the kicker will kick to the right less frequently. Responding to the fact that more shots are being struck to the left, the goalie chooses that side in greater proportion in his mixture. The point of improving your weakness is that you don't have to use it so often.

You can verify this by recalculating the kicker's mixture in response to this change; you will see that the proportion of Left in his mixture goes up from 38.3 percent to 47.1 percent.

And the goalie's work on his right-side skill does yield a benefit: the average rate of goal scoring in the equilibrium goes down from 79.6 percent to 76.5 percent.

Upon reflection, the seeming paradox has a very natural game-theoretic logic after all. What is best for you depends not only on what you do but what other players do. That is what strategic interdependence is, and should be, all about.

CASE STUDY: JANKEN STEP GAME*

The scene is a sushi bar in downtown Tokyo. Takashi and Yuichi are sitting at the bar drinking sake while waiting for their sushi. Each has ordered the house specialty, uni sashimi (sea urchin). Unfortunately, the chef reports that he has only one serving of uni left. Who will defer to the other?

In America, the two might flip a coin. In Japan, the two would more likely play the Janken game, better known in the West as Rock Paper Scissors. Of course, by now you are experts in RPS, so to make the problem a little more challenging, we introduce a variant called the Janken step game.

The Janken step game is played on a staircase. As before, the two players simultaneously cast rock, paper, or scissors. But now, the winner climbs up the staircase: five steps if the winner played paper (five fingers); two steps for winning with scissors (two fingers); one step for winning with rock (no fingers). Ties are replayed. Normally, the winner is the first to the top of the stairs. We simplify the game slightly by assuming that each player's goal is just to get as far ahead of the other player as possible.

What is the equilibrium mixture of strategies for this version of the Janken step game?

Case Discussion

Since each step puts the winner further ahead and the loser that much further behind, we have a zero-sum game. Considering all possible pairs of moves leads to the following game table. The payoffs are measured in terms of steps ahead.

---

* This case first appeared in the Japanese edition of *Thinking Strategically*. It is the result of a project undertaken by Takashi Kanno and Yuichi Shimazu while students at Yale School of Management. They were also the book's Japanese translators.

		Yuichi's choice		
		Rock	Paper	Scissors
Takashi's choice	Rock	0 0	5 −5	−1 1
	Paper	−5 5	0 0	2 −2
	Scissors	1 −1	−2 2	0 0

How can we find the equilibrium mixture of throwing Paper, Scissors, and Rock? Earlier we showed you some simple numerical calculations and graphical methods that are useful when each side has only two alternatives, like forehand and backhand. But in the Janken step game, there are three alternatives.

The first question to ask is what strategies will be part of the equilibrium mixture. Here the answer is that all three are essential. To confirm this, imagine that Yuichi never plays Rock. Then Takashi would never play Paper, in which case Yuichi would never use Scissors. Continuing along this line implies that Takashi would never use Rock, thus Yuichi would never use Paper. The assumption that Yuichi never uses Rock eliminates all of his strategies and so, must be false. A similar argument demonstrates that the other two strategies are indispensable to Yuichi's (and Takashi's) mixing equilibrium.

We now know that all three strategies must be used in the equilibrium mixture. The question becomes when will all three strategies be used. The players are interested in maximizing their payoffs, not mixing for mixing's sake. Yuichi is willing to randomize between Rock, Paper, and Scissors if and only if all three options are equally attractive. (If Rock offered Yuichi a higher payoff than either Paper or Scissors, then he should play Rock exclusively; but that would not be an equilibrium.) Thus, the special case when all three strategies give Yuichi the same expected payoff is what defines Takashi's equilibrium mixture.

Let us suppose that Takashi uses the following mixing rule:

> p = probability that Takashi casts paper;
> q = probability that Takashi casts scissors;
> 1 − (p + q) = probability that Takashi casts rock.

Then if Yuichi plays rock, he will fall behind five steps if Takashi plays paper (p) and win one step if Takashi plays scissors (q), for a net payoff of −5p + q. In the same way, Yuichi would get the following payoffs from each of his strategies:

> Rock: $-5p + 1q + 0(1 - (p + q)) = -5p + q$
> Scissors: $2p + 0q - 1(1 - (p + q)) = 3p + q - 1$
> Paper: $0p - 2q + 5(1 - (p + q)) = -5p - 7q + 5$

Yuichi will find the three options equally attractive only when

$$-5p + q = 3p + q - 1 = -5p - 7q + 5$$

Solving these equations reveals: p = 1/8, q = 5/8, and (1 − p − q) = 2/8.

This defines Takashi's equilibrium mixture. The game is symmetric, so Yuichi will randomize according to the same probabilities.

Note that when both Yuichi and Takashi use their equilibrium mixture, their expected payoff from each strategy is zero. While this is not a general feature of mixed strategy outcomes, it is always true for symmetric zero-sum games. There is no reason why Yuichi should be favored over Takashi, or vice versa.

In chapter 14, "Fooling All the People Some of the Time: The Las Vegas Slots" offers another case study on choice and chance.

CHAPTER 6

Strategic
Moves

CHANGING THE GAME

Millions of people make at least one New Year's resolution every year. A Google search for the phrase "New Year's resolutions" produces 2.12 million pages. According to a U.S. government web site, the most popular of these resolutions is "lose weight." This is followed by "pay off debt," "save money," "get a better job," "get fit," "eat right," "get a better education," "drink less alcohol," and "quit smoking."[1]

Wikipedia, the free online encyclopedia, defines a New Year's resolution as "a commitment that an individual makes to a project or a habit, often a lifestyle change that is generally interpreted as advantageous." Note the word "commitment." Most people have an intuitive understanding of it, in the sense of a resolve, a pledge, or an act of binding oneself. We will soon make the concept more precise in its game-theoretic usage.

What happens to all these wonderful life-improving plans? A CNN survey reports that 30 percent of the resolutions are not even kept into February, and only 1 in 5 stays on track for six months or longer.[2] Many reasons contribute to this failure: people set themselves excessively ambitious goals, they do not have good methods for measuring their progress, they lack the time, and so on. But by far the most important cause of failure is that, like Oscar Wilde, most people can resist anything except temptation. When they see and smell those steaks, french fries, and desserts, their diets are doomed. When those new electronic gadgets beckon, the resolution to keep the credit card in the wallet falters. When they are sitting comfortably in their armchairs watching sports on TV, actually exercising seems too much like hard work.

Many medical and lifestyle advisers offer tips for success in keeping resolutions. These include basics such as setting reasonable and measurable goals, working toward them in small steps, setting up a regime of healthy food and exercise that is varied to prevent boredom, not getting discouraged, and not giving up after any setbacks. But the advice also includes strategies for creating the right incentives, and an important feature of these is a support system. People are advised to join groups that diet and exercise together and to publicize their resolutions among their family and friends. The feeling that one is not alone in this endeavor surely helps, but so does the shameful prospect of public failure.

This shame factor was powerfully harnessed by one of us (Nalebuff) in an ABC *Primetime* program, *Life: The Game*.[3] As described in our opening chapter, the overweight participants agreed to be photographed wearing just a bikini. Anyone who failed to lose 15 pounds in the following two months would have his or her photographs displayed on national television and on the program's web site. The desire to avoid this fate served as a powerful incentive. All but one of the participants lost the 15 pounds or more; one failed but only narrowly.

Where does game theory come in? The struggle to lose weight (or to save more money) is a game of one's current self (who takes a long-run viewpoint and wants to improve health or wealth) against a future short-run self (who is tempted to overeat and overspend). The current self's resolution constitutes a commitment to behave better. But this commitment must be irreversible; the future self should be denied the possibility of reneging. The current self does this by taking an associated action—being photographed in an embarrassing outfit and giving up control over the use of these pictures to the program's producer to display if the weight loss is insufficient. This changes the game by changing the future self's incentives. The temptation to overeat or overspend still exists, but it is countered by the prospect of shameful exposure.

Such actions that change the game to ensure a better outcome for the player taking the actions are called *strategic moves*. In this chapter we will explicate and illustrate many of these moves. There are two aspects to consider: what needs to be done and how to do it. The former is amenable to the science of game theory, while the latter is specific to each situation—thinking up effective strategic moves in each specific context is more art than science. We will equip you with the basics of the science and try to convey some of the art through our examples. But we must leave it to you to further develop the art you will need in the games you play, based on your knowledge of the situations.

For our second example of changing the game, imagine yourself as an American male teenager in the 1950s. You live in a small town. It is a clear Saturday evening. You are with a group of friends, playing games of rivalry to decide who is the alpha male. Tonight's contests start with a game of chicken. As you race toward a head-on collision, you know that the one who swerves first is the loser, or chicken. You want to win.

This is a dangerous game. If both of you attempt to win, both may end up in the hospital, or worse. We analyzed this game in chapter 4 from the perspective of Nash equilibrium (and in the context of the Stone Age hunters Fred and Barney) and found that it had two Nash equilibria, one where you go straight and your rival swerves, and the other where you swerve while your rival continues straight. Of course you prefer the first to the second. Here we take the analysis to a higher level. Can you do anything to achieve your preferred outcome?

You could establish a reputation as someone who never swerves. However, to do that you must have won similar games in the past, so the question just transfers itself to what you could have done in those games.

Here is a fanciful but effective device. Suppose you disconnect your steering wheel from the shaft and throw it out of the window in a manner that makes it very visible to your rival. He now knows that you *can't* swerve. The whole onus of avoiding the collision is on him. You have changed the game. In the new game you have just one strategy, namely to go straight. And then your rival's best (actually, least bad) response is to swerve. You are helpless as a driver, but that very helplessness makes you a winner in the game of chicken.

The way you have changed this game in your favor is surprising at first sight. By losing your steering wheel, you have restricted your own freedom of action. How can it be beneficial to have fewer choices? Because in this game, freedom to swerve is merely freedom to become the chicken; freedom to choose is freedom to lose. Our study of strategic moves will produce other seemingly surprising lessons.

This example also serves to motivate a fair warning about strategic moves. Their success is not guaranteed, and sometimes they can be outright dangerous. In reality, there are delays in action and observation. In chicken, what if your rival gets the same idea, and each of you simultaneously sees the other's steering wheel flying through the air? Too late. Now you are headed helplessly toward a crash.

So try these devices at your own risk, and don't sue us if you fail.

A LITTLE HISTORY

People and nations have made commitments, threats, and promises for millennia. They have intuitively recognized the importance of credibility in such actions. They have used such strategies and devised counterstrategies against other players' use of them. When Homer's Odysseus tied himself to the mast, he was making a credible commitment not to be lured by the song of the Sirens. Parents understand that while a threat of cold-bloodedly punishing a child for misbehavior is not credible, the threat "Do you want Mommy to get angry?" is much more believable. Kings throughout history have understood that voluntarily exchanging hostages—giving up a beloved child or other relative to live in a rival monarch's family—helps make their mutual promises of peaceful coexistence credible.

Game theory helps us understand and unify the conceptual framework of such strategies. However, in its first decade, game theory focused on characterizing different kinds of equilibria in a *given game*—backward reasoning in sequential-move games, minimax in two-person zero-sum games, and Nash in more general simultaneous-move games—and illustrating them in important contexts like the prisoners' dilemma, assurance, battle of the sexes, and chicken.[4] Thomas Schelling gets the honor and credit for being the first person to develop the idea that one or both players might take actions to *change the game* as a central theme of game theory. His articles in the late 1950s and early 1960s, collected and elaborated on in his books *The Strategy of Conflict* (1960) and *Arms and Influence* (1965),[5] gave us precise formulations of the concepts of commitment, threat, and promise. Schelling clarified just what is needed for credibility. He also analyzed the subtle and risky strategy of brinkmanship, which had previously been much misunderstood.

A more rigorous formal development of the concept of credibility, namely subgame perfect equilibrium, which is a generalization of the backward reasoning equilibrium we discussed in chapter 2, came a few years later from Reinhard Selten, who in 1994 was in the first group of game theorists to receive a Nobel Prize, jointly with John Nash and John Harsanyi.

COMMITMENTS

Of course, you don't have to wait until New Year's Day to make a good resolution. Every night you can resolve to wake up early the next morning to get a good start on the day, or perhaps to go for that five-mile run. But you know that when morning comes, you will prefer to stay in bed for another half-hour or hour (or longer). This is a game of your resolute nighttime self against your own future weak-willed morning self. In the game as structured, the morning self has the advantage of the second move. However, the nighttime self can change the game to create and seize first-mover advantage by setting the alarm clock. This is intended as a commitment to get out of bed when the alarm rings, but will it work? Alarm clocks have snooze buttons, and the morning self can hit the button, repeatedly. (Of course an even earlier self could have searched and bought an alarm clock without a snooze button, but that may not have been possible.) The night self can still make the commitment credible by keeping the alarm clock on the wardrobe across the room instead of on the bedside table; then the morning self will have to get out of bed to shut off the noise. If this is not enough and the morning self stumbles straight back into bed, then the night self will have to think up some other device, perhaps an alarm clock that at the same time starts brewing coffee, so the wonderful smell will induce the morning self out of bed.*

This example nicely illustrates the two aspects of commitments and credibility: what and how. The "what" part is the scientific or game-theoretic aspect—seizing first-mover advantage. The "how" part is the practical aspect or the art—thinking up devices for making strategic moves credible in a specific situation.

We can illustrate the mechanics or science of the commitment of the alarm clock using the tree diagrams of chapter 2. In the original game, where the night self takes no action, the game is trivial:

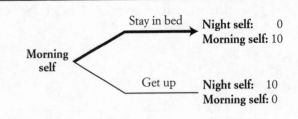

---

* There are some amazing gadgets on the market. Clocky is an alarm clock with wheels. When the buzzer goes off, the clock jumps off your nightstand and scurries away. By the time you capture and silence it, you are fully awake.

The morning self stays in bed and gets its preferred payoff, which we have assigned 10 points, leaving the night self with its worse payoff, to which we have assigned 0 points. The precise number of points does not matter much; all that matters is that for each self, the preferred alternative is assigned more points than the less-preferred one.

The night self can change the game into the following:

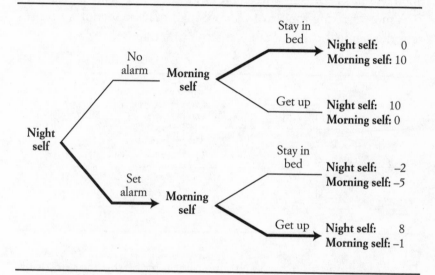

Now the payoff numbers matter a little bit and need more explanation. Along the upper main branch where the night self does not set an alarm, the tree is as before. Along the lower main branch, we have supposed that the night self has a small cost, which we have set at 2 points, of setting the alarm clock. So if the morning self heeds the alarm and gets up, the night self will get 8 points, instead of the 10 in the original game. But if the morning self were to ignore the alarm, the night self would get −2 points since the cost of setting the alarm was wasted. The morning self has an annoyance cost of hearing the alarm; it is only 1 if it gets out of bed to turn off the alarm quickly but would be intolerably large (15 points) if it stayed in bed and the alarm went on and on, converting the pleasure of the bed (10) into a payoff of -5 (= $10 - 15$). If the alarm has been set, the morning self prefers -1 to -5 and gets up. The night self looks ahead to this, and reasons that setting the alarm will give it 8 points in the eventual outcome, which is better than the zero it would get in the original game.* Therefore, in the backward reasoning equilibrium of the game, the morning self does get up if an alarm has been set, and the night self sets the alarm.

A more striking aspect of commitment may be seen if we represent this game in a game table, instead of a tree.

		Morning self		Morning self	
		Stay in bed		Get up	
Night self	No alarm	0	10	10	0
	Set alarm	-2	-5	8	-1

The table shows that for each given strategy of the morning self, the night self's payoff from Set alarm is smaller than that from No alarm: -2 is less than 0, and 8 is less than 10. Therefore for the night self, the strategy Set alarm is dominated by No alarm. Nevertheless, the night self finds it desirable to commit to Set alarm!

How can it be good to choose a dominated strategy and not play a dominant strategy? To understand this, we need to understand the concept of dominance more clearly. No alarm dominates Set alarm from the perspective of the night self because, for each *given* strategy of the morning self, No alarm yields a higher

* If the cost of the action were too high—for example if the night self had to set a timed incendiary device that would start a fire in the bed to get the morning self out of it—then the commitment would not be optimal for the night self to make.

payoff to the night self than does Set alarm. If the morning self chooses Stay in bed, the night self gets 0 from No alarm and –2 from Set alarm; if the morning self chooses Get up, the night self gets 10 from No alarm and 8 from Set alarm. If moves are simultaneous, or if the night self moves second, he cannot affect what the morning self chooses and must accept it as given. But the very purpose of a strategic move is to *alter* the other player's choice, not to take it as given. If the night self chooses Set alarm, the morning self will choose Get up and the night self will have payoff 8; if the night self chooses No alarm, the morning self will choose Stay in bed and the night self's payoff will be 0; and 8 is greater than 0. The payoffs of 10 and –2, and their comparisons with 8 and 0, respectively, become irrelevant. Thus the concept of dominance loses its significance for a first mover in a sequential game.

For most of the examples we give in this chapter, you will be able to get the idea without drawing any such explicit trees or tables, so we will generally offer only verbal statements and reasoning. But you can reinforce your understanding of the game, and of the tree method, by drawing them for yourself if you wish.

THREATS AND PROMISES

A commitment is an *unconditional* strategic move; as the Nike slogan says, you "just do it"; then the other players are followers. The night self simply sets the alarm on the bureau and the timer on the coffee machine. The night self has no further moves in the game; one might even say that the night self ceases to exist in the morning. The morning self is the follower player or second mover, and its best (or least bad) response to the night self's commitment strategy is to get out of bed.

Threats and promises, on the other hand, are more complex *conditional* moves; they require you to fix in advance a *response rule,* stating how you would respond to the other player's move in the actual game. A threat is a response rule that punishes others who fail to act as you would like them to. A promise is an offer to reward other players who act as you would like them to.

The response rule prescribes your action as a response to the others' moves. Although you act as a follower in the actual game, the response rule must put be in place *before* others make their moves. A parent telling a child "No dessert unless you eat your spinach" is establishing such a response rule. Of course, this rule must be in place and clearly communicated before the child feeds her spinach to the dog.

Therefore such moves require you to change the game in more complex ways. You must seize the first-mover status in the matter of putting the response rule in place and communicating it to the other player. You must ensure that your response rule is credible, namely that if and when the time comes for you to make the stated response, you will actually choose it. This may require changing the game in some way to ensure that the choice is in fact best for you in that situation. But in the game that follows, you must then have the second move so you will have the ability to respond to the other's choice. This may require you to restructure the order of moves in the game, and that adds its own difficulties to your making the strategic move.

To illustrate these ideas, we will use the example of the pricing rivalry of the catalog merchants B. B. Lean and Rainbow's End, which we developed as a simultaneous-move game in chapters 3 and 4. Let us recapitulate its basic points. The two are competing over a specific item, a deluxe chambray shirt. Their joint interests are best served if the two collude and charge a monopoly price of $80. In this situation each will make a profit of $72,000. But each has the temptation to undercut the other, and if they both do so, in the Nash equilibrium each will charge only $40 and make a profit of only $40,000. This is their prisoners' dilemma, or a lose-lose game; when each gives way to the temptation to make a bigger profit for itself, both lose.

Now let us see if strategic moves can resolve the dilemma. A commitment by one of them to keep its price high won't do; the other will simply exploit it to the disadvantage of the first. What about conditional moves? Rainbow's End might employ a threat ("If you charge a low price, so will I") or a promise ("If you keep your price at the monopoly level, so will I"). But if the actual game of choosing prices in the catalog has simultaneous moves in the sense that neither can observe the other's catalog before setting its own in print, how can Rainbow's End *respond* to B. B. Lean's move at all? It must change the game so it has the opportunity to choose its price after it knows the other's price.

A clever commonly used device, the meet-the-competition clause, achieves this purpose. In its catalog, Rainbow's End prints the price $80, but with a footnote: "We will meet any lower price charged by any competitor." Now the catalogs are printed and mailed simultaneously, but if B. B. Lean has cheated and printed a price lower than $80, perhaps all the way down to the Nash equilibrium price of $40, then Rainbow's End automatically matches that cut. Any customer who might have a slight preference or loyalty toward Rainbow's End need not switch to B. B. Lean for its lower price, he can simply order from Rainbow's End as usual and pay the lower price listed in the B. B. Lean catalog.

We will return to this example again to illustrate other aspects of strategic moves. For now, just note two distinct aspects: the scientific or "what" aspect (the threat to match any price cut) and the art or the "how" aspect (the meet-the-competition clause that makes the threat possible and credible).

DETERRENCE AND COMPELLENCE

The overall purpose of threats and promises is similar to that of commitments, namely, to induce the others to take actions different than they would otherwise. In the case of threats and promises, it is useful to classify the overall purpose into two distinct categories. When you want to stop the others from doing something they would otherwise do, that is *deterrence*. Its mirror image, namely to compel the others to do something they would not

(MINI) TRIP TO THE GYM NO. 4

Set up the tree for the game of the cold war, and show how the U.S. threat changes the equilibrium outcome of the game.

otherwise do, can then be termed *compellence.*6

When a bank robber holds the employees hostage and establishes a response rule that he will kill them if his demands are rejected, he is making a compellent threat. When, during the cold war, the United States threatened to respond with nuclear weapons if the Soviet Union attacked any NATO country, it made a deterrent threat. The two threats share a common feature: *both* sides will bear an extra cost if the threat has to be carried out. The bank robber compounds the punishment he will face when caught if he adds murder to his original crime of armed robbery; the United States would suffer horribly in a nuclear war when it could have lived with a Soviet-dominated Europe.

Promises can also be compellent or deterrent. A compellent promise is designed to induce someone to take a favorable action. For example, a prosecutor who needs a witness to buttress his case promises one defendant a more lenient sentence if he turns state's evidence against his codefendants. A deterrent promise is designed to prevent someone from taking an action that is against your interests, as when the mobsters promise a confederate they will protect him if he keeps his mouth shut. Like the two kinds of threats, the two promises also share a common feature. After the other player has complied with one's wishes, the promisor no longer needs to pay the cost of delivering the reward and has the temptation to renege. Thus, after the mob bosses on trial are acquitted for lack of evidence, they might kill the confederate anyway to avoid the risk of any future trouble or blackmail.

A QUICK REFERENCE GUIDE

We have thrown many concepts at you thick and fast. To help you remember them, and to refer to them later at a glance, here is a chart:

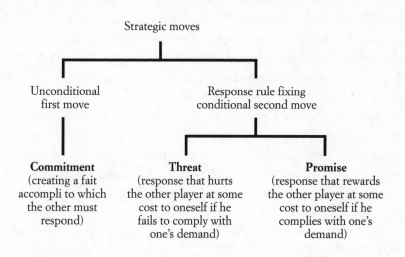

And here is a table summarizing, in the form of pregame statements of the strategic mover, how threats and promises seek to achieve each of the two aims: deterrence and compellence. "If, in the game to follow, you . . .

	Deterrence	Compellence
Threat	...do what I don't want you to do...	...don't do what I want you to do...
	...then I will respond with an action which will hurt you (and will also hurt me).	
Promise	...don't do what I don't want you to do...	...do what I want you to do...
	...then I will respond with an action which will reward you (and will be costly for me)."	

WARNINGS AND ASSURANCES

All threats and promises have a common feature: the response rule requires you to take actions that you would not take in its absence. If instead the rule merely says that you will do what is best at the time, this is as if there is no rule: there is no *change* in others' expectations about your future actions and hence no change in their actions. Still, there is an informational role for stating what will happen, even without any rule; these statements are called *warnings* and *assurances*.

When it is in your interest to carry out a "threat," we call this a *warning.* For example, if the president warns he will veto a bill not to his liking, this is simply an indication of his intentions. It would be a threat if he were willing to sign the bill but strategically committed to veto it in order to induce Congress to offer something even better.

To illustrate this in a business context, let us examine whether B. B. Lean's matching Rainbow's End's price cuts constitutes a threat or a warning. In chapter 4 we considered the best response of B. B. Lean to various prices that Rainbow's End could conceivably charge. We found that it was somewhere between zero and full response. If B. B. Lean were to keep its price unchanged while Rainbow's End cut its price, then B. B. Lean would lose too many customers to its rival. But if B. B. Lean were to match Rainbow's End's price cut dollar for dollar, its own profit margin would be squeezed too far. In the example we developed, B. B. Lean struck the optimal balance between these two considerations by reducing its price by 40 cents for each dollar reduction in Rainbow's End's price.

But if B. B. Lean wants to threaten Rainbow's End to deter it from initiating any price cuts, it may need to threaten a larger response than the 40 cents per dollar that would be optimal in the event of an actual price cut by Rainbow's End. In fact, B. B. Lean may want to threaten a superaggressive response of more than a dollar. It can do so by printing a beat-the-competition clause instead of merely a meet-the-competition clause in its catalog. Such devices are genuine threats in our terminology. B. B. Lean would find it costly to carry out the actions if put to the test by Rainbow's End. Its threat is made credible by printing its policy in the catalog, so its customers can rely upon it as law, and

B. B. Lean cannot renege on it. If B. B. Lean had said in its cata-log: "For every dollar that Rainbow's End's price falls short of $80, we will charge 40 cents less than our catalog price of $80," this would be merely a warning to Rainbow's End; if put to the test, B. B. Lean would want to go through with the stated response anyway.

When it is in your interest to carry out a promise, we call this an assurance. In the shirt pricing example, B. B. Lean may secretly want to tell Rainbow's End that if they hold to the collusive price of $80, so will B. B. Lean. In the game played once, this is not in the interest of B. B. Lean after the fact. Therefore it is a genuine strategic move, namely a promise. If the game was repeated so that continued mutual cooperation was an equilibrium, as we saw in chapter 3, then the statement from B. B. Lean would be an assur-ance, intended merely to inform Rainbow's End that B. B. Lean was quite aware of the nature of the repeated game and how it offered a resolution to their prisoners' dilemma.

To reiterate the point, threats and promises are truly strategic moves, whereas warnings and assurances play more of an informa-tional role. Warnings or assurances do not change your response rule in order to influence another party. Instead, you are simply informing them of how you will want to respond based on their actions. In stark contrast, the sole purpose of a threat or promise is to change your response rule away from what will be best when the time comes, not in order to inform but to manipulate.

Because threats and promises indicate that you will act against your own interest, their credibility becomes the key issue. After others have moved, you have an incentive to break your threat or promise. Some other accompanying change in the game is needed to ensure credibility. Without credibility, other players will not be influenced by mere words. Children who know that their parents get pleasure from giving them toys are not influenced by threats to withhold toys unless the parents take some prior action to make the threat credible.

Strategic moves, therefore, contain two elements: the planned course of action and the associated actions that make this course credible. We will try to give you a better appreciation of both aspects by making two passes through the ideas. In the remainder of this chapter we focus attention on the former, or what needs to be done to make threats and promises. Think of this as a menu of moves. In the next chapter we will turn our focus to the recipes for credibility—that is, how to make threats and promises believable and therefore effective.

OTHER PLAYERS' STRATEGIC MOVES

It is natural to think of the advantages you can get from strategic moves, but you should also think about how such moves made by other players will affect you. In some cases you might even benefit by relinquishing the opportunity to make such a move and purposely allowing someone else to do so. Three such logical possibilities are:

> You may allow someone to make an unconditional move before you respond.
>
> You may wait for a threat before taking any action.
>
> You may wait for a promise before taking any action.

We have already seen examples in which someone who could move first does even better by relinquishing this option, allowing the other side to make an unconditional move. This is true whenever it is better to follow than to lead, as in the chapter 1 tale of the America's Cup race (and in the chapter 14 case study on gambling at the Cambridge May Ball). More generally, if the game, played sequentially, has a second mover advantage, you can benefit by arranging things so that the other player must move first, thereby making an unconditional commitment. While it can be advantageous to give up the initiative, this is not a general rule. Sometimes your goal will be to prevent your opponent from making an unconditional commitment. This was the motivation behind Chinese military strategist Sun Tzu's advice to leave the enemy an escape route—the idea is to prevent the enemy from making a commitment to fight to the death.

It is never advantageous to allow others to threaten you. You could always do what they wanted you to do without the threat. The fact that they can make you worse off if you do not cooperate cannot help, because it limits your available options. But this maxim applies only to allowing threats. If the other side can make promises, then you can both be better off. A simple example is the prisoners' dilemma, where both players can benefit if even one player has some way to make a credible promise to keep quiet. Note that it must be a conditional move, a promise, not an unconditional commitment. If the other player were to make a commitment to keep quiet, you would simply exploit it by confessing, and, knowing this, he would not make such a move.

SIMILARITIES AND DIFFERENCES BETWEEN THREATS AND PROMISES

Sometimes the distinctions between threats and promises are blurred. A friend was mugged in New York City with the following promise: If you "lend" me twenty dollars, I promise I won't hurt you. More relevant was the mugger's implicit threat that if our friend *didn't* lend him the money, he would be hurt.

As this story suggests, the distinction between a threat and a promise depends only on what you call the status quo. The traditional mugger threatens to hurt you if you don't give him some money. If you don't, he starts hurting you, making that the new status quo, and promises to stop once you give him money. A compellent threat is just like a deterrent promise with a change of status quo; likewise, a deterrent threat and a compellent promise differ only in their status quo.

So should you use a threat or a promise? The answer depends on two considerations. The first is the cost. A threat can be less costly; in fact, it is costless if it is successful. If it changes the other player's behavior in the way you want, you don't have to carry out the costly action you had threatened. A promise, if successful, must be fulfilled—if the other player acts as you want him to, you have to deliver the costly action you had promised. If a company could threaten its employees with terrible consequences should their performance fall short of being excellent, it could save a lot of money that it usually pays out to fulfill its promises of incentive bonuses. Indeed, Stalin tried using just sticks instead of carrots—threats of being sent to the Gulag in Siberia instead of promises of better pay or living conditions—to get good performance from Soviet workers. But his system did not work because its methods for judging performance were inaccurate, arbitrary, and corrupt. We will return to this point in the next section.

The second consideration in the choice between a threat and a promise is whether the purpose is deterrence or compellence. The two have different time dimensions. Deterrence does not necessarily have a deadline. It simply involves telling the other player not to do such and such, and credibly communicating the bad consequences that would follow if he takes the forbidden action. So the United States says to the Soviet Union: "Don't invade Western Europe," or God says to Adam and Eve, "Don't eat the apple." "When?" "Ever."* Therefore deterrence can be achieved more simply and better by a threat. You set up a tripwire, and it is up to the other to decide whether to trigger it.

In contrast, compellence must have a deadline. When a mother says to her child, "Clean your room," a time limit such as "before 5:00 P.M. today" must accompany it. Otherwise the child can defeat the purpose by procrastination: "I have soccer practice today; I will do it tomorrow," and when tomorrow comes, some other, more urgent task will come up. If the mother has threatened some dire consequence, she does not want to invoke it for each seemingly tiny delay. The child can defeat her threat "slice by slice," a strategy that Schelling calls *salami tactics.*

Therefore compellence is often better achieved by giving the other player the incentive not to procrastinate. This means that earlier performance must get a better reward or lighter punish-

---

* If the threatener changes his mind, he can always lift the threat. Thus, if the United States eventually got fed up with de Gaulle's antics, it could simply hint to the Soviet Union that it would be okay if they now invaded France.

ment. This is a promise. The mother says: "You will get that special treat for dessert when you have cleaned your room," and the mugger says: "The knife at your throat will go away as soon as you have given me your money."

CLARITY AND CERTAINTY

When making a threat or a promise, you must communicate to the other player quite clearly what actions will bring what punishment (or what reward). Otherwise, the other may form a wrong idea of what is forbidden and what is encouraged and miscalculate the consequences of his actions. Stalin's stick-type "incentives" for workers in the Soviet Union suffered from this crucial flaw. The monitoring system was arbitrary and corrupt, so the worker stood almost as much a risk of going to Siberia if he worked hard as if he shirked. So why work?

But clarity does not have to be a simple either-or choice. In fact, such a stark alternative may be poor strategy. The United States wanted to deter the Soviet Union from invading Western Europe. But threatening all-out nuclear war in the event of the smallest transgression, say a handful of soldiers straying across the border, might be too risky. When a company wants to promise rewards to its workers for improved productivity, a bonus that increases gradually with an increase in output or profit may be better than offering nothing if the performance does not exceed a set target, and a very large sum if it does.

For a threat or promise to have its desired effect, the other player must believe it. Clarity without certainty doesn't cut it. Certainty does not mean a complete lack of risk. When a company offers stock bonuses to its managers, the value of the promised reward is uncertain, influenced by many factors that affect the market and are outside the control of the manager. But the manager should be told just how many shares he will get in relation to the immediately measurable indicator of his performance on which the bonus is based.

Nor does certainty require that everything happens at once.

Threats and promises that work in multiple small steps are especially useful against salami tactics. When we give exams to students, there are always a few who attempt to keep writing after the
time is up, in the hope of getting a few extra points. Grant them an
extra minute and they will go past that, grant another minute and
it becomes five, and so on. The dire punishment of refusing to
accept an exam that is two or three minutes late would not be
credible, but levying a graduated penalty of a few grade points per
minute of delay is perfectly credible.

LARGE THREATS

If a threat is successful, the threatened action does not have to
be carried out. Even though it may be costly for you to carry it out,
since you don't have to do so, the cost is irrelevant. So why not use
a huge threat that would really frighten the other player into
acceding to your wishes? Instead of politely asking your dinner
table neighbor to please pass the salt, why don't you threaten him
with: "If you don't pass the salt, I will smash your head"? Instead
of patiently negotiating with trading partner countries in an
attempt to persuade them to lower their barrier against our
exports, why doesn't the United States threaten that if they don't
buy more of our beef or wheat or oranges, we will nuke them?

This is an obviously horrific idea; the threats are too large to be
useable or believable. In part this is because they would generate
terror and revulsion at the gross violation of all social norms of
behavior. But in part it is also because the assumption that you
would never have to carry out the threatened action is not 100 percent valid. Suppose something goes wrong. Your dinner table
neighbor may be the obstinate kind who revolts at any prospect of
bullying, or a tough guy who enjoys an opportunity for a fight. If
he refuses to comply, you must either go through with the threatened action or back down and face the humiliation and loss of reputation. Similar considerations apply to the United States if it tries
to threaten another country with a harsh military action in an
economic dispute. Even slight risks of such hugely costly errors

provide strong arguments for keeping threats at the smallest level needed to keep them effective.

Very often you don't know the exact size of a threat that is needed to deter or compel your adversary. You want to keep the size as low as possible to minimize the cost to you in the event that things go wrong and you have to go through with the action. So you start small and gradually raise the size of the threat. This is the delicate strategy of brinkmanship.

BRINKMANSHIP

In the book and movie *L.A. Confidential,* the "good cop" Ed Exley is interrogating a suspect, Leroy Fontaine, when the hot-tempered cop Bud White intervenes:

> The door banged open. Bud White stepped in, threw Fontaine against the wall.
>
> Ed froze.
>
> White pulled out his .38, broke the cylinder, dropped shells on the floor. Fontaine shook head to toe; Ed kept freezing. White snapped the cylinder shut, stuck the gun in Fontaine's mouth. "One in six. Where's the girl?"
>
> Fontaine chewed steel; White squeezed the trigger twice: clicks, empty chambers. [So now the risk has risen to one in four.] Fontaine slid down the wall; White pulled the gun back, held him up by his hair. *"Where's the girl?"*
>
> Ed kept freezing. White pulled the trigger—another little click. [So now it is one in three.] Fontaine, bug-eyed. "S-ss-sylvester F-fitch, one-o-nine and Avalon, gray corner house please don' hurt me no—"
>
> White ran out.[7]

Obviously White is threatening Fontaine to compel him to reveal the information. But what is the threat? It is not simply: "If you don't tell me, I will kill you." It is: "If you don't tell me, I will pull the trigger. If the bullet happens to be in the firing chamber, you will die." It is creating a *risk* that Fontaine will be killed. And every time the threat is repeated, the risk is increasing. Finally, at one in three, Fontaine finds the risk too high and spills the information. But there were other possibilities: White may have feared that the information would die with Fontaine, found the risk too high, backed down, and tried something else. Or the thing they both feared—the bullet reaches the firing chamber and Fontaine dies— might have come about.

A similar situation arises in the movie *The Gods Must Be Crazy.* There has been an unsuccessful attempt on the life of the president of a country in Africa. The presidential guards have caught one of the attackers, and he is being interrogated for information about the rest of his group. He stands blindfolded with his back to the open door of a helicopter with its rotors whirring. The officer facing him asks: "Who is your leader? Where is your hideout?" No answer. The officer pushes him out of the door of the helicopter. The scene switches to the outside. We see that the helicopter is actually hovering just a foot off the ground, and the man has fallen on his back. The interrogating officer appears at the door, laughs, and says to the man: "The next time it will be a little bit higher." The scared man gives away the information.

What is the purpose of such threats of increasing risk? We argued in the previous section that there are good reasons for keeping the size of a threat down to the smallest level that will have the desired effect. But you may not know in advance the smallest effective size of a threat. That is why it makes sense to start small, and increase it gradually to find out when it works. As the size of the threatened action grows, the cost of carrying it out grows too. In the examples above, the way to increase the size of the threat is to increase the *risk* of the bad thing happening. The maker and the recipient of the threat are then engaged in a game of exploring each other's tolerance for this cost or the risk. Is a one-in-four chance of Fontaine being killed too large for Fontaine or for White? If not, then try one in three. They continue this eyeball-to-eyeball confrontation until one of them blinks—or until the outcome they both fear transpires.

This is the strategy that Schelling called *brinkmanship.** The term is often interpreted as taking an adversary to the brink of disaster in order to get him to blink first. Standing on the brink, you threaten to push him off if he fails to comply with your wishes. Of course, he will take you down with him. That is why, says Schelling, the pure and simple threat of cold-bloodedly pushing the adversary off the brink is not credible.

> If the brink is clearly marked and provides a firm footing, no loose pebbles underfoot and no gusts of wind to catch one off guard, if each climber is in full control of himself and never gets dizzy, neither can pose any risk to the other by approaching the brink. . . . [W]hile either can deliberately jump off, he cannot credibly pretend that he is about to. Any attempt to intimidate or to deter the other climber depends on the threat of slipping or stumbling. . . . [O]ne can credibly threaten to fall off accidentally by standing near the brink.

> Deterrence has to be understood in relation to this uncertainty. . . . A response that carries some risk of war [through a compounding of actions and reactions, of calculations and miscalculations, of alarms and false alarms] can be plausible, even reasonable, at a time when a final, ultimate decision to have a general war would be implausible or unreasonable.[8]

The 1962 Cuban missile crisis provided perhaps the most famous example of brinkmanship. The Soviet Union, under its mercurial leader Nikita Khrushchev, had begun to install nuclear missiles on Cuba, ninety miles from the American mainland. On October 14, American reconnaissance airplanes brought back photographs of missile sites under construction. After a week of tense discussions within his administration, on October 22 President John F. Kennedy announced a naval quarantine of Cuba. Had the Soviet Union taken up the challenge, the crisis could have escalated to the point of all-out nuclear war between the superpowers. Kennedy himself estimated the probability of this as "between one out of

* Many people say "brinksmanship"—which sounds more like the art of robbing an armored truck.

three and even." But after a few anxious days of public posturing and secret negotiation, Khrushchev looked over the nuclear brink, did not like what he saw, and pulled back. In return for a face-saving compromise involving eventual withdrawal of U.S. missiles from Turkey, he ordered the Soviet missiles in Cuba dismantled and shipped back.[9]

Just where was the brink in the Cuban missile crisis? Had the Soviets tried to defy the blockade, for example, the United States was unlikely to launch its strategic missiles at once. But events and tempers would have heated up another notch, and the risk of Armageddon would have increased perceptibly.

Soldiers and military experts speak of the "fog of war"—a situation in which both sides act with disrupted lines of communication, individual acts of fear or courage, and a great deal of general uncertainty. There is too much going on to keep everything under control. This serves the purpose of creating some risk. Even the president found it difficult to control the operations of the naval blockade of Cuba once put into play. Kennedy tried to move the blockade from 800 miles to 500 miles off the shore of Cuba in order to give Khrushchev more time. Yet evidence based on the first ship boarded, the *Marcula* (a Lebanese freighter under charter by the Soviets), indicates that the blockade was never moved.[10]

The key to understanding brinkmanship is to realize that the brink is not a sharp precipice but a slippery slope, getting gradually steeper. Kennedy took the world some way down this slope; Khrushchev did not risk going farther, and then the two arranged a pullback to the safe ground above.*

The essence of brinkmanship is the deliberate creation of risk. This risk should be sufficiently intolerable to your opponent to induce him to eliminate the risk by following your wishes. The

---

* Of course, it would be a mistake to think of the Cuban missile crisis as a game with only two players, Kennedy and Khrushchev. On each side, there was another game of internal politics, with the civilian and military authorities disagreeing among themselves and with one another. Graham Allison's *Essence of Decision* (Boston: Little, Brown, 1971) makes a compelling case for regarding the crisis as just such a complex multiperson game.

game of chicken, discussed in the preceding chapters, is of this kind. Our earlier discussions supposed that each driver had just two choices, whether to swerve or to go straight. But in reality the choice is not whether to swerve but when to swerve. The longer the two keep on going straight, the greater the risk of a collision. Eventually the cars get so close to each other that even if one of the drivers decides that the danger is too high and swerves, it may be too late to avoid a collision. In other words, brinkmanship is "chicken in real time": a game of increasing risk, just like the interrogation games in the movies.

Once we recognize this, we see brinkmanship everywhere. In most confrontations, for example, between a company and a labor union, a husband and a wife, a parent and a child, and the president and Congress, one or both of the players cannot be sure of the other party's objectives and capabilities. Therefore most threats carry a risk of error, and almost every threat contains an element of brinkmanship. Understanding the potentialities and risks of this strategic move can prove crucial in your life. Use it carefully, and understand that even with the best care it may fail, because the bad thing you and the other player both dread may come to pass while you are raising the stakes. If your assessment is that in this confrontation you will "blink first"—that is, the probability of the bad thing happening will get too large for your own tolerance before the limit of the other player's tolerance is reached—then you may be better advised not to embark on the path of brinkmanship in the first place.

We will return to some aspects of the art of practicing brinkmanship in the next chapter. For now, we end on a cautionary note. With any exercise of brinkmanship, there is always the danger of falling off the brink. While we look back at the Cuban missile crisis as a successful use of brinkmanship, our evaluation would be very different if the risk of a superpower war had turned into a reality. The survivors would have cursed Kennedy for recklessly and unnecessarily flaming a crisis into a conflagration. Yet in any exercise of brinkmanship, the risk of falling off the brink can turn into a reality. The massacre of the Chinese students occupying

Beijing's Tiananmen Square in June 1989 is a tragic example. The students were on a collision course with the hard-liners in their government. One side would have to lose; either the hard-liners would cede power to more reform-minded leaders or the students would compromise on their demands. During the confrontation, there was a continual risk that the hard-liners would overreact and use force to squelch the democracy movement. When two sides are playing a game of brinkmanship and neither side is backing down, there is a chance that the situation will get out of control, with tragic consequences.

In the aftermath of Tiananmen Square, government leaders became more aware of the dangers in brinkmanship—for both sides. Faced with similar democracy protests in East Germany and Czechoslovakia, the communist leaders decided to give in to popular demands. In Romania, the government tried to hold firm against a reform movement, using violent repression to maintain power. The violence escalated almost to the level of a civil war, and in the end President Nicolae Ceauşescu was executed for crimes against his people.

CASE STUDY: TWO WRONGS KEEP THINGS RIGHT

Parents often face a difficult problem in punishing their children for bad behavior. Children have an uncanny sense of when the parents' threat to punish may not be credible. They recognize that the punishment may hurt the parents as much as the children (although for different reasons). The standard parental dodge to this inconsistency is that the punishment is for the child's own good. How can parents do a better job at making their threat to punish bad behavior credible?

Case Discussion

With two parents and one child, we have a three-person game. Teamwork can help the parents make an honest threat to punish a misbehaving child. Say the son misbehaves, and the father is scheduled to carry out the punishment. If the son attempts to rescue himself by pointing out the "irrationality" of his father's actions, the father can respond that he would, given the choice, prefer not to punish his son. But, were he to fail in carrying out the punishment, that would be breaking an agreement with his wife. Breaking that agreement would be worse than the cost of punishing the child. Thus the threat to punish is made credible.

Single parents can play this game, but the argument gets much more convoluted, as the punishment agreement must be made with the child. Now if the son attempts to rescue himself by pointing out the "irrationality" of his father's actions, the father can respond that he would, given the choice, prefer not to punish his son. But, were he to fail in carrying out the punishment, then this would be a misdeed on his part, a misdeed for which he should be punished. Thus, he is punishing his son only to prevent getting punished himself. But who is there to punish him? It's the son! The son replies that were his father to forgive him, he too would forgive his father and not punish his father for failing to punish him. The father responds that were his son to fail to punish him for being lenient, this would be the *second* punishable offense done by the son in the same day! And so on and so forth do they keep each other honest. This may seem a little far-fetched, but it is no less convoluted than most real arguments used to justify punishing kids who misbehave.

A compelling example of how two people can keep each other honest comes from Yale economist Dean Karlan. Dean was keen to lose weight and so wrote a contract with one of his friends that if either of them was ever above 175 pounds, the overweight one would owe the other $1,000 per pound. Dean is a professor, and so that was a large financial penalty looming over his head. The threat worked for him and for his friend, too. But there was always the question of whether the friends would actually take each other's money.

Dean's friend got lazy and creeped up to 190. Dean called him on the scale and took $15,000 of his money. Dean didn't want to take money from his friend, but he knew that by doing so, his friend would then have no hesitation to take the money back should Dean ever fail. Dean engaged in the punishment to ensure that he would be punished if need be. Knowing that this threat is real has worked for Dean. If you'd like, he offers this service to others though his Commitment Store, which we discuss in the next chapter.

This concludes our brief sketch of the "what" of threats and promises. (For more practice, have a look at the "Arms Across the Ocean" case study in chapter 14.) Although we did have to say something about credibility, that was not the focus so far. In the next chapter we turn our attention to the issue of making credible strategic moves. We can offer only a general guide to this; it is largely an art that you must acquire by thinking about and through the dynamics of your own specific situation.

CHAPTER 7

Making Strategies Credible

IN GOD WE TRUST?

Early in Genesis, God explains to Adam the punishment for eating from the tree of knowledge.

> You are free to eat from any tree in the garden; but you must not eat from the tree of the knowledge of good and evil, for when you eat of it you will surely die. (2:16–17)[1]

Would you eat the apple? What would be the point of gaining the knowledge only to die moments later? And yet the wily serpent tempts Eve into having a taste. The serpent suggests that God was bluffing.

> "You will not surely die," the serpent said to the woman. "For God knows that when you eat of it your eyes will be opened, and you will be like God, knowing good and evil." (3:4–5)

As we all know, Adam and Eve do partake, and of course God catches them. Now remember the threat. God should smite them down and start all over again.

And therein lies the problem. It would be costly for God to follow through. He'd have to destroy his creation, made in his own image, and the whole sixth day's good work would be wasted. God comes up with a revised and much less drastic punishment. Adam and Eve are banished from the Garden of Eden. Adam has to till the barren ground. For Eve, childbirth was made painful. Yes, they were punished, but it didn't come anywhere close to getting killed. The snake was right after all.*

This is the genesis of the problem of making a credible threat. If we can't believe a threat from God, whose word can we believe?

Harry Potter? Here we have a hero, a brave young wizard with a heart of gold, who is willing to sacrifice his life in order to defeat He-Who-Must-Not-Be-Named. And yet, in the *Deathly Hallows* finale, Potter promises the goblin Griphook that if he helps Harry break into the vault at Gringotts Wizarding Bank, the Sword of Gryffindor will be Griphook's reward. While Harry does intend to eventually return the sword to the goblins, he first plans to use it to destroy some Horcruxes. Hermione points out that Griphook expects to get the sword right away. Harry is willing to mislead, even cheat, Griphook in order to achieve his larger objective. As it turns out, Griphook does get the sword, but only by nabbing it from Harry during their escape from Gringotts. Even Harry has a credibility problem.

We want to convince others—children, associates, rivals—that they should (or shouldn't) take some action . . . or else. We want to convince others that they should help us because of our promise. But the threat or the promise is often not in our interest to carry out in the end. How do we change the game to make it credible?

Commitments, threats, and promises will not improve your outcome in a game if they are not credible. We emphasized this in the previous chapter and discussed some aspects of credibility. But the focus there was on the more mechanical aspects of strategic moves, namely, what needs to be done to change the game. We split the

---

* For more on this interpretation see David Plotz's Blogging the Bible at www.slate.com/id/2141712/entry/2141714. We recognize that this exposition glosses over some details of the story that are considered important in more mainstream interpretations. (Remember, we're economists, not theologians.) In the common Christian interpretation, God kept his promise: Adam and Eve died spiritually when they ate the apple. This spiritual death was the great fall from grace, restored only by Christ.

topic in this way because the "what" of strategic moves is more amenable to the science of game theory, whereas the "how" aspect is more of an art, which can be conveyed only partially and by suggestion. In this chapter we offer several examples, grouped into categories, to give you some idea of what devices are more likely to succeed in which circumstances. You will have to develop these ideas to suit the context of the games you play, practice the art, and refine it from your own experience. And whereas science often gives clear-cut answers to questions—something either works or it doesn't—success or perfection in art is usually a matter of degree. So don't expect success all the time, and don't be discouraged by the occasional failure either.

THE EIGHTFOLD PATH TO CREDIBILITY

In most situations, mere verbal promises should not be trusted. As Sam Goldwyn put it, "A verbal contract isn't worth the paper it's written on."[2] An incident in *The Maltese Falcon* by Dashiell Hammett, which became a movie classic with Humphrey Bogart as Sam Spade and Sydney Greenstreet as Gutman, further illustrates the point. Gutman gives Sam Spade an envelope containing ten thousand dollars.

> Spade looked up smiling. He said mildly: "We were talking about more money than this."
>
> "Yes sir, we were," Gutman agreed, "but we were talking then. This is actual money, genuine coin of the realm, sir. With a dollar of this you can buy more than with ten dollars of talk."[3]

Indeed, this lesson can be traced all the way back to the eighteenth-century philosopher Thomas Hobbes: "The bonds of words are too weak to bridle men's avarice."[4] Women's too, as King Lear discovered. Words must be backed up by appropriate strategic actions if they are to have an effect on the other players' beliefs and actions.*

---

* If the other players' objectives are perfectly aligned with yours, you can trust their words. For example, in the assurance game of Fred and Barney

We classify the actions that can enhance the credibility of your unconditional and conditional strategic moves and that can help you practice brinkmanship into eight categories, which are based on three broad principles. We will state them first and then illustrate each.

The first principle is to change the payoffs of the game. The idea is to make it in your interest to follow through on your commitment: turn a threat into a warning, a promise into an assurance. This can be done through two broad classes of tactics:

1. Write contracts to back up your resolve.

2. Establish and use a reputation.

Both these tactics make it more costly to break the commitment than to keep it.

A second avenue is to change the game by limiting your ability to back out of a commitment. In this category, we consider three possibilities:

3. Cut off communication.

4. Burn bridges behind you.

5. Leave the outcome beyond your control, or even to chance.

These two principles can be combined: the available actions and their payoffs can both be changed.

If a large commitment is broken down into many smaller ones, then the gain from breaking a little one may be more than offset by the loss of the remaining contract. Thus we have:

6. Move in small steps.

---

planning to meet for a hunt, if one of them can communicate to the other which area he will go to, the other can believe the statement. If the players' interests are partially aligned, some valid inferences can be drawn from statements. This theory of "cheap talk" in games was developed by Vincent Crawford and Joel Sobel and plays an important role in the more advanced levels of game theory. However, in most strategic situations, words are not to be trusted unless backed by actions, so we will focus our attention on such situations.

A third route is to use others to help you maintain commitment. A team may achieve credibility more easily than an individual. Or you may simply hire others to act in your behalf.

7. Develop credibility through teamwork.

8. Employ mandated agents.

We now proceed to illustrate the use of each of these devices. But remember, what we offer is only a basic guide to what is essentially an art.

Contracts

A straightforward way to make your commitment credible is to agree to pay a penalty if you fail to follow through. If your kitchen remodeler gets a large payment up front, he is tempted to slow down the work. But a contract that specifies payment linked to the progress of the work and penalty clauses for delay can make it in his interest to stick to the schedule. The contract is the device that makes the remodeler's promise of completion credible.

Actually, it's not quite that simple. Imagine that a dieting man offers to pay $500 to anyone who catches him eating fattening food. Every time the man thinks of a dessert, he knows that it just isn't worth $500. Don't dismiss this example as incredible; just such a contract was offered by a Mr. Nick Russo—except the amount was $25,000. According to the *Wall Street Journal*, "So, fed up with various weight-loss programs, Mr. Russo decided to take his problem to the public. In addition to going on a 1,000-calorie-a-day diet, he is offering a bounty—$25,000 to the charity of one's choosing—to anyone who spots him eating in a restaurant. He has peppered local eateries . . . with 'wanted' pictures of himself."[5]

But this contract has a fatal flaw: there is no mechanism to prevent renegotiation. With visions of éclairs dancing in his head, Mr. Russo should point out that no one will ever actually get the $25,000 bounty because he will never violate the contract. Hence the contract is worthless to the enforcers. Renegotiation would be in their mutual interest. For example, Mr. Russo might offer to buy a round of drinks in exchange for being released from the contract. The restaurant diners prefer a drink to nothing and let him out of the contract.* For the contracting approach to be successful, the party that enforces the action or collects the penalty must have some independent incentive to do so. In the dieting problem, Mr. Russo's family might also want him to be skinnier and thus would not be tempted by a mere free drink.

The contracting approach is better suited to business dealings. A broken contract typically produces damages, so that the injured party is not willing to give up on the contract for naught. For example, a producer might demand a penalty from a supplier who fails to deliver. The producer is not indifferent about whether the supplier delivers or not. He would rather get his supply than receive the penalty sum. Renegotiating the contract is no longer a mutually attractive option. What happens if the supplier tries the dieter's argument? Suppose he attempts to renegotiate on the grounds that the penalty is so large that the contract will always be honored and the producer will never receive the penalty. This is just what the producer wants, and hence he is not interested in renegotiation. The contract works because the producer is not solely interested in the penalty; he cares about the actions promised in the contract.

In some instances, the contract holder might lose his job if he allows the contract to be rewritten. Thomas Schelling provides a remarkable example of how these ideas have been implemented.[6] In Denver, one rehabilitation center treats wealthy cocaine addicts by having them write a self-incriminating letter that will be made public if they fail random urine analysis. After placing themselves voluntarily in this position, many people will try to buy their way out of the contract. But the person who holds the contract will lose his job if the contract is rewritten; the center will

---

* Even so, Mr. Russo might find it difficult to renegotiate with a large number of people simultaneously. If even one person fails to agree, the renegotiation won't be successful.

lose its reputation if it fails to fire employees who allow contracts to be rewritten.

The ABC *Primetime* program on dieting, which we described in chapter 1, had a similar feature. According to the contract, any dieters who failed to lose the stipulated 15 pounds over two months would have their bikini photos displayed on *Primetime* and on the ABC web site. As it turned out, one woman failed narrowly but was forgiven by the program's producers. She had lost 13 pounds, dropped two dress sizes, and looked great. What mattered was not whether ABC actually broadcast the photos but whether the dieters believed they would.

This act of kindness would seem to have destroyed ABC's credibility to enforce such contracts in a follow-up program. Nonetheless, the show was repeated. The second time around, the dieters were the administrative staff of Bridgeport's minor league baseball team, the Bluefish. Since ABC could no longer be counted on to broadcast the pictures, this time the team agreed to display them on its jumbotron screen at a home game on the night of the weigh-in. Once again, most dieters succeeded, but one woman just missed the 15-pound goal. She asserted that it would cause her great psychological damage if the pictures went up. This implied the threat of a lawsuit, and so ABC and the team backed down. Now participants in any future rounds are unlikely to regard the device as credible, and Barry and ABC will have to think up something else.*

Most contracts specify that some third party will be in charge of enforcing it. A third party does not have any personal interest in whether the contract is upheld. Its incentive to enforce the contract comes from other sources.

Our colleagues Ian Ayres and Dean Karlan have started a company to offer just this sort of third-party contract enforcement. They call it the Commitment Store (www.stickK.com). If you want

---

* How about taking a photo shoot of the ABC producers and their lawyers in Speedos and then give Barry the authority to post the photos on the web if ABC doesn't follow through? Of course, either way, there would be no sequel—after posting the photos, Barry would never work in the business again. Remember, there's always a larger game.

to lose weight, you can go online and sign up for how much you want to lose and what happens if you fail. For example, you can post a $250 bond that will go to a designated charity if you don't reach your goal. (If you succeed, you get your money back.) There is also a pari-mutuel option. You and a friend can wager that you will each lose 15 pounds over the next two months. If you both succeed, the money is returned. But if one fails while the other succeeds, then the loser pays the winner. If you both fail, then the one who loses the most is the winner.

How can you trust the Commitment Store to keep its word? One reason is that they don't have anything to gain. If you fail, the money goes to a charity, not them. Another reason is that they have a reputation to keep. If they are willing to renegotiate, then their service is of no value. And, were they to renegotiate, you might even be able to sue them for breach of contract.

This naturally leads us to the contract-enforcing institution that we know best: the court system. Judges or juries don't stand to gain anything directly whether one side or the other wins a civil case arising from a contract dispute (at least so long as the system is not corrupt). They are motivated to weigh the facts of the case in light of the laws and render an impartial verdict. For the jurors, this is mainly because their education and socialization have taught them to regard this as an important part of a citizen's duties, but also for fear of punishment if they are found to have violated the oath they took when the jury was formed. The judges have their professional pride and ethic that motivates them to be careful and deliver correct verdicts. They have strong career reasons as well: if they make too many errors and are repeatedly overruled on appeal by higher courts, they will not be promoted.

In many countries, alas, the state's courts are corrupt, slow, biased, or simply unreliable. In such situations, other nongovernmental contract-enforcement institutions emerge. Medieval Europe developed a code called Lex Mercatoria, or Merchant Law, for the enforcement of commercial contracts which was applied by private judges at trade fairs.[7]

If the government does not provide contract enforcement as a service to its citizens, someone might do so for a profit. Organized crime often fills the niches of enforcement left unfilled or vacated by formal law.* Diego Gambetta, a professor of sociology at Oxford, conducted a case study on the Sicilian Mafia's role of providing protection to private economic activity, including enforcement of property rights and contracts. He quotes a cattle rancher he interviewed: "When the butcher comes to me to buy an animal, he knows that I want to cheat him [by giving him a low-quality animal]. But I know that he wants to cheat me [by reneging on payment]. Thus we need Peppe [that is, a third party] to make us agree. And we both pay Peppe a percentage of the deal."8 The reason why the rancher and the butcher could not use the formal Italian law was that they were doing informal deals to avoid taxes.

Gambetta's Peppe enforces contracts among his customers using two methods. First, he acts as a store of information about the past behavior of traders in his territory. A trader becomes Peppe's customer by paying him a retainer. When considering a deal with a stranger, the customer asks Peppe what he knows about the trader's past record. If this record is flawed, the customer can refuse the deal. In this role, Peppe is like a credit-rating agency or a Better Business Bureau. Second, Peppe can mete out punishment, typically involving physical violence, to someone who cheats one of his customers. Of course Peppe may collude with the other party to double-cross the customer; the only thing that keeps Peppe honest is concern for his long-run reputation.

Alternative institutions of enforcement, such as the Mafia, get their credibility by developing a reputation. They may also develop expertise, which enables them to evaluate evidence faster or more accurately than the court system can. These advantages can prevail even when the court system is reliable and fair, and the alternative

---

* People who are dissatisfied with the outcome they get from the formal legal system may also take recourse to such extralegal methods for private "justice." At the beginning of the novel and the movie *The Godfather*, the undertaker Amerigo Bonasera comes to the conclusion that American courts are biased against immigrants like him and that only the "godfather's justice" can avenge the dishonor of his daughter.

tribunals coexist with the formal machinery of the law. Many industries have such arbitration panels to adjudicate disputes among their members and between their members and customers. Lisa Bernstein, a professor at the University of Chicago Law School, conducted a now famous study of the tribunal system used by New York diamond traders. She found that this system has some further advantages; it can impose severe sanctions on members who break contracts and then defy the panel's judgment. The panel posts the name and photograph of the miscreant on the bulletin board at the Diamond Traders' Club. This effectively drives the offender out of business. He also faces social ostracism because many of the traders are part of a tightly knit social and religious network.[9]

Thus we have numerous institutions and mechanisms for enforcing contracts. But none of them is proof against renegotiation. The matter comes to the attention and under the adjudication of a third party only when one of the two parties to the contract decides to bring it there. But if the two main parties to the contract have the temptation to renegotiate, they can do so at their joint will and the original contract will not be enforced.

Therefore, contracts alone cannot overcome the credibility problem. Success can be enhanced by using some additional tools for credibility, such as employing parties with independent interests in enforcement or having a sufficiently strong reputation at stake. In fact, if the reputation effect is strong enough, it may be unnecessary to formalize a contract. This is the sense of a person's word being his bond.

A wonderful example of how a strong reputation can obviate the need for a contract comes from the Verdi opera *Rigoletto*. Gambetta quotes:

"Kill the hunchback?! What the devil do you mean?" snaps Sparafucile, opera's prototype of the honorable hit man, at the suggestion that he might kill his client Rigoletto. "Am I a thief? Am I a bandit? What client of mine has ever been cheated? This man pays me and he buys my loyalty."[10]

Sparafucile's agreement with Rigoletto did not need to specify: "It is hereby agreed that the party of the first part shall not under any circumstances kill the party of the second part."

Reputation

If you try a strategic move in a game and then back off, you may lose your reputation for credibility. In a once-in-a-lifetime situation, reputation may be unimportant and therefore of little commitment value. But you typically play several games with different rivals at the same time, or the same rivals at different times. Future rivals will remember your past actions and may hear of your past actions in dealing with others. Therefore you have an incentive to establish a reputation, and this serves to make your future strategic moves credible.

Gambetta, in his study of the Sicilian Mafia, examines how its members can create and maintain a reputation for toughness to lend credibility to their threats. Which devices work, and which ones don't? Wearing dark glasses won't work. Anyone can do that; it does not serve to differentiate a truly tough person. A Sicilian accent won't help; in Sicily almost everyone has a Sicilian accent, and even elsewhere it is just as likely to be an accident of birth as a mark of toughness. No, says Gambetta, the only thing that really works is a record of committing acts of toughness, including murder. "Ultimately, the test consists of the ability to use violence both at the outset of one's career and later, when an established reputation is under attack from authentic and bogus rivals alike."[11] In most business contexts we merely talk of "cutthroat competition"; Mafiosi practice it!

Sometimes a public declaration of your resolve can work by putting your reputation on the line in a public way. During the tense period of the cold war in the early 1960s, President John F. Kennedy made several speeches to create and uphold just such a public reputation. The process began with his inaugural address: "Let every nation know, whether it wishes us well or ill, that we shall pay any price, bear any burden, meet any hardship, support any friend, oppose any foe, to assure the survival and the success of

liberty." During the Berlin crisis in 1961, he explained the importance of the U.S. reputation in terms that illustrate the idea of
strategic reputation: "If we do not meet our commitments to
Berlin, where will we later stand? If we are not true to our word
there, all that we have achieved in collective security, which relies
on these words, will mean nothing." And perhaps most famously
during the Cuban missile crisis, he declared: "any nuclear missile
launched from Cuba against any nation in the Western Hemi-
sphere [would be regarded] as an attack on the United States,
requiring a full retaliatory response against the Soviet Union."12

However, if a public official makes such a declaration and then
acts contrary to it, his reputation can suffer irreparable damage. In
his campaign for the presidency in 1988, George H. W. Bush
famously declared: "Read my lips: no new taxes." But economic
circumstances compelled him to raise taxes a year later, and this
contributed importantly to his defeated bid for reelection in 1992.

Cutting Off Communication

Cutting off communication succeeds as a credible commitment
device because it can make an action truly irreversible. An extreme
form of this tactic arises in the terms of a last will and testament.
Once the party has died, renegotiation is virtually impossible. (For
example, it took an act of the British parliament to change Cecil
Rhodes's will in order to allow female Rhodes Scholars.) In general, where there is a will, there is a way to make your strategy
credible.

One need not die trying to make commitments credible. Irreversibility stands watch at every mailbox. Who has not mailed a
letter and then wished to retrieve it? And it works the other way.
Who has not received a letter he wishes he hadn't? But you can't
send it back and pretend you've never read it once you've opened
the letter. In fact, merely signing for a certified delivery letter acts
as a proof presumptive that you have read the letter.

The movie *Dr. Strangelove*, which is full of clever and not-so-
clever strategic moves, starts out with a good example of the use of
irreversibility. The scene is set in the early 1960s, a high point in the

cold war, with serious fears of a nuclear war between the United States and the Soviet Union. The Air Force Strategic Air Command (SAC) had several wings of bombers constantly in the air, ready to fly to their targets in the Soviet Union if and when the command from the president came. In the movie, General Jack D. Ripper,* who commanded a base housing SAC aircraft, hijacked a provision (Plan R) whereby a lower-echelon commander could order an attack if the president and the rest of the chain of command over him had been knocked out by a preemptive Soviet strike. He ordered a wing of his planes to attack their targets and hoped that the president, presented with this fait accompli, would launch an all-out attack before the inevitable Soviet retaliation was launched.

To make his move irreversible, Ripper did several things. He sealed off the base, cut off the staff's communications with the outside world, and impounded all radios on the base so no one would realize that there was no real emergency. He waited to send the go-code authorizing the attack until the planes were already at their fail-safe points near the boundaries of Soviet air space so that they would not need a further authorization to proceed. He kept the only countercommand the pilots were supposed to obey, the recall code, secret. In fact, later in the movie he killed himself (the ultimate irreversible commitment) rather than risk revealing it under torture. Finally, he sent a phone message to the Pentagon telling them what he had done and was then unavailable for further discussions or questions. An officer read from the transcript of Ripper's message at the resulting meeting at the Pentagon:

> They are on their way in, and no one can bring them back. For the sake of our country and our way of life, I suggest you get the rest of SAC in after them. Otherwise, we will be totally destroyed by Red retaliation. My boys will give you the best kind of start, 1400 megatons worth, and you sure as hell won't stop

---

* Ripper was supposedly modeled after the cigar-chewing U.S. Air Force General Curtis LeMay, famous for his bombing strategy against Japan in World War II and his advocacy of the most hawkish policies and strategies during the cold war.

them now. So let's get going. There's no other choice. God willing, we will prevail in peace and freedom from fear and in true
health through the purity and essence of our natural fluids. God
bless you all.[13]

The officer incredulously concludes: "Then he hung up!" Ripper's
hanging up was meant to be the final act that made his move irreversible. Even the commander in chief, the president of the United
States, would not be able to reach him and order him to recall his
attack.

But Ripper's attempt to achieve a U.S. commitment did not work.
The president did not follow his advice; he instead ordered a nearby
army unit to attack Ripper's base, which they did successfully and
quickly. The president contacted the Soviet premier and even gave
the Soviets details about the attacking planes so they could shoot
them down. The base was not perfectly sealed: an exchange-
program British officer, Lionel Mandrake, discovered a working
radio playing music, and later a pay phone (and a Coke machine to
supply coins) to phone the Pentagon. Most importantly, Ripper's
obsessive doodling enabled Mandrake to guess the recall code.

However, one plane, commanded by a Texan captain with a lot
of initiative, got through. All of this conveys an important practical
lesson in strategy. Theory often makes it sound as if the various
moves being discussed are either 100 percent effective or not at all.
Reality is almost always somewhere in between. So do the best you
can with your strategic thinking, but don't be surprised if something unexpected—an "unknown unknown," as former Defense
Secretary Donald Rumsfeld would say—nullifies your efforts.[14]

There is a serious difficulty with the use of cutting off communication as a device to maintain commitment. If you are incommuni-
cado, it may be difficult, if not impossible, to make sure that the
rival has acceded to your wishes. You must hire others to ensure
that your stipulation is being honored. For example, wills are carried out by trustees, not the deceased. A parental rule against
teenage smoking may be exempt from debate while the parents are
away, but also unenforceable.

Burning Bridges behind You

Armies often achieve commitment by denying themselves an opportunity to retreat. Although Xenophon did not literally burn his bridges behind him, he did write about the advantages of fighting with one's back against a gully.[15] Sun Tzu recognized the reverse strategy, namely the advantage of leaving an opponent an escape route to reduce his resolve to fight. The Trojans, however, got it all backward when the Greeks arrived in Troy to rescue Helen. The Trojans tried to burn the *Greek* ships. They did not succeed, but if they had succeeded, that would simply have made the Greeks all the more determined opponents.

The strategy of burning bridges (or boats) was used by several others. William the Conqueror's army, invading England in 1066, burned its own ships, thus making an unconditional commitment to fight rather than retreat. Hernán Cortés followed the same strategy in his conquest of Mexico, giving orders upon arrival that all but one of his ships be burned or disabled. Although his soldiers were vastly outnumbered, they had no choice but to fight and win. "Had [Cortés] failed, it might well seem an act of madness.... Yet it was the fruit of deliberate calculation. There was no alternative in his mind but to succeed or perish."[16]

The strategy of burning one's ship plays out in *The Hunt for Red October*, where Russian Captain Marko Ramius plans to defect and bring the latest Soviet submarine technology to the United States. Although his officers are loyal, he wants them to have no doubt about their new course. After revealing his plan to them, Ramius explains that just prior to departure he mailed a letter to Admiral Yuri Padorin detailing his intention to defect. Now the Russians will try to sink the sub. There is no turning back. Their only hope is to reach New York harbor.

In the world of business, this strategy applies to attacks on land as well as by sea. For many years, Edwin Land's Polaroid Corporation purposefully refused to diversify out of the instant photography business. With all its chips in instant photography, it was committed to fight against any intruder in the market. On April

20, 1976, after twenty-eight years of a Polaroid monopoly on the instant photography market, Eastman Kodak entered the fray. It announced a new instant film and camera. Polaroid responded aggressively, suing Kodak for patent infringement. Edwin Land, founder and chairman, was prepared to defend his turf: "This is our very soul we are involved with. This is our whole life. For them it's just another field. . . . We will stay in our lot and protect that lot."[17] On October 12, 1990, Polaroid was awarded a $909.4 million judgment against Kodak, which was forced to withdraw its instant film and camera from the market.*

Sometimes, building rather than burning bridges can serve as a credible source of commitment. In the December 1989 reforms in Eastern Europe, building bridges meant knocking down walls. Responding to massive protests and emigration, East Germany's leader, Egon Krenz, wanted to promise reform but didn't have a specific package. The population was skeptical. Why should they believe that his vague promise of reform would be genuine and far-reaching? Even if Krenz was truly in favor of reform, he might fall out of power. Dismantling parts of the Berlin Wall helped the East German government make a credible commitment to reform without having to detail all the specifics. By (re)opening a bridge to the West, the government forced itself to reform or risk an exodus. Since people would still be able to leave in the future, the promise of reform was both credible and worth waiting for. Reunification was to be less than a year away.

Leaving the Outcome beyond Your Control or to Chance

Returning to *Dr. Strangelove*, President Merkin Muffley invites the Soviet ambassador into the Pentagon war room to let him see the situation with his own eyes and be convinced that this was not a

---

* Although Polaroid restored its dominance over the instant photography market, it later lost ground to competition from portable videocassette recorders and minilabs that developed and printed conventional film in one hour and, later still, to digital photography. Lacking bridges, Polaroid began to feel trapped on a sinking island. With a change in philosophy, the company began to branch out into these other areas but without great success.

general U.S. attack on his country. The ambassador explains that even if just one plane succeeded, it would set off the Doomsday Machine, a large number of buried nuclear devices that would pollute the atmosphere and destroy "all human and animal life on earth." The president asks: "Is the [Soviet] premier threatening to explode this device?" The ambassador replies: "No, sir. It is not anything a sane man would do. The Doomsday Machine is designed to trigger itself automatically. . . . It is designed to explode if any attempt is ever made to untrigger it." The president asks his nuclear expert, Dr. Strangelove, how this was possible, and is told: "It is not only possible—it is essential. That is the whole idea of this machine, you know. Deterrence is the art of producing in the mind of the enemy the fear to attack. And so, because of the automated and irrevocable decision making process which rules out human meddling, the Doomsday Machine is terrifying. It's simple to understand. And completely credible and convincing."

The device is such a good deterrent because it makes aggression tantamount to suicide. Faced with an American attack, Soviet premier Dimitri Kissov might refrain from retaliating and risking mutually assured destruction. As long as the Soviet premier has the freedom not to respond, the Americans might risk an attack. With the doomsday device in place, the Soviet response is automatic and the deterrent threat is credible. In the real cold war, the real-life Soviet premier, Khrushchev, attempted to use a similar strategy, threatening that Soviet rockets would fly *automatically* in the event of armed conflict in Berlin.[18]

However, this strategic advantage does not come without a cost. There might be a small accident or unauthorized attack, after which the Soviets would not want to carry out their dire threat but have no choice, as execution is out of their control. This is exactly what happened in *Dr. Strangelove*. To reduce the consequences of errors, you want a threat that is just strong enough to deter the rival. What do you do if the action is indivisible, as a nuclear explosion surely is? You can make the threat milder by creating a risk, but not a certainty, that the dreadful event will occur. This is where brinkmanship comes in.

The device that creates a risk of mutual disaster in brinkmanship is just as automatic as the Doomsday Machine. If the opponent defies you, you no longer control whether to explode the device. But the automatic explosion is not a certainty. It is only a probability. This is just like Russian Roulette. One bullet is loaded into a revolver, the chamber is spun, and the trigger pulled. The shooter no longer controls whether the firing chamber contains the bullet. But he controls the size of the risk beforehand—one in six. Thus brinkmanship is a controlled loss of control: the threatener controls the size of the risk but not the outcome. If he finds an empty chamber and decides to pull the trigger again, he is raising the risk to one in five, just as Bud White did in *L.A. Confidential.* How far he chooses to do this depends on his tolerance for the risk. All the time he is hoping that the opponent has a lower tolerance for risk and will give in, and that the mutually undesirable explosion will not occur before either concedes.

No wonder brinkmanship is such a delicate strategy, fraught with dangers. Practice it at your own peril. We recommend trying it out in relatively innocuous situations before you try it on a really important occasion. Try to control the behavior of your kids, where the bad outcome is merely a messy room or a tantrum, before you try to play negotiation roulette with your spouse, where the bad outcome may be a messy divorce or a court fight.

Moving in Steps

Although two parties may not trust each other when the stakes are large, if the problem of commitment can be reduced to a small enough scale, then the issue of credibility will resolve itself. The threat or promise is broken up into many pieces, and each one is solved separately. Honor among thieves may be restored if they have to trust each other only a little bit at a time. Consider the difference between making a single $1 million payment to another person for a kilogram of cocaine and engaging in 1,000 sequential transactions with this other party, with each transaction limited to $1,000 worth of cocaine. While it might be worthwhile to double-cross your "partner" for $1 million, the gain of $1,000 is too small, since it brings a premature end to a profitable ongoing relationship. Whenever a large degree of commitment is infeasible, one should make do with a small amount and reuse it frequently.

This is also true for homeowners and contractors, who are mutually suspicious. The homeowner is afraid of paying up front and finding incomplete or shoddy work. The contractors are afraid that after they have completed the job, the homeowner may refuse to pay. So at the end of each day (or each week), contractors are paid on the basis of their progress. At most, each side risks losing one day's (or one week's) worth of work or money.

As with brinkmanship, moving in small steps reduces the size of the threat or promise and correspondingly the scale of commitment. There is just one feature to watch out for. Those who understand strategic thinking will reason forward and look backward, and they will worry about the last step. If you expect to be cheated on the last round, you should break off the relationship one round earlier. But then the penultimate round will become the final round, and so you will not have escaped the problem. To avoid the unraveling of trust, there should be no clear final step. As long as there remains a chance of continued business, it will never be worthwhile to cheat. So when a store has a "going out of business" sale with massive price reductions, be especially cautious about the quality of what you are buying.

Teamwork

Often others can help us achieve credible commitment. Although people may be weak on their own, they can build resolve by forming a group. The successful use of peer pressure to achieve commitment has been made famous by Alcoholics Anonymous (AA) and diet centers. The AA approach changes the payoffs if you break your word. It sets up a social institution in which pride and self-respect are lost when commitments are broken. Sometimes teamwork goes far beyond social pressure and employs strong-arm tactics to force us to keep true to our promises. Consider the problem for the front line of an advancing army. If everyone else charges forward, one soldier who hangs back ever so slightly will increase his chance of survival without significantly lowering the probability that the attack will be successful. If every soldier thought the same way, however, the attack would become a retreat.

Of course it doesn't happen that way. A soldier is conditioned through honor to his country, loyalty to fellow soldiers, and belief in the million-dollar wound—an injury that is serious enough to send him home, out of action, but not so serious that he won't fully recover.[19] Those soldiers who lack the will and the courage to follow orders can be motivated by penalties for desertion. If the punishment for desertion is certain and ignominious death, the alternative—advancing forward—becomes much more attractive. Of course soldiers are not interested in killing their fellow countrymen, even deserters. How can soldiers who have difficulty committing to attack the enemy make a credible commitment to kill their countrymen for desertion? The ancient Roman army made falling behind in an attack a capital offense. As the army advanced in a line, any soldier who saw the one next to him falling behind was ordered to kill the deserter immediately. To make this order credible, failing to kill a deserter was also a capital offense. Even though a soldier would rather get on with the battle than go back after a deserter, failing to do so could cost him his own life.*

The tactics of the Roman army live on today in the honor code enforced at West Point, Princeton, and some other universities. Exams are not monitored, and cheating is an offense that leads to expulsion. But, because students are not inclined to "rat" on their classmates, failure to report observed cheating is also a violation of the honor code, and leads to expulsion as well. When the honor code is violated, students report crimes because they do not want to become guilty accomplices by their silence. Similarly, criminal law provides penalties for those who fail to report a crime as an accessory after the fact.

* The motive for punishing deserters is made even stronger if the deserter is able to gain clemency by killing those in line next to him who fail to punish him. Thus if a soldier fails to kill a deserter, there are now two people who can punish him: his neighbor and the deserter, who could save his own life by punishing those who failed to punish him.

Mandated Negotiating Agents

If a worker says he cannot accept any wage increase less than 5 percent, why should the employer believe that he will not subsequently back down and accept 4 percent? Money on the table induces people to try negotiating one more time. The worker's situation can be improved if he has someone else negotiate for him. When the union leader is the negotiator, his position may be less flexible. He may be forced to keep his promise or lose support from his electorate. The union leader may secure a restrictive mandate from his members or put his prestige on the line by declaring his inflexible position in public. In effect, the labor leader becomes a mandated negotiating agent. His authority to act as a negotiator is based on his position. In some cases he simply does not have the authority to compromise; the workers, not the leader, must ratify the contract. In other cases, compromise by the leader would result in his removal.

The device of using mandated negotiating agents becomes especially useful if you are negotiating with someone with whom you share other bonds of friendship or social links that you are reluctant to break. In such situations, you may find it difficult to adhere firmly to negotiating positions and may concede more than you should for the sake of the relationship. An impersonal agent is better able to avoid falling into this trap and can get you a better deal. Members of professional sports teams employ agents partly for this reason, as do authors in their dealings with editors and publishers.

In practice we are concerned with the means as well as the ends of achieving commitment. If the labor leader *voluntarily* commits his own prestige to a certain position, should you (do you) treat his loss of face as you would if it were externally imposed? Someone who tries to stop a train by tying himself to the railroad tracks may get less sympathy than someone else who has been tied there against his will.

A second type of mandated negotiating agent is a machine. Very few people haggle with vending machines over the price; even

fewer do so successfully.* That is why many store clerks and government bureaucrats are required to follow rules mechanically.
The store or the government makes its policy credible; even the
employees benefit by being able to say that negotiating or bending
the rule is "above their grade."

UNDERMINING YOUR OPPONENT'S CREDIBILITY

If you stand to gain by making your strategic moves credibly,
then similarly you will benefit by preventing other players from
making their strategic moves credible. Right? No, not so fast. This
thinking is the relic of the idea that games must be win-lose or
zero-sum, an idea that we have consistently criticized. Many games
can be win-win or positive-sum. In such games, if another player's
strategic move can achieve an outcome that is better for both, then
you benefit by enhancing the credibility of such a move.

For example, in a prisoners' dilemma, if the other player is in a
position to make you a promise to reciprocate your choice of cooperation, then you should try to enable him to make that promise
credibly. Even a threat, deployed mutually, may be in the combined interests of the players. In the previous chapter we saw how
our two catalog merchants, Rainbow's End and B. B. Lean, can use
meet-the-competition or beat-the-competition clauses to make
threats of retaliation against a rival's price cut. When they both use
such a strategy, each removes the temptation for the other firm to
make a price cut in the first place and thereby helps both firms
keep their prices high. Each firm should want the other to have the
ability to make its strategy credible, and if one of them thinks of
the device, it should suggest to the other that both should use it.

That said, there are many situations where the other player's
strategic move can hurt you. The other player's threats do often

---

* According to the U.S. Defense Department, over a five-year period seven
servicemen or dependents were killed and 39 injured by soft-drink machines that
toppled over while being rocked in an attempt to dislodge beverages or change
(*International Herald Tribune*, June 15, 1988).

work against your interests; so make some commitments. In such circumstances, you want to try to prevent the other from making that move credible. Here are a few suggestions for practicing that art. Once again we give them with the caution that they are tricky and even risky, and you should not expect perfect success.

Contracts: Mr. Russo in our story had two selves, one before chocolate éclairs appear on the dessert trolley (BCE) and the other after (ACE). The BCE self sets up the contract to defeat ACE's temptation, but the ACE self can render the contract ineffective by proposing a renegotiation that will benefit all the parties that are present at that point. The BCE self would have refused ACE's proposal, but BCE is no longer there.

If all the distinct parties to the original contract are still present, then to get around a contract you have to propose a new deal that will be in the interests of everyone at that point. Gaining unanimous consent is difficult, but not impossible. Suppose you are playing a repeated prisoners' dilemma game. An explicit or implicit contract says that everyone should cooperate until someone cheats; after that, cooperation will break down and everyone will choose the selfish action. You can try to get away with cheating once by pleading that it was just an innocent error and that all the available gains from future cooperation should not be wasted just because the contract says so. You cannot hope to pull this trick too often, and even the first time others may be suspicious. But it does seem like children get away with "it won't happen again" time after time.

Reputation: You are a student trying to get a deadline extension from your professor. He wants to maintain his reputation and tells you: "If I do this for you, I will have to do the same for everyone in the future." You can come back with: "No one will ever know. It is not in my interest to tell them; if they write better assignments by getting extensions, my grade will suffer because the course is graded on a curve." Similarly, if you are a retailer negotiating a lower price with your supplier, you can credibly promise not to reveal this to your rival retailers. A reputation is valuable only to the extent that it gets publicized; you can make it ineffective by maintaining secrecy.

Communication: Cutting off communication may protect the player making a strategic move by making his action irreversible. But if the other player is unavailable to receive the information about the opponent's commitment or threat in the first place, the strategic move is pointless. A parent's threat—"If you don't stop crying you won't get dessert tonight"—is ineffective against a child who is crying too loudly to hear it.

Burning Bridges: Recall the advice of Sun Tzu: "When you surround an enemy, leave an outlet free."[20] One leaves an outlet free not so that the enemy may actually escape but so that the enemy may believe there is a road to safety.* If the enemy does not see an escape outlet, he will fight with the courage of desperation. Sun Tzu aimed to deny the enemy an opportunity to make his own very credible commitment of fighting to the death.

Moving in Steps: The credibility of mutual promises can be enhanced by breaking large actions into a sequence of small ones. But you can try to destroy the credibility of an opponent's threat by going against his wishes in small steps. Each step should be so small in relation to the threatened costly action that it is not in the interests of the other to invoke it. As previously discussed, this method is called salami tactics; you defuse the threat one slice at a time. The best example comes from Schelling: "Salami tactics, we can be sure, were invented by a child. . . . Tell a child not to go in the water and he'll sit on the bank and submerge his bare feet; he is not yet 'in' the water. Acquiesce, and he'll stand up; no more of him is in the water than before. Think it over, and he'll start wading, not going any deeper; take a moment to decide whether this is different and he'll go a little deeper, arguing that since he goes back and forth it all averages out. Pretty soon we are calling to him not to swim out of sight, wondering whatever happened to all our discipline."[21] Smaller nations understand this, just as small

---

* In a footnote, Sun Tzu suggests that the retreating army should be ambushed. That, of course, only works if the opposing army hasn't read Sun Tzu.

children do. They defy the wishes of superpowers in little ways—vote independently in the United Nations, violate some clauses of trade agreements, and even take successively tiny steps toward acquiring nuclear technology—that are too small to invoke serious retaliation.

Mandated Agents: If the other player aims to achieve credibility for an inflexible negotiating position by using a mandated agent, you might simply refuse to deal with the agent and demand to speak directly to the principal. A channel of communication must be open between those two; after all, the agent has to report the outcome to the principal. Whether the principal agrees to deal directly with you then depends on his reputation or other aspects of his resolve. As an example, suppose you are attempting to negotiate the price of an item in a department store, and the clerk tells you that he has no authority to offer a discount. You can ask to see the manager, who may have such authority. Whether you attempt this depends on your judgment about the likelihood of success, on how badly you want the item, and on your valuation of the humiliation if you fail and have to settle for the listed price.*

This completes our array of illustrative examples of devices for making your own strategic moves credible and for dealing with other players' strategic moves. In practice, any particular situation may require more than one of these devices. And even in combination, they may not work 100 percent of the time. Remember, nothing is perfect. (And, Billy Wilder would have us believe, "Nobody's perfect.") We hope our little guide serves to engage your interest and gives you a starting point for developing these skills in the games you play.

CASE STUDY: A TEXTBOOK EXAMPLE OF CREDIBILITY

The size of the U.S. college textbook market is $7 billion (including course packets). To put that in perspective, the revenue

---

* If you think this is a fool's errand, you should know that even megastores such as Home Depot and Best Buy are "quietly telling their salespeople that negotiation is acceptable" (*New York Times*, March 23, 2008).

from the movie industry is about $10 billion and all of professional sports is $16 billion. Textbooks may not have Heisman trophies or Academy Awards, but it is a big business all the same. Perhaps that might not be so surprising when you consider that textbooks routinely cost more than $150—that's about the price for Samuelson and Nordhaus's *Economics* (18th ed.) or *Thomas' Calculus* (11th ed.)—and students might buy eight or so a year.

Congress has proposed a way out. It wants college bookstores to guarantee that they will buy back the used text. At first glance, this might seem to cut the student's cost in half. If you can sell a $150 book back at the end of the semester for $75, then the real cost is cut in half. Is that right?

Case Discussion

Let's take a step back and look at the world from the publisher's perspective. If a typical textbook will be resold twice on the used market, then the publisher gets to make only one sale rather than three. If they were looking to make $30 profit per student, now they have to make $90 on the first sale to come out even. This is what leads publishers to raise the list price of the text all the way up to $150, which allows them to get their $90 profit up front. And once the books are sold, they have every incentive to eliminate competition from the stock of used textbooks by coming out with a new edition as quickly as they can.

Compare this to a world where the publisher promises not to bring out a revision and the students promise not to resell their used books. Over three years, the publisher could sell three books at $50 each and make the same amount of money. Actually, that ignores the extra printing costs (not to mention the environmental cost of the trees cut down), so we'll bump the price up to $60. In this world, the publishers are just as happy, professors spend less time making unnecessary revisions, and students get a better deal. They buy the book for $60 and get to keep it as a reference, rather than paying $150 and hoping to resell it for $75 (for a net price of $75).

There's one group of students who fare really badly under the present system: the ones who buy a new text in the final year of an edition. A new edition will be assigned for the following year, so they can't sell their used copy back to the store. These unlucky students end up paying the full $150.*

Students aren't dumb. They don't want to be left holding a hot potato. After a text has been on the market for two or three years, they realize that a revision is due. Students anticipate that the effective cost of the book is going to be much higher, and so they respond by not buying the text.[22] (As faculty, we were surprised to learn that somewhere around 20 percent of students don't buy the required texts.)

Eliminating the used textbook market would be an improvement for students, faculty, and publishers. The losers would be the bookstores; they make more money from the status quo. On the $150 text that is resold twice, the store makes $30 on the original sale and then $37.50 twice more each time they buy it back for half-price and resell it at three-quarters of list. They'd make much less selling a new book three times for $60 each.

Forcing the bookstores to buy back used texts doesn't solve the problem. It would only lead them to pay less, as they would anticipate getting stuck with all obsolete texts. Better than forcing the store to buy back the text is having students promise not to resell and thereby eliminating the used market. But how can that promise be credible? Banning used book sales isn't practical.

One answer is to have students lease or rent the textbooks. Students could put down a deposit on the book that is refunded when they return the book (to the publisher, not the bookstore). This is just like having the publishers promise to buy back the text whether there is a new edition or not. Even simpler, publishers could sell licenses to the text, just like a software license, to each student in the class.[23] This would give each student access to a

---

* It is a puzzle why the price of new and used books doesn't change over the revision cycle. One might have guessed that in the year prior to a revision, the publisher would sell the new text for $75 rather than $150. The buy-back price for a used book could be two-thirds list after the first year and one-third list after the second go-around.

copy of the text. The university would pay for the license and bill the students. Now that the publisher is making all of its profits via the license, the books can be sold for a price near production cost, and so there is little motivation to resell the books.

In general, when there is a problem of commitment, one way around the issue is to rent rather than sell the product. That way no one has an incentive to take advantage of the used book stockpile, because there isn't any.

There are two more cases on making strategies credible; see "But One Life to Lay Down for Your Country" and "*United States v. Alcoa*" in chapter 14.

EPILOGUE TO PART II: A NOBEL HISTORY

Game theory was pioneered by John von Neumann. In the early years, the emphasis was on games of pure conflict (zero-sum games). Other games were considered in a cooperative form, that is, the participants were able to choose and implement their actions jointly. For most games played in reality, people choose actions separately but their effects on others are not ones of pure conflict. The breakthrough that allowed us to study general games combining conflict and cooperation is due to John Nash. We explained his concept of (what is now known as) Nash equilibrium in chapter 4.

In our presentation of the equilibrium concept, we assumed that all the people in the game understood the preferences of the other players. They might not know what the other players would do, but they understood each other's objectives. John Harsanyi, who shared the 1994 Nobel Prize with Nash, showed that Nash equilibrium could be extended to games where players are unsure about the other players' preferences.

Another challenge in the application of Nash equilibrium is the potential for multiple solutions. The work of 2005 Nobel laureate Robert Aumann shows that this challenge becomes even greater in repeated games. Almost anything can be a Nash equilibrium in a game that is repeated often enough. Fortunately, there are some tools that help us select one equilibrium over another. Reinhard Selten demonstrated that the Nash equilibrium concept could be refined, thereby eliminating some of the multiplicity, by introducing the idea there is a small possibility that a player will make a mistake when moving. This forces players to ensure that their strategy is optimal even when the game takes an unexpected turn. It turns out that this is much like the idea of look forward, reason backward, but applied to games where people move simultaneously.

When we recognize that players in a game may not have perfect information, it becomes important, even essential, to specify who knows what. I might know that you prefer one outcome over another or that you are lying to me, but if you don't know that I know that, then that changes the game.* Another of Robert Aumann's contributions was bringing the concept of common knowledge to game theory. When two players have common knowledge of something, not only do they both know it, but they each know that the other knows it, knows that the other side knows they know it, and so on to infinity.

Lack of common knowledge is the more common case. One or more of the players in these games lack some crucial piece of information that the other has. The better informed player may want to conceal or distort the information or sometimes may want to convey the truth to a skeptical adversary; the less informed player generally wants to find the truth. This makes the true game between them one of manipulating information. Concealing, revealing, and interpreting information each require their own special strategies.

During the last thirty years, the ideas and theories of information manipulation have revolutionized economics and game theory and have had a huge impact on other social sciences and on evolutionary biology. We've discussed the contributions of 2005 laureate Thomas Schelling, who developed the ideas of commitment and strategic moves. Three other Nobel Prizes in Economics have honored pioneers of these theories and their applications, and there are probably more to come. The very first, given in 1996, honored James Mirrlees and William Vickrey, who developed theories of how to design a game that will achieve truthful revelation of the other player's private information. *The Economist* succinctly characterized their contribution as answering the question: "How do you deal with someone who knows more than you do?"[1]

---

* The movie *Mystery Men* offers a fine illustration of knowing who knows what about whom. Captain Amazing (CA) confronts his nemesis, Captain Frankenstein (CF), who has just escaped from the asylum:

> CF: Captain Amazing! What a surprise!
> CA: Really? I'm not so sure about that. Your first night of freedom and you blow up the asylum. Interesting choice. I knew you couldn't change.
> CF: I knew you'd know that.
> CA: Oh, I know that, and I knew you'd know I know you knew.
> CF: But I didn't. I only knew you'd know that I knew. Did you know that?
> CA: Of course.

Mirrlees did this in the context of designing an income tax system when the government does not know people's income-producing potential, and Vickrey analyzed strategies for selling by auction.

In 2001, the Nobel Prize went to George Akerlof, whose model of the private used car market illustrated how markets can fail when one party has private information; Michael Spence, who developed the "signaling" and "screening" strategies that are used for coping with such information asymmetries; and Joseph Stiglitz, who developed the application of these ideas to insurance, credit, labor, and many other kinds of markets, yielding some startling insights on the limitation of markets.

The 2007 prize was also for information economics. Screening is just one strategy available to gain information about others. More generally, a player (often called the principal) can devise a contract that creates incentives for the other players to reveal their information, directly or indirectly. For example, when A cares what B does but can't monitor B's actions directly, then A can devise an incentive payment that induces B to take an action closer to what A desires. We take up the topic of incentive schemes in chapter 13. The general theory of the design of such mechanisms was developed during the 1970s and 1980s. The 2007 Nobel Prize honored three of the most eminent pioneers of this research, Leonid Hurwicz, Eric Maskin, and Roger Myerson. Hurwicz, at ninety, became the oldest recipient ever of the Economics Nobel Prize; Maskin at fifty-six and Myerson at fifty-seven were among the youngest ever. Information economics and game theory are truly for all ages.

We will present many of these Nobel ideas in the next chapters. You will learn about Akerlof's market for lemons, about Spence's job-market signaling, the Vickrey auction, and Myerson's revenue-equivalence theorem. You will learn how to bid in an auction, run an election, and design an incentive scheme. One of the wonderful aspects of game theory is that it is possible to understand the contributions of Nobel laureates without having to spend years in graduate school. Indeed, some of the ideas may even seem obvious. We think that is true, but only in hindsight, and that is the mark of a truly brilliant insight.

Part III

CHAPTER 8

Interpreting
and Manipulating
Information

THE MARRYING KIND?

True story: Our friend, whom we'll call Sue, was in love. Her beau was an extremely successful executive. He was smart, single, and straight. He professed his love to her. It was a happily-ever-after fairy tale. Well, almost.

The problem was that, at age thirty-seven, Sue wanted to get married and have kids. He was on board with the plan, except that his kids from a previous marriage weren't ready for him to remarry. These things take time, he explained. Sue was willing to wait, so long as she knew that there would be a light at the end of the tunnel. How could she know whether his words were sincere or not? Unfortunately, any public demonstration was out of bounds, as the kids would surely find out.

What she wanted was a credible signal. This is the cousin of a commitment device. In the previous chapter, we emphasized strategies that guaranteed that the person would carry out what he said he'd do. Here, we are looking for something weaker. What Sue wanted was something that would help her understand whether he was truly serious about their relationship.

After much thought, Sue asked him to get a tattoo, a tattoo with her name. A small, discreet tattoo would be just fine. No one else would ever have to see it. If he was in this for the long run, then having Sue's name indelibly inked would be a fitting tribute to their love. But, if commitment wasn't part of his plan, this would be an embarrassing artifact for his next conquest to discover.

He balked, so Sue left. She found a new love and is now happily married with kids. As for her ex, he is still on the runway, on permanent ground delay.

Tell It Like It Is?

Why can't we just rely on others to tell the truth? The answer is obvious: because it might be against their interests.

Much of the time, people's interests and communications are aligned. When you order a steak medium rare, the waiter can safely assume that you really want the steak medium rare. The waiter is trying to please you and so you do best by telling the truth. Things get a bit trickier when you ask for a recommended entrée or advice on wine. Now the waiter might want to steer you to a more expensive item and thereby increase the likely tip.

The British scientist and novelist C. P. Snow attributes just such strategic insight to the mathematician G. H. Hardy: "If the Archbishop of Canterbury says he believes in God, that's all in the way of business, but if he says he doesn't, one can take it he means what he says."[1] Similarly, when the waiter points you to the less expensive flank steak or bargain Chilean wine, you have every reason to believe him. The waiter might also be right when recommending the expensive entrée, but it is harder to know.

The greater the conflict, the less the message can be trusted. Recall the soccer penalty kicker and the goalie from chapter 5. Suppose that, just as he is getting ready to take his shot, the kicker says: "I am going right." Should the goalie believe him? Of course not. Their interests are totally opposed, and the kicker stands to lose by making his intentions known truthfully in advance. But does this mean that the goalie should assume that the kicker will kick to the left? Again, no. The kicker might be trying a second-level deception—lying by telling the truth. The only rational reaction to an assertion made by another player whose interests are totally opposed to yours is to ignore it completely. Don't assume it to be true, but don't assume its opposite to be true either. (Instead, think about the equilibrium of the actual game ignoring what the other side has said and play accordingly; later in this chapter, we explain just how to do this using the example of bluffing in poker.)

Politicians, advertisers, and children are all players in their own strategic games with their own interests and incentives. And what they are telling us serves their own agendas. How should you interpret information that comes from such sources? And conversely, how can you make your claims credible, knowing that others will regard what you say with due suspicion? We start our exploration with perhaps the most famous example of divining the truth from interested parties.

KING SOLOMON'S DILEMMA

Two women came before King Solomon, disputing who was the true mother of a child. The Bible takes up the story in 1 Kings (3:24–28):

> Then the king said, "Bring me a sword." So they brought a sword for the king. He then gave an order: "Cut the living child in two and give half to one and half to the other." The woman whose son was alive was filled with compassion for her son and said to the king, "Please, my lord, give her the living baby! Don't kill him!" But the other said, "Neither I nor you shall have him. Cut him in two!" Then the king gave his ruling: "Give the living baby to the first woman. Do not kill him; she is his mother." When all Israel heard the verdict the king had given, they held the king in awe, because they saw that he had wisdom from God to administer justice.

Alas, strategic experts cannot leave a good story alone. Would the king's device have worked if the second woman, the false claimant, had understood what was going on? No.

The second woman made a strategic blunder. It was her answer in favor of dividing the child that distinguished her from the true mother. She should have simply repeated whatever the first woman said; with both women saying the same thing, the king would not have been able to say which one was the true mother.

The king was more lucky than wise; his strategy worked only because of the second woman's error. As for what Solomon should have done, we offer that as a case study in chapter 14.

DEVICES FOR MANIPULATING INFORMATION

The kinds of problems faced by Sue and Solomon arise in most strategic interactions. Some players know more than others about something that affects the payoffs for them all. Some who possess extra information are keen to conceal it (like the false claimant); others are equally keen to reveal the truth (like the true mother). Players with less information (like King Solomon) typically want to elicit it truthfully from those who know.

Pretending wisdom greater than that of Solomon, game theorists have examined several devices that serve these purposes. In this chapter we will illustrate and explain them in simple terms.

The general principle governing all such situations is: Actions (including tattoos) speak louder than words. Players should watch what another player does, not what he or she says. And, knowing that the others will interpret actions in this way, each player should in turn try to manipulate actions for their information content.

Such games of manipulating behavior to manipulate others' inferences, and seeing through others' manipulation of our inferences, go on every day in all of our lives. To borrow and twist a line from *The Love Song of J. Alfred Prufrock*, you must constantly "prepare a face to meet the faces that you meet." If you do not recognize that your "face," or more generally your actions, are being interpreted in this way, you are likely to behave in a way that works to your own disadvantage, often quite seriously so. Therefore the lessons of this chapter are among the most important you will learn in all of game theory.

Strategic game players who possess any special information will try to conceal it if they will be hurt when other players find out the truth. And they will take actions that, when appropriately interpreted, reveal information that works favorably for them. They know that their actions, like their faces, leak information. They will choose actions that promote favorable leakage; such strategies are called *signaling*. They will act in ways that reduce or eliminate unfavorable leakage; this is *signal jamming*. It typically consists of mimicking something that is appropriate under different circumstances than the ones at hand.

If you want to elicit information from someone else, you should set up a situation where that person would find it optimal to take one action if the information was of one kind, and another action if it was of another kind; action (or inaction) then reveals the information.* This strategy is called *screening*. For example, Sue's request for a tattoo was her screening test. We will now illustrate and explain the working of these devices.

In chapter 1, we argued that poker players should conceal the true strength of their hand by bidding somewhat unpredictably. But the optimal mix of bids is different for hands of different strengths. Therefore, limited information about the probability of a strong hand can be derived from the bids. The same principle holds when someone is trying to convey rather than conceal information: Actions speak louder than words. To be an effective signal, an action should be incapable of being mimicked by a rational liar: it must be unprofitable when the truth differs from what you want to convey.[2]

Your personal characteristics—ability, preferences, intentions—constitute the most important information that you have and others lack. They cannot observe these things, but you can take

---

* Sometimes even actions are hard to observe and interpret. The greatest difficulty arises in judging the quality of a person's work effort. The quantity of effort can easily be measured, but all save the most simple repetitive jobs require some thinking and creativity, and employers or supervisors cannot accurately assess whether an employee is utilizing his or her time well. In such situations, performance has to be judged by outcome. The employer must design a suitable incentive scheme to induce the employee to supply high-quality effort. This is the topic of chapter 13.

actions that credibly signal the information to them. Likewise, they will attempt to infer your characteristics from your actions. Once you become aware of this, you will start seeing signals everywhere and will scrutinize your own actions for their signal content.

When a law firm recruits summer interns with lavish hospitality, it is saying, "You will be well treated here, because we value you highly. You can believe us because if we valued you less then we would not find it in our interest to spend so much money on you." In turn, the interns should realize that it doesn't matter if the food is bad or the entertainment bores them stiff; what's important is the price.

Many colleges are criticized by their alumni for teaching things that proved of no use in their subsequent careers. But such criticism leaves out the signaling value of education. Skills needed to succeed in particular firms and specialized lines of work are often best learned on the job. What employers cannot easily observe but really need to know is a prospective employee's general ability to think and learn. A good degree from a good college acts as a signal of such ability. The graduate is in effect saying, "If I were less able, would I have graduated Princeton with honors?"

But such signaling can turn into a rat race. If the more able get only a little more education, the less able might find it profitable to do likewise, be mistaken for the more able, and be given better jobs and wages. Then the truly more able must get even more education to distinguish themselves. Pretty soon, simple clerical jobs require master's degrees. True abilities remain unchanged; the only people to benefit from the excessive investment in education for signaling are we college professors. Individual workers or firms can do nothing about this wasteful competition; a public policy solution is needed.

IS THE QUALITY GUARANTEED?

Suppose you are in the market to buy a used car. You find two that seem to have the same quality, as far as you can judge. But the first comes with a warranty and the second does not. You surely prefer the first, and are willing to pay more for it. For one thing, you know that if something goes wrong, you will get it fixed free of charge. However, you will still have to spend a lot of time and suffer a lot of inconvenience, and you are not going to be compensated for these hassles. Here another aspect becomes more relevant. You believe that things are less likely to go wrong with the car under warranty in the first place. Why? To answer that, you have to think about the seller's strategy.

The seller has a much better idea of the quality of the car. If he knows that the car is in good condition and not likely to need costly repairs, offering the warranty is relatively costless to him. However, if he knows that the car is in poor condition, he expects to have to incur a lot of cost to fulfill the warranty. Therefore, even after taking into account the higher price that a car under a warranty may fetch, the worse the quality of the car, the more likely the warranty is to be a losing proposition to the seller.

Therefore the warranty becomes an implied statement by the seller: "I know the quality of the car to be sufficiently good that I can afford to offer the warranty." You could not rely on the mere statement: "I know this car to be of excellent quality." With the warranty, the seller is putting his money where his mouth is. The action of offering the warranty is based on the seller's own gain and loss calculation; therefore it is credible in a way that mere words would not be. Someone who knew his car to be of low quality would not offer the warranty. Therefore the action of offering a warranty serves to separate out sellers who merely "talk the talk" from those who can "walk the walk."

Actions that are intended to convey a player's private information to other players are called *signals*. For a signal to be a credible carrier of a specific item of information, *it must be the case that the action is optimal for the player to take if, but only if, he has that specific information*. Thus we are saying that offering a warranty can be a credible signal of the quality of the car. Of course whether it is credible in a specific instance depends on the kinds of things that are potentially likely to go wrong with that kind of car, the cost of fixing them, and the difference in price between a car under a warranty and a similar-looking car without a warranty. For example, if the expected cost of repairs on a good-quality car is $500, while that for a poor-quality car is $2,000, and the price difference with and without a warranty is $800, then you can infer that a seller offering such a warranty knows his car to be of good quality.

You don't have to wait for the seller to think all this through and offer the warranty if he knows his car to be good. If the facts are as we just stated, you can take the initiative and say: "I will pay you an extra $800 for the car if you offer me a warranty." This will be a good deal for the seller if, but only if, he knows his car to be of good quality. In fact you could have offered $600, and he might counter with $1,800. Any price greater than $500 and less than $2,000 for the warranty will serve to induce sellers of good and bad cars to take different actions and thereby reveal their private information, and the two of you might bargain over this range.

Screening comes into play when the less-informed player requires the more-informed player to take such an information-revealing action. The seller might take the initiative and signal the quality of the car by offering the warranty, or the buyer might take the initiative and screen the seller by asking for a warranty. The two strategies can work in similar ways to reveal private information, although there can be technical game-theoretic differences between the resulting equilibria. When both methods are potentially available, which one is used can depend on the historical, cultural, or institutional context of the transaction.

A credible signal has to be against the interests of an owner who knows his car to be of low quality. To drive home the point, how would you interpret a seller's offer to let you get the car inspected by a mechanic? This is not a credible signal. If the mechanic finds some serious flaw and you walk away, the owner is no worse off than before, regardless of the condition of his car. Therefore the owner of a bad car can make the same offer; the action will not serve to convey the information credibly.*

---

* The owner might get the car inspected at his own expense and present a certificate of quality, but you would be rightly suspicious that the mechanic might be in collusion with the owner. To make the signal credible, the owner could

Warranties are credible signals because they have the crucial cost-difference property. Of course the warranty itself has to be credible in the sense that you can enforce its terms when the need arises. Here we see a big difference between a private seller and a car dealership. Enforcement of a warranty given by a private seller is likely to be much harder. Between the time when the car is sold and when the need for a repair arises, a private seller may move, leaving no forwarding address. Or he may lack the money to pay for the repair, and taking him to court and enforcing a judgment may be too costly to the buyer. A dealership is more likely to be in the business for a longer time and may have a reputation to preserve. Of course a dealer can also try to weasel out of payment by claiming that the problem arose because you did not maintain the car properly or drove it recklessly. But on the whole, revelation of the quality of a car (or other consumer durables) through warrantees or other methods is likely to be far more problematic for private transactions than for sales by established dealers.

A similar problem exists for car manufacturers who have not yet established a reputation for high quality. In the late 1990s, Hyundai raised the quality of their cars, but this had not yet been recognized by U.S. consumers. To get its claims of quality across in a dramatic and credible way, in 1999 the company signaled its quality by offering an unprecedented 10-year, 100,000-mile warranty on the power train and 5 years, 50,000 miles on the rest.

A LITTLE HISTORY

George Akerlof chose the used car market as the main example in his classic article showing that information asymmetries can lead to market failures.[3] To illustrate the issue in the simplest way, suppose there are just two types of used cars: lemons (bad quality) and

---

agree to reimburse your inspection cost in the event a mechanic finds a problem. That is more costly for someone with a low-quality car compared to someone selling a high-quality car.

peaches (good quality). Suppose that the owner of each lemon is willing to sell it for $1,000, whereas each potential buyer is willing to pay $1,500 for a lemon. Suppose the owner of each peach is willing to sell it for $3,000, whereas each potential buyer is willing to pay $4,000 for a peach. If the quality of each car were immediately observable to all parties, then the market would work well. All cars would be traded, lemons selling for a price somewhere between $1,000 and $1,500, and each peach between $3,000 and $4,000.

But suppose each seller knows the quality of a car, whereas all that buyers know is that half the cars are lemons and half are peaches. If cars are offered for sale in the same proportion, each buyer would be willing to pay at most

$$\frac{1}{2} \times (\$1,500 + \$4,000) = \$2,750.$$

An owner who knows his car to be a peach is not willing to sell at this price.* Therefore only lemons will be offered for sale. Buyers, knowing this, would offer at most $1,500. The market for peaches would collapse completely, even though buyers are willing to pay a price for provable peaches that sellers are happy to accept. The Panglossian interpretation of markets, namely that they are the best and most efficient institutions for conduct of economic activity, breaks down.

One of us (Dixit) was a graduate student when Akerlof's article first appeared. He and all the other graduate students immediately recognized it as a brilliant and startling idea, the stuff of which sci-

---

* A naïve buyer who offers $2,750 because he thinks that is the average value of a random car will fall victim to the winner's curse. He buys the item but discovers that it isn't worth as much as he thought. This is a problem that arises when the quality of the item being sold is uncertain and your information is only one piece of the puzzle. The very fact that the seller was willing to accept your price says that the missing information wasn't as good as you might have guessed. Sometimes the winner's curse results in a complete breakdown of the market, as in Akerlof's example. In other cases, it just means that you have to bid less in order to avoid losing money. Later, in chapter 10, we show how to avoid getting caught in the snares of the winner's curse.

entific revolutions are made. There was just one problem with it: almost all of them drove used cars, most of which they had bought in private deals, and most of which were not lemons. There must be ways in which market participants cope with the information problems that Akerlof had brought to our attention in such a dramatic example.

There are some obvious ways. Some students have a fair bit of mechanical knowledge about cars, and the rest of them can enlist a friend to inspect a car they are thinking of buying. They can get information about the history of the car from networks of mutual friends. And many owners of high-quality cars are forced to sell them at almost any price, because they are moving far away or even out of the country, or have to switch to bigger cars as their families grow, and so on. Thus there are many practical ways in which markets can mitigate Akerlof's lemons problem.

But we had to wait until Michael Spence's work for the next conceptual breakthrough, namely how strategic actions can communicate information.* He developed the idea of signaling and elucidated the key property—the differences in payoffs from taking an action for players who have different information—that can make signals credible.

The idea of screening evolved from the work of James Mirrlees and William Vickrey but received its clearest statement in the work of Michael Rothschild and Joseph Stiglitz on insurance markets. People have better information about their own risks than do the companies from whom they seek insurance. The companies can require them to take actions, typically to choose from among different plans with different provisions of deductibles and coinsurance. The less risky types will prefer a plan that has a smaller premium but requires them to bear a larger fraction of the risk;

---

* This is a case where the original text is well worth reading: A. Michael Spence, *Market Signaling* (Cambridge, MA: Harvard University Press, 1974). Similar ideas were expressed in the context of psychology in Erving Goffman's classic, *The Presentation of Self in Everyday Life* (New York: Anchor Books, 1959).

this is less attractive to those who know themselves to have higher risk. Thus the choice reveals the insurance applicant's risk type.

This idea of screening by letting people make choices from a suitably designed menu has since become the key to our understanding of many features commonly found in markets, for example, the restrictions on discounted tickets that airlines impose. We will discuss some of these later in this chapter.

The insurance market provided one other input to this topic of information asymmetries. Insurers have long known that their policies selectively attract the worst risks. A life insurance policy that charges the premium of, say, 5 cents for every dollar of coverage will be especially attractive to people whose mortality rate is greater than 5 percent. Of course many people with lower mortality rates will still buy policies, because they need to protect their families, but those at greatest risk will be overrepresented and will buy bigger policies. Raising the price can make matters worse. Now the good risks find the policies too expensive, leaving behind just the worse cases. Once again we have the Groucho Marx effect: anyone willing to buy insurance at those prices is not someone you would want to insure.

In Akerlof's example, potential buyers do not directly know the quality of an individual car and therefore cannot offer different prices for different cars. Thus selling becomes selectively attractive to the owners of lemons. Because the relatively "bad" types are selectively attracted to the transaction, the problem came to be called *adverse selection* in the insurance industry, and the line of research in game theory and economics that deals with problems caused by information asymmetries has inherited that name.

Just as adverse selection is a problem, sometimes the effect can be turned on its head to create "positive selection." Starting from its IPO in 1994, Capital One was one of the most successful companies in America. It had a decade of 40 percent compounded growth—and that is excluding mergers and acquisitions. The key to its success was a clever application of selection. Capital One was a new player in the credit card business. Its big innovation was the transfer of balance option, wherein a customer could bring over an outstanding balance from another credit card and get a lower interest rate (at least for some period).

The reason why this was such a profitable offer comes down to positive selection. Roughly speaking, there are three types of credit card customers, what we will call maxpayers, revolvers, and dead- beats. Maxpayers are the folks who pay their bills in full each month and never borrow on the card. Revolvers are the ones who borrow money on the card and pay it back over time. Deadbeats are also borrowers but, unlike revolvers, are going to default on the loan.

From the credit card issuer's perspective, they obviously lose money on deadbeats. Revolvers are the most profitable of all cus- tomers, especially given the high interest rate on credit cards. It may be surprising, but credit card companies also lose money on maxpayers. The reason is that the fees charged to merchants just barely cover the free one-month loan given to these customers. The small profit doesn't cover billing costs, fraud, and the risk, small but not negligible, that the maxpayer will get divorced (or lose his job) and then default.

Consider who will find the transfer of balance option attractive. Since the maxpayer isn't borrowing money on the card, there is no reason to switch over to Capital One. The deadbeat is not planning to pay the money back, so here, too, there is little interest in switch- ing. Capital One's offer is most attractive to the customers who have large amounts outstanding and are planning to pay the loan back. While Capital One may not be able to identify who the prof- itable customers are, the nature of its offer ends up being attractive just to the profitable type. The offer screens out the unprofitable types. This is the reverse of the Groucho Marx effect. Here, any customer who accepts your offer is one you want to take.

SCREENING AND SIGNALING

You are the chief personnel officer of a company, looking to recruit bright young people who have natural-born talent as man- agers. Each candidate knows whether he or she has this talent, but you don't. Even those lacking the talent look for jobs in your firm, hoping to make a good salary until they are found out. A good manager can generate several million dollars in profits, but a poor one can rack up large losses quickly. Therefore you are on the lookout for evidence of the necessary talent. Unfortunately, such signs are hard to come by. Anyone can come to your interview wearing the right dress and professing the right attitudes; both are widely publicized and easy to imitate. Anyone can get parents, relatives, and friends to write letters attesting to one's leadership skills. You want evidence that is credible and hard to mimic.

What if some candidates can go to a business school and get an MBA? It costs around $200,000 to get one (when you take into account both tuition and foregone salary). College graduates without an MBA, working in an environment where the specialized managerial talent is irrelevant, can earn $50,000 per year. Supposing people need to amortize the expense incurred in earning an MBA over five years, you will have to pay at least an extra $40,000 a year—that is, a total of $90,000 a year—to a candidate with an MBA.

However, this will make no difference if someone who lacks managerial talent can get an MBA just as easily as someone with this talent. Both types will show up with the certificates, expecting to earn enough to pay off the extra expense and still get more money than they could in other occupations. An MBA will serve to discriminate between the two types only if those with managerial talent somehow find it easier or cheaper to earn this degree.

Suppose that anyone possessing this talent is sure to pass their courses and get an MBA, but anyone without the talent has only a 50 percent chance of success. Now suppose you offer a little more than $90,000 a year, say $100,000, to anyone with an MBA. The truly talented find it worthwhile to go and get the degree. What about the untalented? They have a 50 percent chance of making the grade and getting the $100,000 and a 50 percent chance of failing and having to take another job for the standard $50,000. With only a 50 percent chance of doubling their salary, an MBA would net them only $25,000 extra salary on average, so they cannot expect to amortize their MBA expenses over five years. There-fore they will calculate that it is not to their advantage to try for the MBA.

Then you can be assured that anyone with an MBA does have the managerial ability you need; the larger pool of college gradu-ates has sorted itself into two pools in just the right way for you. The MBA serves as a screening device. We emphasize once again that it works because the cost of using the device is less for those you want to attract than for those you want to avoid.

The irony of this is that companies could just as well hire the MBA students on the first day of classes. When the screening device works, only the ones with managerial ability show up. Therefore, firms don't need to wait until the students graduate to know who's talented and who isn't. Of course, if this practice were to become common, then untalented students would start to show up and be the first in line to drop out. The screening only works so long as people spend the two years to make it through.

Thus this screening device comes at a significant cost. If you could identify the talented directly, you could get them to work for you for just over the $50,000 that they could have earned else-where. Now you have to pay the MBAs more than $90,000 to make it worth the while of talented students to incur this extra expense in order to identify themselves. The extra $40,000 per year for five years is the cost of overcoming your informational dis-advantage.

The cost can be attributed to the existence of the untalented in the population. If everyone were a good manager, you would not need to do any screening. Thus the untalented, by their mere exis-tence, are inflicting a negative spillover, or a negative externality in the language of economics, on the rest. The talented initially pay the cost, but the company then has to pay them more, so in the end the cost falls on the company. Such "informational externalities" per-vade all of the examples below, and you should try to pinpoint them in order to understand exactly what is going on in each.

Is it really worth your while to pay this cost, or would you do bet-ter to hire randomly from the whole pool at $50,000 each and take

ONE REASON TO GET AN MBA:

A prospective employer may be concerned about hiring and training a young woman only to find that she leaves the labor force to have children. Whether legal or not, such discrimination still arises. How does an MBA help solve the problem?

An MBA serves as a credible signal that the person intends to work for several years. If she was planning to drop out of the labor force in a year, it would not have made sense to have invested the two years in getting an MBA. She would have done much better to have worked for those two years and one more. Practically speaking, it likely takes at least five years to recover the cost of the MBA in terms of tuition and lost salary. Thus you can believe an MBA when she says that she plans to stick around.

your chances of hiring some untalented people who will cost you money? The answer depends on what proportion of the population is talented, and the size of the losses that each of them can inflict on your firm. Suppose 25 percent of the population of college graduates lacks managerial talent, and each of them can run up losses of a million dollars before they are found out. Then the random hiring policy will cost you $250,000 per hire, on average. That exceeds the $200,000 cost ($40,000 extra salary over five years) of using the MBA to screen out the untalented. Actually, the proportion with managerial talent is probably much smaller, and the potential loss from poor strategies much larger, so the case for using costly screening devices is much stronger. We like to think that the MBA does teach them a few useful skills, too.

Often there are several ways you can identify talent, and you will want to use the cheapest. One way may be to hire people for an in-house training or probationary period. You might let them undertake some small projects under supervision and observe their performance. The cost of this is the salary you have to pay them in the interim, and the risk that the untalented run up some small losses during their probationary period. A second way is to offer contracts with suitably designed backloaded or performance-related compensation. The talented, with confidence in their ability to survive in the firm and generate profits, will be more willing to accept such contracts, while the rest will prefer to take jobs elsewhere that pay a sure $50,000 a year. A third is to observe the performance of managers in other firms and then try to lure away the proven good ones.

Of course, when all firms are doing this, it alters all their calculations of the costs of hiring apprentices, their salary and performance pay structures, etc. Most importantly, competition among firms forces the salaries of the talented above the minimum (for example, $90,000 with the MBA) needed to attract them. In our example, the salaries could not rise above $130,000.* If they did, those lacking managerial talent will also find it pays to go for the MBA, and the pool of MBA's will be "contaminated" by the untalented who are lucky enough to pass.

We have thus far looked at the MBA as a screening device—the firm chose it as a condition of hiring and tied the starting pay to the possession of this degree. But it could also work well as a signaling device, initiated by the candidates. Suppose you, the personnel officer, have not thought of this one. You are hiring at random from the pool at $50,000 a year, and the firm is suffering some losses from the activities of the untalented hires. Someone could come to you with an MBA, explain how it identifies his or her talent, and say: "Knowing that I am a good manger raises your expectation of the profit the company will make from my services by a million. I will work for you if you will pay me more than $75,000 a year." So long as the facts about the ability of the business school to discriminate managerial talent are clear, this will be an attractive proposition for you.

Even though different players initiate the two strategies of screening and signaling, the same principle underlies them both, namely, the action serves to discriminate between the possible types of players or to indicate the specialized information possessed by one of the players.

Signaling via Bureaucracy

In the United States, the government runs a health insurance system called Workers' Compensation to cover the treatment of

---

* Half of the time the untalented will get the degree and, at $130,000, will net an extra $80,000, or $40,000 on average, which is just enough to cover the cost of the degree over five years.

work-related injuries or illnesses. The aims are laudable, but the outcomes have problems. It is difficult for those administering the system to know or judge the severity of an injury (or in some cases even its existence) and the cost of treating it. The workers themselves and the doctors treating them have better information but are also subject to severe temptations to overstate the problems and collect larger sums than are warranted. It has been estimated that 20 percent or more of the claims under Workers' Compensation involve cheating. According to Stan Long, CEO of Oregon's state-owned Workers' Compensation insurer, "If you run a system where you give money to everybody who asks, you are going to get a lot of people asking for money."[4]

The problem can be tackled to some extent using surveillance. The claimants, or at least those suspected of filing false claims, are watched surreptitiously. If they are found doing things incompatible with their claimed injuries—for example, someone with a claim for a severe back injury is seen lifting heavy loads—their claims are denied, and they are prosecuted.

However, surveillance is costly for the scheme, and our analysis of strategies to elicit information suggests some devices to screen those who are truly injured or ill from the false claimants. For example, the claimants could be required to spend a lot of time filling out forms, sitting all day in a bureaucratic office waiting to talk for five minutes to an official, and so on. Those who are actually healthy and can earn good money working all day will have to forgo those earnings and will therefore find this wait too costly. Those who are truly injured and unable to work will be able to spare the time. People often think of bureaucratic delays and inconveniences as proof of the inefficiency of government, but they may sometimes be valuable strategies to cope with informational problems.

Benefits in kind have a similar effect. If the government or an insurance company was giving money to the disabled to buy wheelchairs, people might pretend to be disabled. But if it gave wheelchairs directly, the incentive to pretend would be much less, because someone who didn't need a wheelchair would have to make a lot of effort to sell it on the secondhand market and only get a low price for it. Economists usually argue that cash is superior to transfers in kind, because the recipients can make their own optimal decisions to spend cash in the way that best satisfies their preferences, but in the context of asymmetric information, in-kind benefits can be superior because they serve as screening devices.[5]

Signaling by Not Signaling

"Is there any point to which you would wish to draw my attention?"

"To the curious incident of the dog in the night-time."

"The dog did nothing in the night-time."

"That was the curious incident," remarked Sherlock Holmes.

In the case of Sherlock Holmes in "Silver Blaze," the fact that the dog didn't bark meant that the intruder was familiar. In the case where someone doesn't send a signal, that, too, conveys information. Usually it is bad news, but not always.

If the other player knows that you have an opportunity to take an action that will signal something good about yourself, and you fail to take this action, then the other will interpret that as meaning that you do not have that good attribute. You may have innocently overlooked the strategic signaling role of taking or not taking this action, but that will not do you any good.

College students can take many courses for a letter grade (A to F) or on a pass/fail (P or F) basis. Many students think that a P on their transcript will be interpreted as the average passing grade from the letter scale. With grade inflation as it now exists in the United States, this is at least a B+, more likely an A–. Therefore the pass/fail option looks good.

Graduate schools and employers look at transcripts more strategically. They know that each student has a pretty good estimate of his or her own ability. Those who are so good that they are likely to get an A+ have a strong incentive to signal their ability by taking the course for a letter grade and thereby distinguishing themselves from the average. With many A+ students no longer taking the pass/fail option, the group choosing pass/fail loses much of its upper end. The average grade over this limited pool is no longer an A–, but, say, only a B+. Then those who know they are likely to get an A acquire more of an incentive to distinguish themselves from the herd by taking the course for a letter grade. The pool of pass/fails loses more of its upper end. This process can continue to a point where mostly only those who know they are likely to get a C or worse will choose the pass/fail option. That is how strategic readers of transcripts will interpret a P. Some quite good students who fail to work through this thinking will suffer the consequences of their strategic ignorance.

A friend of ours, John, is brilliant at deal making. He built a worldwide network of classified ad papers through no fewer than 100 acquisitions. When he first sold his company, part of the deal was that he could coinvest with any new acquisition he brought them.* As John explained to the buyer, the fact that he could coinvest would help reassure them that this was a good deal and that they were not overpaying. The buyer understood the reasoning and took it one step further. Did John also understand that if he didn't coinvest, then they would take this as a bad sign and probably wouldn't do the deal? Thus the opportunity to invest would really become a requirement to coinvest. Everything you do sends a signal, including not sending a signal.

Countersignaling

You would think, based on the previous section, that if you have the ability to signal your type, you should. That way, you differentiate yourself from those who can't make the same signal. And yet, some of the people most able to signal refrain from doing so. As Feltovich, Harbaugh, and To explain:

---

* You may have noticed the word "first." The buyer was Cendant, which became the victim of an accounting fraud at one of its acquisitions, CUC. When Cendant's stock tanked, our friend was able to buy back his company at a discount.

The nouveau riche flaunt their wealth, but the old rich scorn such gauche displays. Minor officials prove their status with petty displays of authority, while the truly powerful show their strength through gestures of magnanimity. People of average education show off the studied regularity of their script, but the well educated often scribble illegibly. Mediocre students answer a teacher's easy questions, but the best students are embarrassed to prove their knowledge of trivial points. Acquaintances show their good intentions by politely ignoring one's flaws, while close friends show intimacy by teasingly highlighting them. People of moderate ability seek formal credentials to impress employers and society, but the talented often downplay their credentials even if they have bothered to obtain them. A person of average reputation defensively refutes accusations against his character, while a highly respected person finds it demeaning to dignify accusations with a response.[6]

Their insight is that in some circumstances, the best way to signal your ability or type is by not signaling at all, by refusing to play the signaling game. Imagine that there are three types of potential mates: the gold digger, the question mark, and the true love. One partner asks the other to sign a prenuptial with the following argument: I know you say that you love me. Signing the prenup is cheap if you are in this for the love and quite expensive if you are in this relationship for the money.

That is correct. But the partner could well respond: "I know that you can distinguish true loves from gold diggers. It is the question marks that have you confused. You sometimes confuse gold diggers with question marks and other times confuse question marks with true loves. Therefore, if I were to sign the prenup, that would be saying that I felt the need to distinguish myself from the gold diggers. Hence it would be saying that I was a question mark. So I am going to help you realize that I am a true love rather than a question mark by not signing."

Is this really an equilibrium? Imagine that the gold digger and the true love types don't sign and the question marks do sign. As a result, anyone who signs would be viewed as a question mark. This is worse than the position of the true loves. There is no confusion about those who don't sign—the only ones are the gold diggers and true loves, and the partner can tell those apart.

What would happen if the question marks also decided not to sign? Seeing them not sign, their partner would interpret this to mean they must be either a gold digger or a true love. Depending on how likely it is that the question mark will be mistaken for one rather than the other determines whether this would be a good idea or not. If a question mark is more likely to be seen as a gold digger, then not signing is a bad idea.

The larger point is simple. We have ways to figure out people's types besides what they signal. The very fact that they are signaling is a signal that they are trying to differentiate themselves from some other type that can't afford to make the same signal. In some circumstances, the most powerful signal you can send is that you don't need to signal.*

Sylvia Nasar offers the follow perspective on John Nash: "Fagi Levinson, the [MIT math] department's den mother, said in 1996: 'For Nash to deviate from convention is not as shocking as you might think. They were all prima donnas. If a mathematician was mediocre he had to toe the line and be conventional. If he was good, anything went.'"[7]

Prof. Rick Harbaugh, Ph.D., and Ted To did some further investigation into countersignaling. They listened to voicemail messages across the twenty-six University of California and California State University systems, and they found that fewer than 4 percent of economists at schools with a Ph.D. program used a title on their voicemail message, as compared to 27 percent of their colleagues at universities without a doctoral program.[8] In all cases the faculty had a Ph.D., but reminding the caller of the degree or title suggests that you feel the need of a credential in order to distinguish yourself. The truly impressive faculty could show they were

* Just once in our experience, an assistant professor candidate turned up for his job talk wearing jeans. Our first thought was: only a genius would dare not to wear a suit. Only later did we discover that the airline had lost his luggage.

so famous that they didn't need to signal. Hey, just call us Avinash and Barry.

A Quiz: Now you know enough about the manipulation and interpretation of information to take a quiz. We do not call this a Trip to the Gym. It requires no special calculation or math. But we leave it as a quiz instead of offering any discussion of our own, because the correct answers will be highly specific to the situation of each reader. For the same reason, we ask you to grade yourself.

Signal Jamming

If you are buying a used car from the previous owner, you will want to find out how well he cared for it. You might think that its current condition will serve as a signal, that if the car is washed and polished, and its interior is clean and carpets are vacuumed, it is likely to have been well looked after. However, these are signals that even careless owners can mimic when they offer the car for sale. Most importantly, it costs no more for a careless owner than for a careful owner to get the car cleaned. Therefore the signal does not serve to distinguish between the types. As we saw above in the example of the MBA as a signal of managerial talent, this cost difference is essential if the signal is to be effective in making this distinction.

Actually, some small cost differences do exist. Perhaps those who always take good care of their cars take some pride in the fact and may even enjoy washing, polishing, and cleaning the car. Perhaps the careless are very busy and find it hard to spare the time to do these things or get them done. Can small cost differences between the types suffice for the signal to be effective?

The answer depends on the proportions of the two types in the

> **A TRIP TO THE BAR**
>
> You are on a first date with someone you find attractive. You want to make a good first impression—you won't get a second chance. But you expect your date to be aware that impressions can be faked, so you must devise credible signals of your quality. At the same time, you want to screen your date, to see if your immediate attraction has a more durable basis and decide whether you want to continue the relationship. Find some good strategies for your signaling and screening.

population. To see why, begin by thinking of how prospective buyers will interpret a car's cleanness or dirtiness. If everyone gets the car cleaned prior to putting it up for sale, then a prospective buyer learns nothing from observing its cleanness. When he sees a clean car, he interprets it as nothing other than a random draw from the population of possible owners. A dirty car would be a sure indicator of a careless owner.

Now suppose the proportion of careless owners in the population is quite small. Then a clean car would convey quite a favorable impression: the buyer will think that the probability of the owner being careful is quite high. He will be more likely to buy the car or to pay a higher price for it. For the sake of this benefit, even the careless owners will clean their cars prior to selling. This situation, where all types (or all people possessing different types of information) take the same action, and therefore the action is completely uninformative, is called a *pooling* equilibrium of the signaling game—the different types end up in the same pool of signals. By contrast, the kind of equilibrium where one type signals and the other does not, so that the action accurately identifies or separates the types, is a *separating* equilibrium.

Next suppose the proportion of careless owners is large. Then if everyone cleans his car, a clean car does not convey a favorable impression, and a careless owner does not find it worth his while to incur the cost of cleaning the car. (The careful owners always have clean cars.) Thus we cannot get a pooling equilibrium. But if no careless owner is cleaning the car, a single one who does so will get mistaken for a careful owner, and will find it worth his while to incur the small cost. Therefore we cannot get a separating equilibrium either. What happens is somewhere in between: each careless owner follows a mixed strategy, cleaning his car with a positive probability but not certainty. The resulting population of clean cars on the market has a mixture of careful and careless owners. The prospective buyers know the mixture and can infer back to the probability that the owner of a particular clean car is careful. Their willingness to pay will depend on this probability. In turn, the willingness to pay should be such that each careless owner is indifferent between cleaning his car at the small cost and leaving it dirty and thereby being identified as a careless owner, saving the cost but getting a lower price for the car. The mathematical calculation of all this gets somewhat intricate.

It requires a formula, known as Bayes' Rule, for inferring the probabilities of types on the basis of observation of their actions. A simple example of using this rule is illustrated below in the context of betting in poker, but the general features are simple to describe. Because the action now conveys only partial information to distinguish the two types, the outcome is called *semi-separating*.

BODYGUARD OF LIES

Espionage in wartime provides particularly good examples of strategies to confuse the signals of the other side. As Churchill famously said (to Stalin at the 1943 Tehran Conference) "In wartime, truth is so precious that she should always be attended by a bodyguard of lies."

There is a story of two rival businessmen who meet in the Warsaw train station. "Where are you going?" says the first. "To Minsk," replies the other. "To Minsk, eh? What a nerve you have! I know that you are telling me that you are going to Minsk because you want me to believe that you are going to Pinsk. But it so happens that I know you really *are* going to Minsk. So why are you lying to me?"[9]

Some of the best lies arise when someone speaks the truth in order not to be believed. On June 27, 2007, Ashraf Marwan died in London after a suspicious fall from the balcony of his fourth-story flat in Mayfair, London. Thus ended the life of a man who was either the best-connected spy for Israel or a brilliant Egyptian double agent.[10]

Ashraf Marwan was the son-in-law of Egyptian President Abdel Nasser and his liaison to the intelligence service. He offered his services to the Israeli Mossad, who determined his goods were real. Marwan was Israel's guide to the Egyptian mindset.

In April 1973, Marwan sent the code "Radish," which meant

that a war was imminent. As a result, Israel called up thousands of reservists and wasted tens of millions on what turned out to be a false alarm. Six months later, Marwan signaled "Radish" again. It was October 5. The warning was that Egypt and Syria would simultaneously attack the next day, on the Yom Kippur holiday, at sunset. This time, Marwan's alarm was no longer trusted. The head of military intelligence thought Marwan was a double agent and took his message as evidence that war was not imminent.

The attack came at 2:00 P.M. and almost overran the Israeli army. General Zeira, Israel's intelligence head, lost his job over the fiasco. Whether Marwan was a spy for Israel or a double agent remains uncertain. And if his death wasn't an accident, we don't know if it was the Israelis or the Egyptians who are to blame.

When playing mixed or random strategies, you can't fool the opposition every time. The best you can hope for is to keep them guessing and fool them some of the time. You can know the likelihood of your success but cannot say in advance whether you will succeed on any particular occasion. In this regard, when you know that you are talking to a person who wants to mislead you, it may be best to ignore any statements he makes rather than accept them at face value or to infer that exactly the opposite must be the truth.

Actions do speak a little louder than words. By seeing what your rival does, you can judge the relative likelihood of matters that he wants to conceal from you. It is clear from our examples that you cannot simply take a rival's statements at face value. But that does not mean that you should ignore what he does when trying to discern where his true interests lie. The right proportions to mix one's equilibrium play depend on one's payoffs. Observing a player's move gives some information about the mix being used and is valuable evidence to help infer the rival's payoffs. Betting strategies in poker provide a prime example.

Poker players are well acquainted with the need to mix their plays. John McDonald gives the following advice: "The poker hand must at all times be concealed behind the mask of inconsistency. The good poker player must avoid set practices and act at random, going so far, on occasion, as to violate the elementary principles of correct play."[11] A "tight" player who never bluffs seldom wins a large pot; nobody will ever raise him. He may win many small pots, but invariably ends up a loser. A "loose" player who bluffs too often will always be called, and thus he too goes down to defeat. The best strategy requires a mix of the two.

Suppose you know that a regular poker rival raises two-thirds of the time and calls one-third of the time when he has a good hand. If he has a poor hand, he folds two-thirds of the time and raises the other third of the time. (In general, it is a bad idea to call when you are bluffing, since you do not expect to have a winning hand.) Then you can construct the following table for the probabilities of his actions.

To avoid possible confusion, we should say that this is not a table of payoffs. The columns do not correspond to the strategies of any player but are the possible workings of chance. The entries in the cells are probabilities, not payoffs.

		Action		
		Raise	Call	Fold
Quality of hand	Good	2/3	1/3	0
	Poor	1/3	0	2/3

Suppose that before your rival bids, you believe that good and poor hands are equally likely. Because his mixing probabilities depend on his hand, you get additional information from the bid. If you see him fold, you can be sure he had a poor hand. If he calls, you know his hand is good. But in both these cases, the betting is over. If he raises, the odds are 2:1 that he has a good hand. His bid does not always perfectly reveal his hand, but you know more than when you started. After hearing a raise, you increase the chance that his hand is good from one-half to two-thirds.

The estimation of probabilities conditional on hearing the bid is made using Bayes' Rule. The probability that the other player has a good hand conditional on hearing the bid "X" is the chance that this person would both have a good hand and bid X divided by the chance that he ever bids X. Hearing "fold" implies that his hand must be bad, since a person with a good hand never folds. Hearing "call" implies that his hand must be good, since the only time a player calls is when his hand is good. After hearing "raise," the calculations are only slightly more complicated. The odds that a player both has a good hand and raises is $(1/2)(2/3) = 1/3$, while the chance that the player both has a bad hand and raises—that is, bluffs—is $(1/2)(1/3) = 1/6$. Hence the total chance of hearing a raise is $1/3 + 1/6 = 1/2$. According to Bayes' Rule, the probability that the hand is good conditional on hearing a raise is the fraction of the total probability of hearing a raise that is due to the times when the player has a strong hand: in this case that fraction is $(1/3)/(1/2) = 2/3$.

PRICE DISCRIMINATION BY SCREENING

The application of the concept of screening that most impinges on your life is price discrimination. For almost any good or service, some people are willing to pay more than others—either because they are richer, more impatient, or just have different tastes. So long as the cost of producing and selling the good to a customer is less than what the customer is willing to pay, the seller would like to serve that customer and get the highest possible price. But that would mean charging different prices to different customers—for example, giving discounts to those who are not willing to pay so much, without giving the same low price to those who would pay more.

That is often difficult. The sellers do not know exactly how much each individual customer is willing to pay. Even if they did, firms would have to try to avoid situations where one customer with a low value buys the item at a low price and then resells it to a high-value customer who was being charged a high price. Here we don't worry about the issue of resale. We focus on the information issue, the fact that firms don't know which customers are which when it comes to who has a high willingness to pay and who doesn't.

To overcome this problem, the trick that sellers commonly use is to create different versions of the same good and price the versions differently. Each customer is free to select any version and pay the price set by the seller for that version, so there is no overt discrimination. But the seller sets the attributes and prices of each version so that different types of customers will choose different versions. These actions implicitly reveal the customers' private information, namely their willingness to pay. The sellers are screening the buyers.

When a new book is published, some people are willing to pay more; these are also likely to be the readers who want to get and read the book immediately, either because they need the information at once or because they want to impress their friends and colleagues with their up-to-date reading. Others are willing to pay less and are content to wait. Publishers take advantage of this inverse relationship between willingness to pay and willingness to wait by publishing the book initially in hardcover at a higher price and then a year or so later issuing a paperback edition at a lower price. The difference in the costs of printing the two kinds of books is much smaller than the price difference; the "versioning" is just a ploy to screen the buyers. (Question: In what format are you reading this book: hardcover or paperback?)

Producers of computer software often offer a "lite" or "student" version that has fewer features and sells at a substantially lower price. Some users are willing to pay the higher price, perhaps because their employers are the ones paying it. They may also want all the features, or want to have them available just in case they are needed later. Others are willing to pay less and will settle for the basic features. The cost of serving each new customer is very small: just the cost of burning and mailing a CD, or even less in the case of Internet downloads. So the producers would like to cater to those willing to pay less, while charging more to those who are willing to pay more. They do this by offering different versions with different features at different prices. In fact they often produce the lite version by taking the full version and disabling some features. Thus it is somewhat more costly to produce the lite version, even though its price is lower. This seemingly paradoxical situation has to be understood in terms of its purpose, namely to allow the producers to practice price discrimination by screening.

IBM offered two versions of its laser printer. The E version printed at 5 pages per minute, while for $200 more you could get the fast version that printed at 10 pages per minute. The only difference between the two was that IBM added a chip in the firmware of the E version that added some wait states to slow down the printing.[12] If they hadn't done this, then they would have had to sell all their printers at one price. But with the slowed down version, they could offer a lower price to home users who were willing to wait longer for their printouts.

The Sharp DVE611 DVD player and their DV740U unit were both made in the same Shanghai plant. The key difference was that the DVE611 lacked the ability to play DVDs formatted to the European standard (called PAL) on television sets that use the American standard (called NTSC). However, it turns out that the functionality was there all along, just hidden from the customer. Sharp had shaved down the system switch button and then covered it with the remote control faceplate. There were some ingenious users who figured this out and shared their discovery on the web. You could restore full functionality simply by punching a hole in the faceplate at the appropriate spot.[13] Companies often go through great effort to create damaged versions of their goods, and customers often go to great lengths to restore the product.

Airline pricing is probably the example of price discrimination most familiar to readers, so we develop it a little further to give you an idea of the quantitative aspects of designing such a scheme. For this purpose, we introduce Pie-In-The-Sky (PITS), an airline running a service from Podunk to South Succotash. It carries some business passengers and some tourists; the former type is willing to pay a higher price than the latter. To serve the tourists profitably without giving the same low price to the business travelers, PITS has to develop a way of creating different versions of the same flight and price the versions in such a way that each type will choose a different version. First class and economy class might be one way to do this, and we will take that as our example; another common distinction is that between unrestricted and restricted fares.

Suppose that 30 percent of the customers are businesspeople and 70 percent are tourists; we will do the calculation on the basis of "per 100 customers." The table shows the maximum price each type is willing to pay for each class of service (technically referred to as the *reservation price*) and the costs of providing the two types of service.

Type of service	PITS's cost	Reservation price		PITS's potential profit	
		Tourist	Business	Tourist	Business
Economy	100	140	225	40	125
First	150	175	300	25	150

Begin by setting up a situation that is ideal from PITS's point of view. Suppose it knows the type of each customer, for example, by observing their dress as they come to make their reservations. Also suppose that there are no legal prohibitions or resale possibilities. Then PITS can practice what is called perfect price discrimination. To each businessperson it could sell a first-class ticket at $300 for a profit of $300 − 150 = $150, or an economy ticket at $225, for a profit of $225 − 100 = $125. The former is better for PITS. To each tourist, it could sell a first-class ticket at $175 for a profit of $175 − 150 = $25, or an economy ticket at $140 for a profit of $140 − 100 = $40; the latter is better for PITS. Ideally, PITS would like to sell only first-class tickets to business travelers and only economy-class tickets to tourists, in each case at a price equal to the maximum willingness to pay. PITS's total profit per 100 customers from this strategy will be

$$(140 - 100) \times 70 + (300 - 150) \times 30 = 40 \times 70$$
$$+ 150 \times 30 = 2800 + 4500 = 7300.$$

Now turn to the more realistic scenario where PITS cannot identify the type of each customer, or is not allowed to use the information for purposes of overt discrimination. How can it use the versions to screen the customers?

Most importantly, it cannot charge the business travelers their full willingness to pay for first-class seats. They could buy economy-class seats for $140 when they are willing to pay $225; doing so would give them an extra benefit, or "consumer surplus" in the jargon of economics, of $85. They might use it, for example, for better food or accommodation on their trip. Paying the maximum $300 that they are willing to pay for a first-class seat would give them no consumer surplus. Therefore they would switch to economy class, and screening would fail.

The maximum that PITS can charge for first class must give business travelers at least as much extra benefit as the $85 they can get if they buy an economy-class ticket, so the price of first-class tickets can be at most $300 – 85 = $215. (Perhaps it should be $214 to create a definite positive reason for business travelers to choose first class, but we will ignore the trivial difference.) PITS's profit will be

$$(140 - 100) \times 70 + (215 - 150) \times 30 = 40 \times 70$$
$$+ 65 \times 30 = 2800 + 1950 = 4750.$$

So, as we see, PITS can successfully screen and separate the two types of travelers based on their self-selection of the two types of services. But PITS must sacrifice some profit to achieve this indirect discrimination. It must charge the business travelers less than their full willingness to pay. As a result, PITS's profit per 100 passengers drops from the $7,300 it could achieve if it could discriminate overtly with direct knowledge of each customer's type to the $4,750 it achieves from the indirect discrimination based on self-selection. The difference, $2,550, is precisely 85 times 30, where 85 is the drop in the first-class fare below the business travelers' full willingness to pay for this service and 30 is the number of business travelers.

PITS has to keep the first-class fare sufficiently low to give the business travelers enough incentive to choose this service and not "defect" to making the choice that PITS intends for the tourists.

Such a requirement, or constraint, on the screener's strategy is called an *incentive compatibility constraint.*

The only way PITS could charge more than $215 to business travelers without inducing their defection would be to increase the economy-class fare. For example, if the first-class fare is $240 and the economy-class fare is $165, then business travelers get equal extra benefit (consumer surplus) from the two classes: $300 – 240 from first class and $225 – 165 from economy class, or $60 from each, so they are (only just) willing to buy first-class tickets.

But at $140 the economy-class fare is already at the limit of the tourists' willingness to pay. If PITS raised it to even $141, it would lose these customers altogether. This requirement, namely that the customer type in question remains willing to buy, is called that type's *participation constraint.* PITS's pricing strategy is thus squeezed between the participation constraint of the tourists and the incentive compatibility constraint of the businesspeople. In this situation, the screening strategy above, charging $215 for first class and $140 for economy class, is in fact the most profitable for PITS. It takes a little mathematics to prove that rigorously, so we merely assert it.

TRIP TO THE GYM NO. 5

There is also a participation constraint for business travelers and an incentive compatibility constraint for the tourists. Check that these are automatically satisfied at the stated prices.

Whether this strategy is optimal for PITS depends on the specific numbers in the example. Suppose the proportion of business travelers were much higher, say 50 percent. Then the sacrifice of $85 on each business traveler may be too high to justify keeping the few tourists. PITS may do better not to serve them at all—that is, violate their participation constraint—and raise the price of first-class service for the business travelers. Indeed, the strategy of discrimination by screening with these numbers of travelers yields

$$(140 - 100) \times 50 + (215 - 150) \times 50 = 40 \times 50$$
$$+ 65 \times 50 = 2000 + 3250 = 5250,$$

while the strategy of serving only business travelers in first class at $300 would yield

$$(300 - 150) \times 50 = 150 \times 50 = 7500.$$

If there are only a few customers with low willingness to pay, the seller might find it better not to serve them at all than to offer sufficiently low prices to the mass of high-paying customers to prevent their switching to the low-priced version.

Now that you know what to look for, you will see screening for price discrimination everywhere. And if you look in the research literature, you will see analyses of strategies for screening by self-selection equally frequently.[14] Some of these strategies are quite complicated, and the theories need a lot of mathematics. But the basic idea driving all these instances is the interplay between the twin requirements of incentive compatibility and participation.

CASE STUDY: GOING UNDERCOVER

Another friend of ours, Tanya, is an anthropologist. While most anthropologists travel to the ends of the earth to study some unusual tribe, Tanya did her fieldwork in London. Her subject was witches.

Yes, witches. Even in modern-day London there are still a surprisingly large number of people who gather together to trade spells and study witchcraft. Not that being a modern witch is easy; it requires a certain amount of rationalization to be a witch riding the tube. Often anthropologists have trouble gaining their subject's confidence. But Tanya's group was especially welcoming. When she told them she was an anthropologist, they saw this as a clever ruse: she was really a witch with a great cover story.

One of the unusual features of the witches' meetings is that they took place in the nude. Why might that be?

Case Discussion

Any outsider group has to worry that its members will be observers rather than participants. Are you sitting there making fun of the whole process, or are you being a part of it? If you are sitting there in the nude, it is pretty hard to say that you are just watching and making fun of the others. You are well into it.

Thus the nudity is a credible screening device. If you truly believe in the coven, then it is relatively costless to be there in the nude. But if you are a skeptic, then being there in the nude is hard to explain, both to others and to yourself.* For the same reason, gang initiation rites often involve taking actions that are relatively cheap if you are truly interested in gang life (tattoos, committing crimes) but quite costly if you are an undercover cop trying to infiltrate the gang.

For more cases on interpreting and manipulating information, see "The Other Person's Envelope Is Always Greener," "But One Life to Lay Down for Your Country," "King Solomon's Dilemma Redux," and "The King Lear Problem" in chapter 14.

---

* In the film *Gray's Anatomy*, the monologist Spalding Gray told a similar tale of his challenging experience in a Native American sweat lodge.

CHAPTER 9

Cooperation
and Coordination

FOR WHOM THE BELL CURVE TOLLS

In the 1950s the Ivy League colleges were faced with a problem. Each school wanted to produce a winning football team. The colleges found themselves overemphasizing athletics and compromising their academic standards in order to build a championship team. Yet no matter how often they practiced or how much money they spent, at the end of the season the standings were much as they had been before. The average win-loss record was still 50:50. The inescapable mathematical fact is that for every winner there had to be a loser. All the extra work canceled itself out.

The excitement of college sports depends as much on the closeness and intensity of the competition as on the level of skill. Many fans prefer college basketball and football to the professional versions; while the level of skill is lower, there is often more excitement and intensity to the competition. With this idea in mind, the colleges got smart. They joined together and agreed to limit spring training to one day. Although there were more fumbles, the games were no less exciting. Athletes had more time to concentrate on their studies. Everyone was better off, except some alumni who wanted their alma maters to excel at football and forget about academic work.

Many students would like to have a similar agreement with their fellow students before examinations. When grades are based on a traditional bell curve, one's relative standing in the class matters more than the absolute level of one's knowledge. It matters not how much you know, only that others know less than you. The way to gain an advantage over the other students is to study more. If they all do so, they all have more knowledge, but the relative standings and therefore the bottom line—the grades—are largely unchanged. If only everyone in the class could agree to limit spring studying to one (preferably rainy) day, they would get the same grades with less effort.

The feature common to these situations is that success is determined by *relative* rather than *absolute* performance. When one participant improves his own ranking, he necessarily worsens everyone else's ranking. But the fact that one's victory requires someone else's defeat does not make the game zero-sum. In a zero-sum game it is not possible to make everyone better off. Here, it is. The scope for gain comes from reducing the inputs. While there might always be the same number of winners and losers, it can be less costly for everyone to play the game.

The source of the problem of why (some) students study too much is that they do not have to pay a price or compensation to the others. Each student's studying is akin to a factory's polluting: it makes it more difficult for all the other students to breathe. Because there is no market for buying and selling studying time, the result is a rat race: each participant strives too hard, with too little to show for his efforts. But no one team or student is willing to be the only one, or the leader, in reducing the effort. This is just like a prisoners' dilemma with more than two prisoners. An escape from the horns of this dilemma requires an enforceable collective agreement.

As with the Ivy League or OPEC, the trick is to form a cartel to limit competition. The problem for high-school students is that the cartel cannot easily detect cheating. For the collectivity of students, a cheater is one who studies more to sneak an advantage over the others. It is hard to tell if some are secretly studying until after they have aced the test. By then it is too late.

In some small towns, high-school students do have a way to enforce "no-studying" cartels. Everyone gets together and cruises Main Street at night. The absence of those home studying is noticed. Punishment can be social ostracism or worse.

To arrange a self-enforcing cartel is difficult. It is all the better if an outsider enforces the collective agreement limiting competition. This is just what happened for cigarette advertising, although not intentionally. In the old days, cigarette companies used to spend money to convince consumers to "walk a mile" for their product or to "fight rather than switch." The different campaigns made advertising agencies rich, but their main purpose was defensive— each company advertised because the others did, too. Then, in 1968, cigarette advertisements were banned from TV by law. The companies thought this restriction would hurt them and fought against it. But, when the smoke cleared, they saw that the ban helped them all avoid costly advertising campaigns and thus improved all their profits.

THE ROUTE LESS TRAVELED

There are two main ways to commute from Berkeley to San Francisco. One is driving over the Bay Bridge, and the other is taking public transportation, the Bay Area Rapid Transit train (BART). Crossing the bridge is the shortest route, and with no traffic, a car can make the trip in 20 minutes. But that is rarely the case. The bridge has only four lanes and is easily congested.* We suppose that each additional 2,000 cars (per hour) causes a 10-minute delay for everyone on the road. For example, with 2,000 cars the travel time rises to 30 minutes; at 4,000 cars, to 40 minutes.

The BART train makes a number of stops, and one has to walk to the station and wait for the train. It is fair to say that the trip takes closer to 40 minutes along this route, but the train never

---

* Sometimes, after earthquakes, it is closed altogether.

fights traffic. When train usage rises, they put on more cars, and the commuting time stays roughly constant.

If, during rush hour, 10,000 commuters want to go from Berkeley to San Francisco, how will the commuters be distributed over the two routes? Each commuter will act selfishly, choosing the route that minimizes his own transportation time. Left to their own devices, 40 percent will drive and 60 percent will take the train. The commuting time will be 40 minutes for everyone. This outcome is the equilibrium of a game.

We can see this result by asking what would happen if the split were different. Suppose only 2,000 drivers took the Bay Bridge. With less congestion, the trip would take less time (30 minutes) along this route. Then some of the 8,000 BART commuters would find out that they could save time by switching, and would do so. Conversely, if there were, say, 8,000 drivers using the Bay Bridge, each spending 60 minutes, some of them would switch to the train for the faster trip it provides. But when there are 4,000 drivers on the Bay Bridge and 6,000 on the train, no one can gain by switching: the commuters have reached an equilibrium.

We can show the equilibrium using a simple chart, which is quite similar in spirit to the one in chapter 4 describing the classroom experiment of the prisoners' dilemma. In this chart, we are holding the total number of commuters constant at 10,000, so that when there are 2,000 cars using the bridge, it implies that 8,000 commuters are using BART. The rising line shows how the trip time on the Bay Bridge increases as the number of drivers on it increases. The flat line shows the constant time of 40 minutes for the train. The lines intersect at E, showing that the trip times on the two routes are equal when the number of drivers on the Bay Bridge is 4,000. This graphic depiction is a useful tool to describe the equilibrium, and we will use it often in this chapter.

Is this equilibrium good for the commuters as a whole? Not really. It is easy to find a better pattern. Suppose only 2,000 take the Bay Bridge. Each of them saves 10 minutes. The 2,000 who switch to the train are still spending the same time as they did before, namely 40 minutes. So are the 6,000 who were already

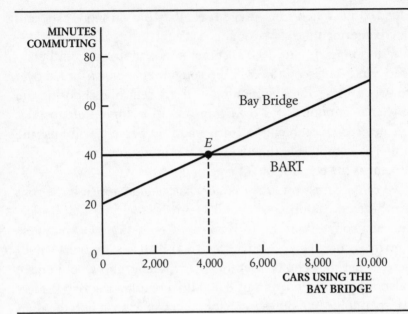

taking the train. We have just saved 20,000 person-minutes (or almost two weeks) from the total travel time.

Why is this saving possible? Or, in other words, why were the drivers left to themselves not guided by an invisible hand to the best mix of routes? The answer again lies in the cost that each user of the Bay Bridge inflicts on the others. When an extra driver takes this road, the travel time of all the other users goes up by a little bit. But the newcomer is not required to pay a price that reflects this cost. He takes into account only his own travel time.

What traffic pattern is best for the group of drivers as a whole? In fact, the one we constructed, with 2,000 cars on the Bay Bridge and a total time saving of 20,000 minutes, is best. To see this, try a couple of others. If there are 3,000 cars on the Bay Bridge, the travel time is 35 minutes, with a saving of 5 minutes each, or 15,000 minutes in all. With only 1,000 cars, the travel time is 25 minutes, and each saves 15 minutes, but the total saving is again only 15,000 minutes. The intermediate point with 2,000 drivers, each saving 10 minutes, is best.

How can the best pattern be achieved? Devotees of central planning will think of issuing 2,000 licenses to use the Bay Bridge. If they are worried about the inequity of allowing those with licenses to travel in 30 minutes while the other 8,000 must take the train and spend 40 minutes, they will devise an ingenious system of rotating the licenses among the population every month.

A market-based solution charges people for the harm they cause to others. Suppose each person values an hour of time at $12, that is, each would be willing to pay $12 to save an hour. Then charge a toll for driving on the Bay Bridge; set the toll $2 above the BART fare. By our supposition, people regard an extra $2 cost as equivalent to 10 minutes of time. Now the equilibrium commuting pattern will have 2,000 cars on the Bay Bridge and 8,000 riders on BART. Each user of the Bay Bridge spends 30 minutes plus an extra $2 in commuting costs; each BART rider spends 40 minutes. The total effective costs are the same, and no one wants to switch to the other route. In the process we have collected $4,000 of toll revenue (plus an additional 2,000 BART fares), which can then go into the county's budget, thus benefiting everyone because taxes can be lower than they would otherwise be.

A solution even closer to the spirit of free enterprise would be to allow private ownership of the Bay Bridge. The owner realizes that people are willing to pay for the advantage of a faster trip on a less congested road. He charges a price, therefore, for the privilege. How can he maximize his revenue? By maximizing the total value of the time saved, of course.

The invisible hand guides people to an optimal commuting pattern only when the good "commuting time" is priced. With the profit-maximizing toll on the bridge, time really is money. Those commuters who ride BART are selling time to those who use the bridge.

Finally, we recognize that the cost of collecting the toll sometimes exceeds the resulting benefit of saving people's time. Creating a marketplace is not a free lunch. The toll booths may be a primary cause of the congestion. If so, it may be best to tolerate the original inefficient route choices.

CATCH-22?

Chapter 4 offered the first examples of games with many equilibria. Where should two strangers meet in New York City: Times Square or the Empire State Building? Who should return a disconnected phone call? In those examples it was not important which of the conventions was chosen, so long as everyone agreed on the same convention. But sometimes one convention is much better than another. Even so, that doesn't mean it will always get adopted. If one convention has become established and then some change in circumstances makes another one more desirable, it can be especially hard to bring about the change.

The keyboard design on most typewriters is a case in point. In the late 1800s, there was no standard pattern for the arrangement of letters on the typewriter keyboard. Then in 1873 Christopher Scholes helped design a "new, improved" layout. The layout became known as QWERTY, after the letter arrangement of the first six letters in the top row. QWERTY was chosen to *maximize* the distance between the most frequently used letters. This was a good solution in its day; it deliberately slowed down the typist, and reduced the jamming of keys on manual typewriters. By 1904, the Remington Sewing Machine Company of New York was mass-producing typewriters with this layout, and it became the de facto industry standard. But with the advent of electric typewriters and, later, computers, this jamming problem became irrelevant. Engineers developed new keyboard layouts, such as DSK (Dvorak's Simplified Keyboard), which reduced the distance typists' fingers traveled by over 50 percent. The same material can be typed in 5 to 10 percent less time using DSK than QWERTY.[1] But QWERTY is the established system. Almost all keyboards use it, so we all learn it and are reluctant to learn a second layout. Keyboard manufacturers continue, therefore, with QWERTY. The vicious circle is complete.[2]

If history had worked differently, and if the DSK standard had been adopted from the outset, that would have been better for today's technology. However, given where we are, the question of whether or not we should switch standards involves further considerations. There is a lot of inertia, in the form of machines, key-boards, and trained typists, behind QWERTY. Is it worthwhile to retool?

From the point of view of society as a whole, the answer would seem to be yes. During the Second World War, the U.S. Navy used DSK typewriters on a large scale, and retrained typists to use them. It found that the cost of retraining could be fully recouped in only ten days of use.

However, this study and the overall advantage of DSK has been called into question by Professors Stan Liebowitz and Stephen Margolis.[3] It appears that an interested party, one Lieutenant Commander August Dvorak, was involved in conducting the original study. A 1956 General Services Administration study found that it took a month of four-hour-a-day training for typists to catch up to their old QWERTY speed. At that point, further training on the Dvorak keyboard was less effective than providing training to QWERTY typists. To the extent that DSK is superior, the biggest gain is when typists learn this system from the start.

If the typist becomes so good that he or she almost never has to look at the keyboard, then learning DSK makes sense. With today's software, it is a relatively simple matter to reassign the keys from one layout to another. (On a Mac, it is a simple switch on the keyboard menu.) Thus the keyboard layout almost doesn't matter. Almost. The problem is: how does one learn to touch type on a mislabeled keyboard? Anyone who wants to reassign the layout from QWERTY to DSK but cannot yet touch type must look at the keyboard and mentally convert each key to its DSK value. This is not practical. Therefore beginners have to learn QWERTY any-way and that greatly reduces the gains from also learning DSK.

No individual user can change the social convention. The unco-ordinated decisions of individuals keep us tied to QWERTY. The problem is called a bandwagon effect and can be illustrated using the following chart. On the horizontal axis we show the fraction of typists using QWERTY. The vertical axis details the chance that a new typist will learn QWERTY as opposed to DSK. As drawn, if 85 percent of typists are using QWERTY, then the chances are

95 percent that a new typist will choose to learn QWERTY and only 5 percent that the new typist will learn DSK. The way the curve is drawn is meant to emphasize the superiority of the DSK layout. A majority of new typists will learn DSK rather than QWERTY provided that QWERTY has anything less than a 70 percent market share. In spite of this handicap, it is possible for QWERTY to dominate in equilibrium. (Indeed, this possibility is just what has happened in the prevailing equilibrium.)

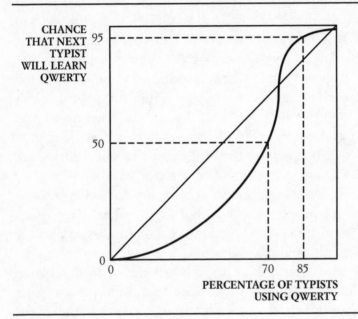

The choice of which keyboard to use is a strategy. When the fraction using each technology is constant over time, we are at an equilibrium of the game. Showing that this game converges to an equilibrium is not easy. The random choice of each new typist is constantly disrupting the system. Recent high-powered mathematical tools in the field of stochastic approximation theory have allowed economists and statisticians to prove that this dynamic game does converge to an equilibrium.[4] We now describe the possible outcomes.

If the fraction of typists using QWERTY exceeds 72 percent,

there is the expectation that an even greater fraction of people will learn QWERTY. The prevalence of QWERTY expands until it reaches 98 percent. At that point, the fraction of new typists learning QWERTY just equals its predominance in the population, 98 percent, and so there is no more upward pressure.*

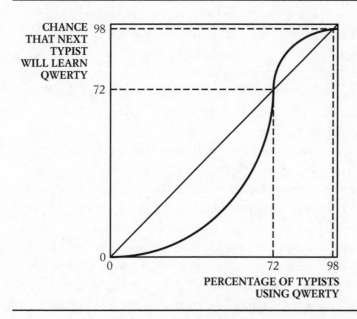

Conversely, if the fraction of typists using QWERTY falls below 72 percent, then there is the expectation that DSK will take over. Fewer than 72 percent of the new typists learn QWERTY, and the subsequent fall in its usage gives new typists an even greater incentive to learn the superior layout of DSK. Once all typists are using DSK there is no reason for a new typist to learn QWERTY, and QWERTY will die out.

* If the number of typists using QWERTY is above 98 percent, the number is expected to fall back to 98 percent. There will always be a small number, somewhere up to 2 percent, of new typists who will choose to learn DSK because they are interested in the superior technology and are not concerned with the compatibility issue.

The mathematics says only that we will end up at one of these two possible outcomes: everyone using DSK or 98 percent using QWERTY. It does not say which will occur. If we were starting from scratch, the odds are in favor of DSK being the predominant keyboard layout. But we are not. History matters. The historical accident that led to QWERTY capturing nearly 100 percent of typists ends up being self-perpetuating, even though the original motivation for QWERTY is long since obsolete.

Since bad luck or the convergence to an inferior equilibrium is self-perpetuating, there is the possibility of making everyone better off. But it requires coordinated action. If the major computer manufacturers coordinate on a new keyboard layout or a major employer, such as the federal government, trains its employees on a new keyboard, this could switch the equilibrium all the way from one extreme to the other. The essential point is that it is not necessary to convert everyone, just a critical mass. Given enough of a toehold, the better technology can take it from there.

The QWERTY problem is but one minor example of a more widespread problem. Our preference for gasoline engines over steam and light-water nuclear reactors over gas-cooled is better explained by historical accidents than by the superiority of the adopted technologies. Brian Arthur, an economist at Stanford and one of the developers of the mathematical tools used to study bandwagon effects, tells the story of how we ended up with gasoline-powered cars.

> In 1890 there were three ways to power automobiles—steam, gasoline, and electricity—and of these one was patently *inferior* to the other two: gasoline. . . . [A turning point for gasoline was] an 1895 horseless carriage competition sponsored by the *Chicago Times-Herald*. This was won by a gasoline-powered Duryea—one of only two cars to finish out of six starters—and has been cited as the possible inspiration for R. E. Olds to patent in 1896 a gasoline power source, which he subsequently mass-produced in the "Curved-Dash Olds." Gasoline thus overcame its slow start. Steam continued to be viable as an automotive power source until 1914, when there was an outbreak of

hoof-and-mouth disease in North America. This led to the withdrawal of horse troughs—which is where steam cars could fill
with water. It took the Stanley brothers about three years to
develop a condenser and boiler system that did not need to be
filled every thirty or forty miles. But by then it was too late. The
steam engine never recovered.[5]

While there is little doubt that today's gasoline technology is
better than steam, that's not the right comparison. How good
would steam have been if it had had the benefit of seventy-five
years of research and development? While we may never know,
some engineers believe that steam was the better bet.[6]

In the United States, almost all nuclear power is generated by
light-water reactors. Yet there are reasons to believe that the alternative technologies of heavy-water or gas-cooled reactors would
have been superior, especially given the same amount of learning
and experience. Canada's experience with heavy-water reactors
allows them to generate power for 25 percent less cost than light-
water reactors of equivalent size in the United States. Heavy-water
reactors can operate without the need to reprocess fuel. Perhaps
most important is the safety comparison. Both heavy-water and
gas-cooled reactors have a significantly lower risk of a meltdown—
the former because the high pressure is distributed over many
tubes rather than a single core vessel, and the latter because of the
much slower temperature rise in the event of coolant loss.[7]

The question of how light-water reactors came to dominate has
been studied by Robin Cowen in a 1987 Stanford University Ph.D.
thesis. The first consumer for nuclear power was the U.S. Navy. In
1949, then Captain Rickover made the pragmatic choice in favor
of light-water reactors. He had two good reasons. It was the most
compact technology at the time, an important consideration for
submarines, and it was the furthest advanced, suggesting that it
would have the quickest route to implementation. In 1954, the first
nuclear-powered submarine, *Nautilus*, was launched. The results
looked positive.

At the same time civilian nuclear power became a high priority.
The Soviets had exploded their first nuclear bomb in 1949. In

response, Atomic Energy Commissioner T. Murray warned, "Once we become fully conscious of the possibility that [energy-poor] nations will gravitate towards the USSR if it wins the nuclear power race, it will be quite clear that this race is no Everest-climbing, kudos-providing contest."[8] General Electric and West-inghouse, with their experience producing light-water reactors for the nuclear-powered submarines, were the natural choice to develop civilian power stations. Considerations of proven reliabil-ity and speed of implementation took precedence over finding the most cost-effective and safe technology. Although light-water was first chosen as an interim technology, this gave it enough of a head start down the learning curve that the other options have never had the chance to catch up.

The adoption of QWERTY, gasoline engines, and light-water reactors are but three demonstrations of how history matters in determining today's technology choices, though the historical rea-sons may be irrelevant considerations in the present. Typewriter-key jamming, hoof-and-mouth disease, and submarine space constraints are not relevant to today's trade-offs between the com-peting technologies. The important insight from game theory is to recognize early on the potential for future lock-in—once one option has enough of a head start, superior technological alterna-tives may never get the chance to develop. Thus there is a poten-tially great payoff in the early stages from spending more time figuring out not only what technology meets today's constraints but also what options will be the best for the future.

FASTER THAN A SPEEDING TICKET

Just how fast should you drive? In particular, should you abide by the speed limit? Again the answer is found by looking at the game where your decision interacts with those of all the other drivers.

If nobody is abiding by the law, then you have two reasons to break it too. First, some experts argue that it is actually safer to drive at the same speed as the flow of traffic.[9] On most highways, anyone who tries to drive at fifty-five miles per hour creates a dangerous obstacle that everyone else must go around. Second, when you tag along with the other speeders, your chances of getting caught are almost zero. The police simply cannot pull over more than a small percentage of the speeding cars. As long as you go with the flow of traffic, there is safety in numbers.*

As more people become law-abiding, both reasons to speed vanish. It becomes more dangerous to speed, since this requires weaving in and out of traffic. And your chances of getting caught increase dramatically.

We show this in a chart similar to the one for commuters from Berkeley to San Francisco. The horizontal axis measures the percentage of drivers who abide by the speed limit. The lines *A* and *B* show each driver's calculation of his benefit from (*A*) abiding by and (*B*) breaking the law. Our argument says that if no one else is

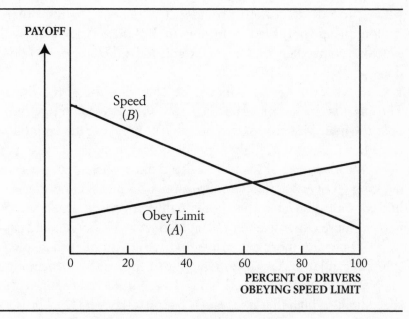

---

* The police love the law-breaking equilibrium because it gives them probable cause to stop any speeder—and anyone who isn't speeding is even more suspicious.

keeping under the limit (the far left end), neither should you (line *B* is higher than line *A*); if everyone else is law-abiding (the far right end), you should be too (line *A* is higher than line *B*). Once again there are three equilibria, of which only the extreme ones can arise from the process of social dynamics as drivers adjust to one another's behavior.

In the case of the commuters choosing between the Bay Bridge and BART, the dynamics converged on the equilibrium in the middle. Here the tendency is toward one of the extremes. The difference arises because of the way interactions work. With commuting, either choice becomes *less* attractive when more of the others follow you, whereas with speeding, additional company makes it *more* attractive.

The general theme of one person's decision affecting the others applies here, too. If one driver speeds up, he makes it a little safer for the others to speed. If no one is speeding, no one is willing to be the first to do so and provide this "benefit" to the others without being "rewarded" for doing so. But there is a new twist: if everyone is speeding, then no one wants to be the only one to slow down.

Can this situation be affected by changing the speed limit? The chart is drawn for a specific speed limit, say 55 m.p.h. Suppose the limit is raised to 65. The value of breaking the limit falls, since beyond a point, higher speeds do become dangerous, and the extra advantage of going 75 instead of 65 is less than the gain of going 65 over 55. Furthermore, above 55 miles an hour, gasoline consumption goes up exponentially with speed. It may be 20 percent more expensive to drive at 65 than at 55, but it could easily be 40 percent more expensive to drive at 75 rather than at 65.

What can lawmakers learn from this if they want to encourage people to drive at the speed limit? It is not necessary to set the speed limit so high that everyone is happy to obey it. The key is to get a critical mass of drivers obeying the speed limit. Thus a short phase of extremely strict enforcement and harsh penalties can change the behavior of enough drivers to generate the momentum toward full compliance. The equilibrium moves from one extreme (where everyone speeds) to the other (where everyone complies). With the new equilibrium, the police can cut back on enforcement, and the compliance behavior is self-sustaining. More generally, what this suggests is that short but intense enforcement can be significantly more effective than the same total effort applied at a more moderate level for a longer time.[10]

A similar logic applies to fuel-economy standards. For many years, the vast majority of Americans supported a large increase in the Corporate Average Fuel Economy (CAFE) standards. Finally, in 2007 President Bush signed legislation mandating an increase from 27.5 mpg to 35 mpg for cars (and similar increases for trucks) to be phased in gradually starting in 2011 before taking full effect in 2020. But if most people want higher fuel economy, nothing prevents them from buying a fuel-efficient car. Why is it that folks who want higher fuel standards keep on driving gas-guzzling SUVs?

One reason is that people are concerned that fuel-efficient cars are lighter and thus less safe in the event of an accident. Light cars are especially unsafe when hit by a Hummer. Folks are more willing to drive a light car when they know that the other cars on the road are light as well. As speeding leads to speeding, the more heavy cars there are out there, the more everyone needs to drive an SUV in order to be safe. Just like people, cars have become 20 percent heavier over the last two decades. The end result is that we end up with low fuel economy and no one is safer. A move to higher CAFE standards is the coordination device that could help shift enough people from heavy to light cars so that (almost) everyone would be happier driving a light car.[11] Perhaps even more important than a technological advance is the coordination change that would shift the mix of cars and thereby allow us to improve fuel economy right away.

Arguments in favor of collective, rather than individual, decisions are not the preserve of liberals, left-wingers, and any remaining socialists. The impeccably conservative economist Milton Friedman made the same logical argument about redistribution of wealth in his classic *Capitalism and Freedom*:

I am distressed by the sight of poverty; I am benefited by its alleviation; but I am benefited equally whether I or someone else
pays for its alleviation; the benefits of other people's charity
therefore partly accrue to me. To put it differently, we might all of
us be willing to contribute to the relief of poverty, *provided* everyone else did. We might not be willing to contribute the same
amount without such assurance. In small communities, public
pressure can suffice to realize the proviso even with private charity. In the large impersonal communities that are increasingly
coming to dominate our society, it is much more difficult for it to
do so. Suppose one accepts, as I do, this line of reasoning as jus-
tifying governmental action to relieve poverty . . .[12]

WHY DID THEY LEAVE?

American cities have few racially integrated neighborhoods. If
the proportion of black residents in an area rises above a critical
level, it quickly increases further to nearly 100 percent. If it falls
below a critical level, the expected course is for the neighborhood
to become all white. Preservation of racial balance requires some
ingenious public policies.

Is the de facto segregation of most neighborhoods the product
of widespread racism? These days, a large majority of urban Americans would regard mixed neighborhoods as desirable.* The more
likely difficulty is that segregation can result as the equilibrium of a
game in which each household chooses where to live, even when
they all have a measure of racial tolerance. This idea is due to
Thomas Schelling.[13] We shall now outline it, and show how it
explains the success of the Chicago suburb Oak Park in maintaining an integrated community.

Racial tolerance is not a matter of black or white; there are
shades of gray. Different people, black or white, have different
views about the best racial mix. For example, very few whites insist

---

* Of course the fact that people have *any* preferences about the racial mix
of their neighbors is a form of racism, albeit a less extreme one than total
intolerance.

on a neighborhood that is 99 or even 95 percent white; yet most will feel out of place in one that is only 1 or 5 percent white. The majority would be happy with a mix somewhere in between.

We can illustrate the evolution of neighborhood dynamics using a chart similar to the one from the QWERTY story. On the vertical axis is the probability that a new person moving into the neighborhood will be white. This is plotted in relationship to the current racial mix, shown on the horizontal axis. The far right end of the curve shows that once a neighborhood becomes completely segregated (all white), the odds are overwhelming that the next person who moves into the neighborhood will also be white. If the current mix falls to 95 percent or 90 percent white, the odds are still very high that the next person to move in will also be white. If the mix changes much further, then there is a sharp drop-off in the probability that the next person to join the community will be white. Finally, as the actual percentage of whites drops to zero, meaning that the neighborhood is now segregated at the other extreme, the probability is very high that the next person to move in will be black.

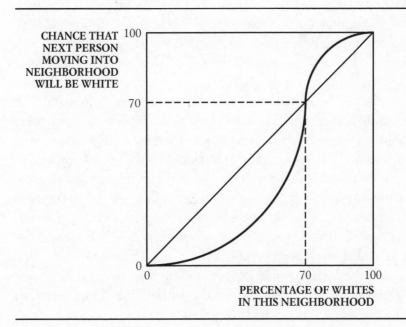

CHANCE THAT NEXT PERSON MOVING INTO NEIGHBORHOOD WILL BE WHITE

100

70

0

0   70   100

PERCENTAGE OF WHITES IN THIS NEIGHBORHOOD

In this situation, the equilibrium will be where the racial mix of the population exactly equals the mix of new entrants to the community. Only in this event are the dynamics stable. There are three such equilibria: two at the extremes where the neighborhood is all white and all black, and one in the middle where there is a mix. The theory so far does not tell us which of the three equilibria is the most likely. In order to answer this question, we need to examine the forces that move the system toward or away from an equilibrium—that is, the social dynamics of the situation.

Social dynamics will always drive the neighborhood to one of the extreme equilibria. Schelling labeled this phenomenon "tipping" (an idea later popularized by Malcolm Gladwell's book *The Tipping Point*). Let us see why it occurs. Suppose the middle equilibrium has 70 percent whites and 30 percent blacks. By chance, let one black family move out and be replaced by a white family. Then the proportion of whites in this neighborhood becomes slightly above 70 percent. Looking at the chart, the probability that the next entrant will also be white is then above 70 percent. The upward pressure is reinforced by the new entrants. Say the racial mix shifts to 75:25 percent. The tipping pressure continues. The chance that a new entrant will be white is above 75 percent, so the expectation is that the neighborhood will become increasingly segregated. This goes on until the mix of new entrants is the same as the mix in the neighborhood. As drawn, that occurs again only when the neighborhood is all white. If the process had started with one white family moving out and one black family moving in, there would have been a chain reaction in the opposite direction, and the odds are that the neighborhood would have become all black.

The problem is that the 70:30 percent mix is not a stable equilibrium. If this mix is somehow disrupted, as it will surely be, there is a tendency to move toward one of the extremes. Sadly, from the extremes there is no similar tendency to move back toward the middle. Although segregation is the predicted equilibrium, that does not mean that people are better off at this outcome. Everyone might prefer to live in a mixed neighborhood. But they rarely exist and, even when found, tend not to last.

Once again, the source of the problem is the effect of one household's action on the others. Starting at a 70:30 percent mix, when one white family replaces a black family, this may make the neighborhood a little less attractive for future blacks to move in. But it is not assessed a fine for this. Perhaps there should be a neighborhood departure tax analogous to road tolls. But that would be counter to a more basic principle, namely the freedom to live where one chooses. If society wants to prevent tipping, it must look for some other policy measures.

If we cannot fine a departing family for the damage it causes, both to those who remain and those who now might choose not to come, we must take measures that will reduce the incentives for others to follow suit. If one white family leaves, the neighborhood should not become less attractive to another white family. If one black family leaves, the neighborhood should not become less attractive to another black family. Public policy can help prevent the tipping process from gathering momentum.

The racially integrated Chicago suburb of Oak Park provides an ingenious example of policies that work. It uses two tools: first, the town banned the use of "For Sale" signs in front yards, and, secondly, the town offers insurance that guarantees homeowners that they will not lose the value of their house and property because of a change in the racial mix.

If by chance two houses on the same street are for sale at the same time, "For Sale" signs would spread this news quickly to all neighbors and prospective purchasers. Eliminating such signs makes it possible to conceal the news that would be interpreted as bad; nobody need know until after a house has been sold that it was even up for sale. The result is that panics are avoided (unless they are justified, in which case they are just delayed).

By itself, the first policy is not enough. Homeowners might still worry that they should sell their house while the going is good. If you wait until the neighborhood has tipped, you've waited too long and may find that you've lost most of the value of your home, which is a large part of most people's wealth. Once the town provides insurance, this is no longer an issue. In other words, the insurance removes the *economic* fear that accelerates tipping. In fact, if the guarantee succeeds in preventing tipping, property values will not fall and the policy will not cost the taxpayers anything.

Tipping to an all-black equilibrium has been the more common problem in urban America. But in recent years gentrification, which is just tipping to an all-rich equilibrium, has been on the rise. Left unattended, the free market will often head to these unsatisfactory outcomes. But public policy, combined with an awareness of how tipping works, can help stop the momentum toward tipping and preserve the delicate balances.

IT CAN BE LONELY AT THE TOP

Top law firms generally choose their partners from among their junior associates. Those not chosen must leave the firm, and generally move to a lower-ranked one. At the mythical firm Justin-Case, the standards were so high that for many years no new partners were selected. The junior associates protested about this lack of advancement. The partners responded with a new system that looked very democratic.

Here is what they did. At the time of the annual partnership decision, the abilities of the ten junior associates were rated from 1 to 10, with 10 being the best. The junior associates were told their rating privately. Then they were ushered into a meeting room where they were to decide by majority vote the cutoff level for partnership.

They all agreed that everyone making partner was a good idea and certainly preferable to the old days when nobody made partner. So they began with a cutoff of 1. Then some high-rated junior associate suggested that they raise the cutoff to 2. He argued that this would improve the average quality of the partnership. Eight junior associates agreed with him. The sole dissenting vote came from the least able member, who would no longer make partner.

Next, someone proposed that they raise the standard from 2 to 3. Eight people were still above this standard, and they all voted for this improvement in the quality of the partnership. The person ranked 2 voted against, as this move deprived him of partnership. What was surprising was that the lowest-rated junior associate was in favor of this raising of the standards. In neither case would he make partner. But at least in the latter he would be grouped with someone who had ability 2. Therefore, upon seeing that he was not selected, other law firms would not be able to infer his exact ability. They would guess that he was either a 1 or a 2, a level of uncertainty that would be to his advantage. The proposal to raise the standard to 3 passed 9:1.

With each new cutoff level someone proposed raising it by one. All those strictly above voted in favor so as to raise the quality of the partnership (without sacrificing their own position), while all those strictly below joined in support of raising the standard so as to make their failure less consequential. Each time there was only one dissenter, the associate right at the cutoff level who would no longer make partner. But he was outvoted 9:1.

And so it went, until the standard was raised all the way up to 10. Finally, someone proposed that they raise the standard to 11 so that *nobody* would make partner. Everybody rated 9 and below thought that this was a fine proposal, since once more this improved the average quality of those rejected. Outsiders would not take it as a bad sign that they didn't make partner, as nobody made partner at this law firm. The sole voice against was the most able junior associate, who lost his chance to make partner. But he was outvoted 9:1.

The series of votes brought everybody back to the old system, which they all considered worse than the alternative of promotion for all. Even so, each resolution along the way passed 9:1. There are two morals to this story.

When actions are taken in a piecemeal way, each step of the way can appear attractive to the vast majority of decision makers. But the end is worse than the beginning for everyone. The reason is that voting ignores the intensity of preferences. In our example, all those in favor gain a small amount, while the one person against loses a lot. In the series of ten votes, each junior associate has nine small victories and one major loss that outweighs all the combined gains. Similar problems tend to plague bills involving reforms of taxes or trade tariffs; they get killed by a series of amendments. Each step gets a majority approval, but the end result has enough fatal flaws so that it loses the support of a majority.

Just because an individual recognizes the problem does not mean an individual can stop the process. It is a slippery slope, too dangerous to get onto. The group as a whole must look ahead and reason back in a coordinated way, and set up the rules so as to prevent taking the first steps on the slope. There is safety when individuals agree to consider reforms only as a package rather than as a series of small steps. With a package deal, everyone knows where he will end up. A series of small steps can look attractive at first, but one unfavorable move can more than wipe out the entire series of gains.

In 1989, Congress learned this danger first-hand in its failed attempt to vote itself a 50 percent pay raise. Initially, the pay raise seemed to have wide support in both houses. When the public realized what was about to happen, they protested loudly to their representatives. Consequently, each member of Congress had a private incentive to vote against the pay hike, provided he or she thought that the hike would still pass. The best scenario would be to get the higher salary while having protested against it. Unfortunately (for them), too many members of Congress took this approach, and suddenly passage no longer seemed certain. As each defection moved them further down the slippery slope, there was all the more reason to vote against it. If the pay hike were to fail, the worst possible position would be to go on record supporting the salary hike, pay the political price, and yet not get the raise. At first, there was the potential for a few individuals to selfishly improve their own position. But each defection increased the incentive to follow suit, and soon enough the proposal was dead.

There is a second, quite different moral to the Justin-Case story. If you are going to fail, you might as well fail at a difficult task. Failure causes others to downgrade their expectations of you in the future. The seriousness of this problem depends on what you attempt. Failure to climb Mt. Everest is considerably less damning than failure to finish a 10K race. The point is that when other people's perception of your ability matters, it might be better for you to do things that *increase* your chance of failing in order to reduce its consequence. People who apply to Harvard instead of the local college or ask the most popular student to prom instead of a more realistic prospect are following such strategies.

Psychologists see this behavior in other contexts. Some individuals are afraid to recognize the limits of their own ability. In these cases they take actions that increase the chance of failure in order to avoid facing their ability. For example, a marginal student may not study for a test so that if he fails, the failure can be blamed on his lack of studying rather than intrinsic ability. Although perverse and counterproductive, there is no invisible hand to protect you in games against yourself.

POLITICIANS AND APPLE CIDER

Two political parties are trying to choose their positions on the liberal-conservative ideological spectrum. First the incumbent takes a stand, then the challenger responds.

Suppose the voters range uniformly over the spectrum. For concreteness, number the political positions from 0 to 100, where 0 represents radical left and 100 represents arch-conservative. If the incumbent chooses a position such as 48, slightly more liberal than the middle of the road, the challenger will take a position between that and the middle—say, 49. Then voters with preferences of 48 and under will vote for the incumbent; all others, making up just over 51 percent of the population, will vote for the challenger. The challenger will win.

If the incumbent takes a position above 50, then the challenger will locate between that and 50. Again this will get him more than half the votes.

By the principle of looking ahead and reasoning backward, the incumbent can figure out that his best bet is to locate right in the middle. (As with highways, the position in the middle of the road is called the median.) When voters' preferences are not necessarily uniform, the incumbent locates at the position where 50 percent of the voters are located to the left and 50 percent are to the right. This median is not necessarily the average position. The median position is determined by where there are an equal number of voices on each side, while the average gives weight to how far the voices are away. At this location, the forces pulling for more conservative or more liberal positions have equal numbers. The best the challenger can do is imitate the incumbent. The two parties take identical stands, so each gets 50 percent of the votes if issues are the only thing that counts. The losers in this process are the voters, who get an echo rather than a choice.

In practice, parties do not take identical hard positions, but each fudges its stand around the middle ground. This phenomenon was first recognized by Columbia University economist Harold Hotelling in 1929. He pointed out similar examples in economic and social affairs: "Our cities become uneconomically large and the business districts within them are too concentrated. Methodist and Presbyterian churches are too much alike; cider is too homogeneous."[14]

Would the excess homogeneity persist if there were three parties? Suppose they take turns to choose and revise their positions, and have no ideological baggage to tie them down. A party located on the outside will edge closer to its neighbor to chip away some of its support. This will squeeze the party in the middle to such an extent that when its turn comes, it will want to jump to the outside and acquire a whole new and larger base of voters. This process will then continue, and there will be *no* equilibrium. In practice, parties have enough ideological baggage, and voters have enough party loyalty, to prevent such rapid switches.

In other cases, locations won't be fixed. Consider three people all looking for a taxi in Manhattan. Though they start waiting at the same time, the one at the most uptown position will catch the first taxi going downtown, and the one located farthest downtown will catch the first uptown cab. The one in the middle is squeezed out. If the middle person isn't willing to be usurped, he will move either uptown or downtown to preempt one of the other two. Until the taxi arrives, there may not be an equilibrium; no individual is content to remain squeezed in the middle. Here we have yet another, and quite different, failure of an uncoordinated decision process; it may not have a determinate outcome at all. In such a situation, society has to find a different and coordinated way of reaching a stable outcome.

A RECAPITULATION

In this chapter we described many instances in which the games people play have more losers than winners. Uncoordinated choices interact to produce a poor outcome for society. Let us summarize the problems briefly, and you can then try out the ideas on the case study.

First we looked at games in which each person had an either-or choice. One problem was the familiar multiperson prisoners' dilemma: everyone made the same choice, and it was the wrong one. Next we saw examples in which some people made one choice while their colleagues made another, but the proportions were not optimal from the standpoint of the group as a whole. This happened because one of the choices involved spillovers—that is, effects on others—which the choosers failed to take into account. Then we had situations in which either extreme—everyone choosing one thing or everyone choosing the other—was an equilibrium. To choose one, or make sure the right one was chosen, required social conventions, penalties, or restraints on people's behavior. Even then, powerful historical forces might keep the group locked into the wrong equilibrium.

Turning to situations with several alternatives, we saw how the group could voluntarily slide down a slippery path to an outcome it would collectively regret. In other examples, we found a tendency toward excessive homogeneity. Sometimes there might be an equilibrium held together by people's mutually reinforcing expectations about what others think. In still other cases, equilibrium might fail to exist altogether, and another way to reach a stable outcome would have to be found.

The point of these stories is that the free market doesn't always get it right. There are two fundamental problems. One is that history matters. Our greater experience with gasoline engines, QWERTY keyboards, and light-water nuclear reactors may lock us in to continued use of these inferior technologies. Accidents of history cannot necessarily be corrected by today's market. When one looks forward to recognize that lock-in will be a potential problem, this provides a reason for government policy to encourage more diversity before the standard is set. Or if we seem stuck with an inferior standard, public policy can guide a coordinated change from one standard to another. Moving from measurements in inches and feet to the metric system is one example; coordinating the use of daylight saving time is another.

Inferior standards may be behavioral rather than technological. Examples include an equilibrium in which everyone cheats on his taxes, or drives above the speed limit, or even just arrives at parties an hour after the stated time. The move from one equilibrium to a better one can be most effectively accomplished via a short and intense campaign. The trick is to get a critical mass of people to switch, and then the bandwagon effect makes the new equilibrium self-sustaining. In contrast, a little bit of pressure over a long period of time would not have the same effect.

The other general problem with laissez-faire is that so much of what matters in life takes place outside the economic marketplace. Goods ranging from common courtesy to clean air are frequently unpriced, so there is no invisible hand to guide selfish behavior. Sometimes creating a price can solve the problem, as with congestion on the Bay Bridge. Other times, pricing the good changes its nature. For example, donated blood is typically superior to blood that is purchased, because the types of individuals who sell their blood for money are likely to be in a much poorer state of health. The coordination failures illustrated in this chapter are meant to show the role for public policy. But before you get carried away, check the case below.

CASE STUDY: A PRESCRIPTION FOR ALLOCATING DENTISTS

In this case study, we explore the coordination problem of how the invisible hand allocates (or misallocates) the supply of dentists between cities and rural areas. In many ways the problem will seem closely related to our analysis of whether to drive or take the train from Berkeley to San Francisco. Will the invisible hand guide the right numbers to each place?

It is often argued that there is not so much a shortage of dentists as a problem of misallocation. Just as too many drivers, left to their own resources, would take the Bay Bridge, is it the case that too many dentists choose the city over the countryside? And if so, does that mean society should place a toll on those who want to practice city dentistry?

For the purposes of this case study, we greatly simplify the dentists' decision problem. Living in the city or in the countryside are considered equally attractive. The choice is based solely on financial considerations—they go where they will earn the most money. Like the commuters between Berkeley and San Francisco, the decision is made selfishly; dentists maximize their individual payoffs.

Since there are many rural areas without enough dentists, this suggests that there is room for an increased number of dentists to practice in rural areas without causing any congestion. Thus rural dentistry is like the train route. At its best, being a rural dentist is not quite as lucrative as having a large city practice, but it is a more certain route to an above-average income. Both the incomes and the value to society of rural dentists stay roughly constant as their numbers grow.

Being a city practitioner is akin to driving over the Bay Bridge— it is wonderful when you are alone and not so great when the city gets too crowded. The first dentist in an area can be extremely valuable and maintain a large practice. But with too many dentists around, there is the potential for congestion and price competition. As the number increases, city dentists will be competing for the same patient pool, and their talents will be underutilized. If the population of city dentists grows too much, they may end up earning less than their rural counterparts. In short, as the number of city practices increases, the value of the marginal service that they perform falls, as does their income.

We depict this story in a simple chart, again quite similar to the driving versus train example. Suppose there are 100,000 new dentists choosing between city and rural practices. Thus, if the number of new city dentists is 25,000, then there will be 75,000 new rural dentists.

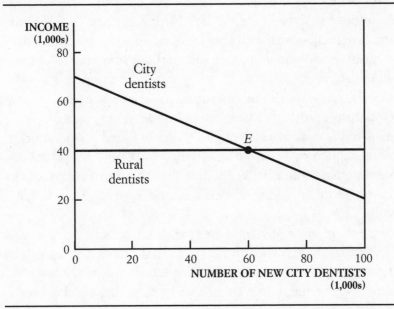

The falling line (city dentists) and the flat line (rural dentists) represent the financial advantages of taking the respective paths. At the far left, where everyone chooses rural practices, city dentists' incomes are above the incomes of those with rural practices. This is reversed at the far right, where everyone chooses city dentistry.

The equilibrium for career choices is at E, where the two options provide the same financial rewards. To verify this, suppose that the distribution of career choice results in only 25,000 new city dentists. Since city dentists' incomes are higher there than rural dentists' incomes, we expect that more new dentists will choose city over rural practices. This will move the distribution of city vs. rural to the right. The reverse adjustment would take place if we started to the right of *E*, where city dentists are the lower paid of the two. Only when *E* is reached will next year's career choices broadly replicate those of this year, and the system will settle down to an equilibrium.

Is this outcome the best for society?

Case Discussion

As in the case of the commuters, the equilibrium does not maximize the combined income of dentists. *But society cares about the consumers of dentistry as well as the practitioners.* In fact, left alone, the market solution at *E* is the best for society as a whole. The reason is that there are two side effects created when one more person decides to be a city dentist. The additional city dentist lowers all other dentists' incomes, imposing a cost on the existing city dentists. But this reduction in price is a benefit to consumers. The two side effects exactly cancel each other out. The difference between this story and our commuting example is that no one benefited from the extra commuting time when the Bay Bridge became congested. When the side effect is a change in price (or income), then the purchasers benefit at the producers' cost. There is zero net effect.

From society's viewpoint, a dentist should not worry about lowering colleagues' incomes. Each dentist should pursue the highest-paying practice. As each person makes a selfish choice, we are invisibly led to the right distribution of dentists between city and rural areas. And, the two careers will have equal incomes.*

Of course the American Dental Association may look at this differently. It may place more weight on the loss to city dentists' incomes than on the saving to consumers. From the dental profession's perspective there is indeed a misallocation, with too many

---

* Or, to the extent that living in a city is worth more than living in a rural area, this differential will be reflected in income differences.

dentists practicing in the city. If more dentists took rural practices, then the potential advantages of a city practice would not be "wasted" by competition and congestion. Taken as a whole, the income of dentists would rise if it were possible to keep the number of city dentists below the free-market level. Although dentists cannot place a toll on those who want to practice in the city, it is in the profession's self-interest to create a fund that subsidizes dental students who commit to establish a rural practice.

For some more case studies on cooperation and coordination, see "Here's Mud in Your Eye," "A Burqa for Prices," and "The King Lear Problem" in chapter 14.

CHAPTER 10

Auctions, Bidding, and Contests

IT WASN'T THAT long ago that the typical image of an auction was that of an auctioneer with a snooty British accent calling out to a hushed room of bejeweled art collectors sitting in Louis XIV chairs and tugging at their ears to bid. With eBay, auctions have become just a touch more democratic.

The most familiar auction is where an item is put up for sale and the high bidder wins. At Sotheby's, it is a painting or an antique. On eBay, it is a Pez dispenser, a used drum set, or almost anything (except a kidney). On Google and Yahoo!, auctions for ad positions next to keyword searches bring in well over $10 billion. In Australia, even houses are sold via auctions. The common denominator is that we have one seller and many buyers. The buyers compete against each other to gain the object and the high bidder wins.

The view of an auction as a way to sell something is too narrow. Auctions are also used to buy items. A good illustration is where a local government wants to build a road and takes bids to determine who will build the road. Here the winning bidder is the one who makes the *lowest* bid, as the government wants to buy the paving service as cheaply as possible. This is a called a procurement auction. There's one buyer and many sellers to get the buyer's business.*

Bidding in an auction requires a strategy—though many people think all they need is a paddle. That leads to problems when people bid based on emotion or excitement. They live to regret it. To do well in an auction setting requires a strategy. Should you bid early or wait until the auction is almost over and then jump in? If you value an item at $100, how high should you bid? How do you avoid winning the auction but then regretting that you've overpaid? As we have discussed before, this phenomenon is known as the winner's curse; here we'll explain how to avoid it.

Should you even bid in the auction? The house auction market in Australia illustrates the buyer's dilemma. Imagine that you are interested in a house that is due to be auctioned on July 1. But there's a house you like even more that will be auctioned off a week later. Do you wait to bid in the second auction and risk ending up with neither?

Our plan is to start with a description of some basic auction types and then discuss how game theory can help you bid—and know when not to.

---

* Procurement auctions are more complicated because the bidders aren't using identical currencies. In a normal auction, when Avinash bids $20 and Barry bids $25, the seller knows that $25 is a better bid. But, in a procurement auction, it isn't clear that Avinash's offer to build a road for $20 is better than Barry's offer of $25—the quality of work may be different. This explains why reverse auctions wouldn't work well on eBay. Imagine you wanted to buy a Pearl Export drum set. This is a fairly common item on eBay and there are usually a dozen or more sets for sale at any time. To run a procurement auction, you would need all of the sellers to bid against each other. At the auction close, you'd then buy the drum set at the lowest bid (assuming it was below your reservation number). The problem is that you might care about the color or age of the drums or the reputation of the seller for reliability and prompt delivery. The lowest bid isn't necessarily the best one. But if you aren't always going to pick the lowest bid, then the sellers don't know how low they need to bid in order to win your business. A solution, which often works better in theory than in practice, is to impose performance standards. The problem is that bidders who would perform above the minimum standard are often not rewarded for doing so in their bid. Because procurement auctions are more complicated in this way, we focus our attention on regular auctions.

ENGLISH AND JAPANESE AUCTIONS

The most famous type of auction is known as the English or ascending auction. In this format, the auctioneer stands at the front of the room calling out ever-increasing bids:

Do I hear 30? 30 from the lady in the pink hat.

40? Yes, 40 from the gentleman on my left.

Will someone bid 50? 50, anyone?

40 going once, going twice, sold.

Here the optimal bidding strategy, although it hardly merits the term strategy, is simple. You bid until the price exceeds your value and then you drop out.

There is often a bit of trickiness over the issue of bidding increments. Imagine that the bidding goes up in units of 10 and your value is 95. Then you would stop bidding at 90. Of course, knowing that, you might want to think about whether you should be the high bidder at 70 or 80, recognizing that 90 will be your last bid. In the discussion that follows, we will assume that the bidding increments are very small, say a penny, so that these endgame issues are not important.

The only hard part is determining what is meant by your "value." What we mean by your value is your walkaway number. It is the highest price at which you still want to win the item. At a dollar more you would rather pass, and at a dollar less you are willing to pay the price, but just barely. Your value might include a premium you put on not having the item fall into a rival's hands. It could include the excitement of winning the bidding. It could include the expected resale value in the future. When all of the components are put together, it is the number such that if you had to pay that price, you no longer care if you've won or lost the auction.

Values come in two flavors, private and common. In a world of private values, your value for the item doesn't depend at all on what others think it is worth. Thus your value of a personalized signed copy of *The Art of Strategy* doesn't depend on what your neighbor might think it is worth. In a common value situation, the bidders understand that the item has the same value for all of them, although each might have a different view as to what that common value is. A standard example is bidding for an offshore oil lease. There is some amount of oil underground. While that amount might be unknown, it is the same whether Exxon or Shell wins the bidding.

In truth, the value of an item usually has elements of both private and common components. Thus one oil company might be better at extracting the oil than the other, which then adds a private value element to something that is mostly common.

In situations with a common value, your best guess as to the value of an item might depend on who else or how many others are bidding and when they drop out. An English auction keeps that information hidden, as you never know who else is willing to bid but hasn't yet moved. Nor are you sure when someone has dropped out. You know their last bid, but not how high they would have gone.

There is a variation of the English auction that is more transparent. In something called a Japanese auction, all of the bidders start with their hands raised or buttons pressed. The bidding goes up via a clock. The clock might start at 30 and then proceed to 31, 32, . . . and upwards. So long as your hand is raised, you are in the bidding. You drop out by lowering your hand. The catch is that once you lower your hand, you can't put it back up again. The auction ends when only one bidder remains.

An advantage of the Japanese auction is that it is always clear how many bidders are active. In an English auction, someone can remain silent even though they were willing to bid all along. The person can then make a surprise entry late in the contest. In the Japanese auction, you know exactly how many competitors there are and even the prices at which each drops out. The Japanese auction is thus just like an English auction where everyone has to reveal their hand.

The outcome of a Japanese auction is easy to predict. Since bidders drop out when the price hits their value, the last person remaining will be the one with the highest valuation. The price the winner will pay will equal the second highest valuation. The reason is that the auction ends at the moment the second-to-last bidder drops out. The last price is the valuation of the second highest bidder.

Thus the item is sold to the person with the highest valuation, and the seller receives a payment equal to the second highest valuation.

VICKREY AUCTION

In 1961, Columbia University economist and future Nobel laureate William Vickrey developed a different type of auction. He called it a second-price auction, though we now call it a Vickrey auction in his honor.*

In a Vickrey auction, all the bids are placed in a sealed envelope. When the envelopes are opened to determine the winner, the highest bid wins. But there's a twist. The winner doesn't pay his or her bid. Instead, the winner only has to pay the second highest bid.

What is remarkable, even magical, about this auction is that all the bidders have a dominant strategy: bid their true valuation. In a regular sealed-bid auction where the high bidder wins and pays his actual bid, bidding strategy is a complicated problem. What you should bid depends on how many other bidders are in the game, what you think their value is for the item, even what you think they think your value is. The result is a complicated game where everyone has to consider what everyone else is doing.

In a Vickrey auction, all you have to do is figure out what the item is worth to you and then write down that amount. You don't

---

* His seminal paper was "Counterspeculation, Auctions, and Competitive Sealed Tenders," *Journal of Finance* 16 (1961): 8–37. While Vickrey was the first to study the second-price auction, its use goes back at least to the nineteenth century, when it was used by stamp collectors. There is even evidence that Goethe employed a second-price auction in 1797 when selling his manuscript to a publisher. See Benny Moldovanu and Manfred Tietzel, "Goethe's Second-Price Auction," *Journal of Political Economy* 106 (1998): 854–59.

need to hire a game theorist to help you bid, alas. Actually, we like that result. Our goal is to be strategic when designing a game so that the players don't have to be strategic when they play it.

The reason why your bidding strategy is so simple is that it is a dominant strategy. A dominant strategy is your best play no matter what others are doing. Thus you don't need to know how many others there are or what they are thinking or doing. Your best strategy doesn't depend on what anyone else bids.

This brings us to the question of how we know that bidding your valuation is a dominant strategy. The following example is the basis of the general argument.

You are in a Vickrey auction and your true value of the item is $60. But instead of bidding $60, you bid $50. To show that this is a bad idea, we put on our consequentialist hat. When does bidding $50 rather than $60 lead to a different outcome? Actually, it is easier to turn this question around. When does bidding $50 or $60 lead to the same result?

TRIP TO THE GYM NO. 6

Imagine that you could find out how much the other bidders were submitting in a Vickrey auction before you had to put in your bid. Ignoring the ethical issues for a moment, how much would this be worth to you?

If someone else bids $63 or $70 or anything else above $60, then both $50 and $60 are losing bids. Hence there is no difference between them. In both cases, you lose the auction and walk away with nothing.

The $50 and $60 bids also lead to identical (but this time happier) outcomes if the highest other bid is below $50, say $43. If you bid $60, then you win and pay $43. If you had bid $50, you would also have won and paid $43. The reason is that in both cases you are the highest bidder and what you pay is the second highest bid, which is $43. Bidding $50 doesn't save you any money (compared to bidding $60) when the second highest bid is $43 or anything below $50.

We've looked at the cases in which the two bids lead to exactly the same result. Based on these cases, there is no reason to prefer one bid over the other. What is left is where the paths diverge. This is how we can judge which bid leads to a better result.

There's no difference if any rival bid is above $60 or all are below $50. The only remaining case is where the highest competitive bid is between $50 and $60, say $53. If you bid $60, then you will win and pay $53. If you were to have bid $50, then you would lose. Since your value is $60, you would rather win and pay $53 than lose.

Thus the only time a bid of $50 leads to a different outcome than one of $60 is when you lose the auction but you wished you had won. This demonstrates that you never want to bid $50 or anything below your true value. A similar argument shows that you never want to bid more than your true value.

REVENUE EQUIVALENCE

At this point, you may have figured out that the Vickrey auction gets you to the same outcome as the English (or Japanese) auction, all in one step. In both cases, the person with the highest value ends up winning the auction. In both cases, what the winning bidder has to pay is the second highest valuation.

In the English (or Japanese) auction, everyone bids up to his or her value, so the auction ends when the bidding gets up to the second highest valuation. The remaining bidder is the person with the highest value. And subject to the vagaries of the bidding interval, what the winning bidder pays is the bid at which the penultimate bidder drops out, namely the second highest valuation.

In the Vickrey auction, everyone bids his or her true value. Thus the person with the highest valuation is the winning bidder. According to the rules, that person only has to pay the second highest bid, which is just the second highest valuation.

Thus it appears that the two auctions get to exactly the same place. The same person wins and the winner pays the same price. Of course, there is always the issue of the bidding interval in that the person with a valuation of 95 might drop out at 90 if the bidding increment comes in 10s. But with small enough increments, the person will drop out just at the valuation.

There is one subtle difference between the two auctions. In the English auction, a bidder learns something about what others think the item is worth by seeing some of their bids. (There are many potential bids that are not seen.) In the Japanese variant, the bidders learn even more. Everyone sees where everyone drops out. In contrast, in the Vickrey auction, the winning bidder doesn't get a chance to learn anything about the other bids until the auction is all over. Of course, in a private value auction, a bidder doesn't care about what others think the item is worth. Thus the extra information is irrelevant. This allows us to conclude that in a private value setting the seller would make the same amount of money by employing a Vickrey auction or an English (or Japanese) auction.

It turns out that this is part of a much more general result. In many cases a change in the rules doesn't lead to any more or less revenue to the seller.

Buyer's Premium

If you win an auction at Sotheby's or Christie's, you might be surprised to learn that you owe more than you bid. This isn't just sales tax we are talking about. The auction houses tack on a 20 percent buyer's premium. If you win the auction with a $1,000 bid, you will be expected to write them a check for $1,200.

Who pays the buyer's premium? The obvious answer is the buyer. But if the answer were really that obvious, we surely wouldn't have asked the question—or would we, just to keep you honest?

Okay, it isn't the buyer who pays—it's the seller. To get this result all we need to assume is that the buyer is aware of this rule and takes it into account when bidding. Put yourself in the position of a collector who is willing to pay $600. How high will you bid? Your top bid should be $500—as you can anticipate that saying $500 really means that you have to pay $600 after the buyer's premium.

You can think of the buyer's premium as being nothing more than a currency conversion or a code. When you say $100, you really mean $120.* Each bidder scales back his bid accordingly.

---

* One way to think of this game is that it is as if the auction is conducted in New York, but the bidding is in euros. Thus when a bidder says "€500," he or

If your winning bid is $100, you have to write a check for $120. You don't care that the $120 is divided up $100 to the seller and $20 to the auction house. You only care that the painting costs $120. From your perspective, you can just as well imagine that the buyer receives the full $120 and then turns over $20 to the auction house.

Our point is that the winning bidder still pays the same amount. The only difference is that now the auction house gets some percentage of the total. Thus the cost is borne entirely by the seller, not the buyer.

The larger takeaway here is that you may change the rules of the game, but the players will adapt their strategies to take those new rules into account. In many cases, they will precisely offset what you've done.

ONLINE AUCTIONS

While the Vickrey auction may go all the way back to Goethe, it was still relatively uncommon until recently. It has now become the standard for online auctions. Take the case of eBay. You don't bid directly in an eBay auction. Instead, you do something called a proxy bid. You authorize eBay to bid up to your proxy. Thus if you give them a proxy bid of $100 and the current high bid is $12, eBay will first bid $13 for you. If that bid is high enough to win, that's where they stop. But if someone else has put a proxy bid of $26, then eBay will bid up to $26 for that person and your proxy will go all the way up to $27.

It would seem that this is just like a Vickrey auction. Think of the proxy bids as being like the bids in a Vickrey auction. The person with the highest proxy bid ends up being the winner and the amount the person pays is equal to the second highest proxy bid.

---

she expects to pay $600. It should be clear that changing the currency used for bidding shouldn't bring in more money to the auction. If Sotheby's said that Monday's auction would be conducted in euros, everyone could do the calculation to translate their bid into dollars (or yen, for that matter). They understand the true cost of bidding "100" whatever the units.

To make this concrete, imagine there are three proxy bids:

> A: $26
> B: $33
> C: $100

A's proxy will drop out once the bidding gets to $26. B's proxy will force the bidding up to that level. And C's proxy will push the bidding all the way up to $34. Thus C will win the auction and pay the second highest proxy bid.

If everyone had to submit their proxy bids at the same time and once and for all, the game truly would be the same as a Vickrey auction and we could advise everyone to play it straight and bid their true value. Bidding the truth would be a dominant strategy.

But the game isn't quite played that way, and these little hiccups lead people to get fancy with their bids. One complication is that eBay will often have several similar items for sale all at the same time. Thus if you want to buy a used Pearl Export drum set, you have a choice of ten or so to choose from at any one time. You might like to bid on whichever one is cheapest up to $400. While you are willing to pay up to $400 for any one of the sets, you wouldn't bid $300 on one version while another might be bought at $250. You might also prefer to bid in an auction that is closing sooner over one that ends in a week so you don't have to wait to know whether or not you've won.

What this comes down to is that your value of the item being sold depends on what else is up for sale, both now and in the future. Thus you can't place a valuation independent of the auction.

Sniping

Let's take a case where multiple items and issues of timing aren't a concern. Consider an auction for a one-of-a-kind item. Now is there any reason not to play it straight and enter your true value in your proxy bid?

As an empirical matter, people don't play it straight. They often wait until the last minute or even second before entering their best proxy bid. The name for this is *sniping*. Indeed, there are Internet services such as Bidnapper and others that will do the sniping for you so you don't have to wait around for the auction to end before submitting your bid.

Why snipe? We've shown that bidding your true value is a dominant strategy in a Vickrey auction. Sniping must arise because of the subtle differences between proxy bidding and a Vickrey auction. The key difference is that the other bidders might get to learn something from your proxy bid before the auction is over. If what they learn influences how they bid, then you can have an incentive to keep your bid, even your proxy bid, hidden.

An early proxy bid might give away valuable information. For example, if a furniture dealer bids on a specific Bauhaus chair, you might (quite reasonably) infer that the piece is authentic and of historical interest. If the dealer is willing to buy the chair at a price of $1,000, then you would be happy to pay $1,200, a better price than you could hope to get from that same dealer. Thus the dealer doesn't want others to know how high he or she is willing to go. That leads the dealer to wait until the very end before putting in a bid. At that point, it is too late for you or others to react. By the time you've discovered the dealer is bidding, the auction is over. Of course, that implies that the bidder's true identity is known to others and that the bidder can't come up with an alias.* Since sniping is so common, that suggests there are other explanations.

We think the best explanation for sniping is that many bidders don't know their own value. Take the case of a vintage Porsche 911. The bidding starts at $1. Of course, we don't value the car at $1. We value the car at $100, even at $1,000. Provided the bidding is below $1,000 we can be confident that this is a great deal. We don't have to look up the Blue Book value or even speak to our

---

* While an alias is easy to create, if the bidder doesn't have a track record, the seller might be reluctant to accept the bid.

spouse about the need for an extra car. The point here is that we are lazy. Figuring out our true value for an item takes work. If we can win the auction without ever having to go through that effort, we would prefer to take the shortcut.

This is where sniping comes into play. Imagine that our expert buyer values the vintage Porsche at $19,000. The buyer would prefer to keep the bidding low for as long as possible. If the buyer enters a $19,000 proxy at the start, then our mindless proxy bid of $1,000 will push the price right up to $1,000. At that point, we will learn that we need to get more information. In the process, our spouse might go along for the ride and let us bid up to $9,000. That could push the final price up to $9,000 or higher if other bidders get a chance to do their homework.

But if the $19,000 proxy bidder keeps his or her powder dry, then the bidding may not escalate above $1,000 until the last moments of the auction, at which point it is too late for us to reenter a higher bid, assuming we were even paying attention and could get the quick okay from the spouse to bid more.

The reason to snipe is to keep others in the dark about their own valuations. You don't want people to learn that their lazy bid doesn't have a chance to win. If they find out early enough, they will have a chance to do their homework, which can only lead to you having to pay more, if you still win.

BIDDING AS IF YOU'VE WON

A powerful idea in game theory is the concept of acting like a consequentialist. By that we mean to look ahead and see where your actions have consequences. You should then assume that situation is the relevant one at the time of your initial play. It turns out that this perspective is critical in auctions and in life. It is the key tool to avoid the winner's curse.

To make this concrete, imagine that you ask someone to marry you. The person can say yes or no. If the answer is no, then there's nothing for you to go through with. But if the answer is yes, then you are on your way to getting hitched. Our point is that you should presume the answer will be yes at the time you pop the question. We are well aware that this is taking an optimistic perspective. Your intended could say no and you will be very disappointed. The reason to assume that your intended will say yes is to be prepared for that outcome. In that case, you should be saying yes as well. If upon hearing that your intended says yes, you then want to reconsider, you shouldn't have asked in the first place.

In a marriage proposal, assuming the answer will be yes is a pretty natural way of going about things. In the case of negotiations and auctions, this is an approach that has to be learned. Try your hand at the following game.

ACME

You are a potential buyer for ACME. Because of your extensive knowledge of game theory, you will be able to increase ACME's value by 50 percent, whatever it is. The problem is that you have some doubts as to the current value. After completing your due diligence, you place the value at somewhere between $2 million and $12 million. The average value is $7 million and your view is that all the options in the $2 to $12 million range are equally likely. The way the bidding is set up, you get to make a single take-it-or-leave-it bid to the owners. They will accept any bid that exceeds the current value and reject otherwise.

Say that you bid $10 million. If it turns out that the company is presently worth $8 million, then you can make it worth $12 million. You will pay $10 million for a company worth $12 million, and so your profit will be $2 million. If the company is only worth $4 million, then you will make it worth $6 million, but have paid $10 million and thus end up $4 million behind.

What is the most you can offer the current owners and still expect to break even? By break even we mean that you might not make money in every situation, but on average you'll neither make nor lose any money. Note that we don't recommend bidding this amount. You should always bid something less that this amount. This is just a way of figuring out a cap on your bids.

When faced with this problem, most people reason as follows:

On average the company is worth $7 million. I can make it worth 50 percent more, or $10.5 million. Thus I can bid up to $10.5 million and still not expect to lose money.

Is $10.5 million where you came out? We hope not.

Think back to the marriage proposal. You've proposed an acquisition here. What if they say yes? Do you still want to go ahead? If you offer $10.5 million and the owners say yes, then you've learned some bad news. You now know that the company is not worth $11 million or $12 million today. When the owners say yes to an offer of $10.5 million, the company will be worth somewhere between $2 million and $10.5 million, or $6.25 million on average. The problem is that even with your 50 percent increase in performance, that only brings the value up to $9.375 million, well below the $10.5 million you offered.

This is a serious problem. It appears that if they say yes, then you no longer want to buy the company. The solution to this problem is to presume that your offer will be accepted. In that case, if you were to offer $8 million, you can predict that when it is accepted the company is worth between $2 million and $8 million, for an average value of $5 million. A 50 percent premium on $5 million only gets you up to $7.5 million, not enough to justify the $8 million offer.

An offer of $6 million just does the trick. You can anticipate that when the seller says yes, the company is worth between $2 million and $6 million, for an average value of $4 million. The 50 percent premium brings the value to you back up to $6 million or breakeven. The fact that the seller says yes is bad news but not fatal to the deal. You have to adjust down your offer to take into account the circumstances under which a seller will say yes to you.

Let's put this all together. If you offer $6 million and you presume that your offer will be accepted, then you anticipate that the company will only be worth $4 million and you won't be disappointed when your offer is accepted.* Quite often your offer will be rejected, in which case you will have underestimated the value of the company, but in those cases you don't end up with the company, so the mistake doesn't matter.

This idea of presuming you've won is a critical ingredient to making the right bid in a sealed-bid auction.

SEALED-BID AUCTIONS

The rules of a sealed-bid auction are simple. Everyone puts his or her bid in a sealed envelope. The envelopes are opened and the high bidder wins and pays his or her bid.

The tricky part of a sealed-bid auction is determining how much to bid. For starters, you should never bid your valuation (or worse, something more). If you do so, you are guaranteed to break even at best. This strategy is dominated by shading your bid to some amount below your valuation. That way, at least you have a chance to come out ahead.† How much you should shade your bid depends on how many others are competing in the auction and what you expect others to bid. But what they bid depends on what they expect you to bid. The key step to cutting through this infinite loop of expectations is to always bid as if you've won. When putting down your bid, you should always assume that all of the other bidders are below you. And then with that assumption, you should

---

* If you are wondering how we came up with the $6 million, here's the calculation to employ. If an offer of $X is accepted, then the seller's value is between 2 and X for an average value of $(2 + X)/2$. You make the company worth an extra 50 percent or 1.5 times the original value. Breakeven implies that your bid of X equals $(3/2) \times (2 + X)/2$ or $4X = 3(2 + X)$ or $X = 6$. It is easier to check that 6 is the right answer than to come up with the number.

† In a procurement auction, this advice gets turned around. Imagine yourself as a bidder for a contract, say construction of a stretch of highway. Your cost (which includes the normal return you require on investment) is $10 million. What bid should you submit? You should never submit a bid that is lower than your cost. For example, suppose you bid $9 million. This makes no difference if you don't win, but if you do win, you will be paid a price that is less than your cost. You will be paving the way to your own bankruptcy.

ask if this is your best bid. Of course, you will often be wrong when making that assumption. But when you're wrong, it won't matter—others will have outbid you and so you won't have won the auction. But when you're right, you'll be the winning bidder and thus have made the correct assumption.

Here's a way of demonstrating that you should always bid as if you've won. Imagine that you have a confederate inside the auction house. The confederate has the ability to adjust your bid downward in the event that you have the highest bid. Unfortunately, he doesn't know the other bids and can't tell you precisely how much to lower your bid. And if you don't have the highest bid, there's nothing he can do to help you.

Would you want to employ his service? You might not because it is unethical. You might not because you are afraid of turning a winning bid into a losing bid. But play along and imagine that you are willing to use his services. Your original bid was $100, and after learning that this was the winning bid, you instruct him to lower the bid to $80.

If this was a good idea, you might as well have bid $80 right from the start. Why? Let's compare the two cases.

Case A	Case B
Bid $100	Bid $80
Lower bid to $80 if $100 was highest	

If $100 would have lost, then there's no difference between bidding $100 or $80. Both would be losing bids. If $100 would have won, then your confederate would lower the bid to $80, in which case you will end up in the same place as if you had bid $80 all along. In short, there is no advantage to bidding $100 and then reducing it to $80 (when you've won) compared to bidding $80 from the start. Since you can get the same result without having the confederate and acting unethically, you might as well bid $80 from the start. What this all says is that when you are thinking about how much to bid, you should pretend that all of the other bidders are somewhere below your bid. Armed with this assumption, you then consider your best bid.

We'll return to figuring out just how much to bid after making a short detour to the Netherlands.

DUTCH AUCTIONS

Stocks are traded on the New York Stock Exchange. Electronics are sold in Akihabara, Tokyo. Holland is where the world goes to buy flowers. At the Aalsmeer Flower Auction, the auction "house" takes up some 160 acres. On a typical day, some 14 million flowers and a million potted plants change hands.

What makes Aalsmeer and other Dutch auctions a bit different from Sotheby's is that the bidding goes in reverse. Instead of starting with a low price and having the auctioneer call out successively higher prices, the auction starts with a high price that declines. Imagine a clock that starts at a hundred and then winds down to 99, 98, . . . The first person to stop the clock wins the auction and pays the price at which the clock was stopped.

This auction is the reverse of the Japanese auction. In the Japanese auction, all of the bidders indicate their participation. The prices keep rising until only one bidder is left. In the Dutch auction, prices start high and fall until the first bidder indicates his or her participation. If you raise your hand in a Dutch auction, the auction stops and you've won.

You don't have to go to the Netherlands to participate in a Dutch auction. You could send an agent to bid for you. Think for a moment about the instructions you might give your agent. You might say to wait until the price of petunias falls to €86.3 and then bid. As you contemplate those instructions, you should anticipate that if the bidding ever gets down to €86.3, then you will be the winning bidder. If you were at the auction house, you'd know that all of the other bidders have yet to act. Armed with this knowledge, you don't want to change your bid. If you wait a moment longer, one of the other bidders might jump in and take your place.

Of course that is true all along. Anytime you wait, another bidder might jump in. The issue is that the longer you wait, the bigger the profit you risk losing. And the longer you wait, the greater the risk that one of the other bidders is about to jump in. At your optimal bid, the savings from paying a lower bid is no longer worth the increased risk of losing the prize.

In many ways this is similar to what you might do in a sealed-bid auction. The instruction you give your bidding agent is akin to what you would write down as your sealed bid. Everyone else does the same. The person who writes down the highest number is the same as the person who first raises his or her hand.

The only difference between a Dutch auction and a sealed-bid auction is when you bid in a Dutch auction, you know you've won. When you write down your bid in a sealed-bid auction, you only find out later if you've won or not. But remember our guidance. In a sealed-bid auction, you are supposed to bid *as if* you've won. You're supposed to pretend that all of the other bidders are somewhere below you. This is exactly the situation you are in when competing in a Dutch auction.

TRIP TO THE GYM NO. 7

How much should you bid in a sealed-bid auction? For simplicity, you can assume that there are only two bidders. You believe that the other bidder has a value that is equally likely to be anything between 0 and 100, and the other bidder has the same belief about you.

Thus the way you bid in the two auctions is identical. Just as an English auction and a Vickrey auction end up in the same place, so do a sealed-bid auction and a Dutch auction. Since participants bid the same amount, the sellers get the same amount. Of course, that doesn't yet tell us how much to bid. It just says that we have two mysteries with the same answer.

The answer for how much to bid comes from one of the most remarkable results in auction theory: the revenue equivalence theorem. It turns out that when the valuations are private and the game is symmetric, the seller makes the same amount of money on average whether the auction type is English, Vickrey, Dutch, or sealed-bid.* What that means is that there is a symmetric equilib-

---

* This result was first established by Roger Myerson. At a fundamental level, it is due to the fact that each bidder is focused on the ends, not the means. A bidder

rium to the Dutch and sealed-bid auctions where the optimal bidding strategy is to bid what you think the next highest person's value is given the belief that you have the highest value.

In a symmetric auction, everyone has the same beliefs about everyone else. For example, everyone might think that each bidder's value is equally likely to be anything between 0 and 100. In this case, whether the auction is Dutch or a sealed-bid, you should bid what you expect the next highest bidder's value to be given that all of the other values are below your own. If your value is 60, you should bid 30 if there is only one other bidder. You should bid 40 if there are two other bidders and 45 if there are three other bidders.*

You can see that this would lead to revenue equivalence. In a Vickrey auction, the person with the highest value wins but only pays the second highest bid, which is the second highest valuation. In a sealed-bid auction, everyone bids what they think the second highest valuation is (given they are the highest). The person with the truly highest valuation will win and the bid will be the same on average as the result in a Vickrey auction.

The larger moral here is that you can write down a set of rules for a game, but the players can undo those rules. You can say that everyone has to pay twice their bid, but that will just lead people to bid half as much. You could say that people have to pay the square of their bids, but that will just lead people to square root what they would otherwise have done. That is ultimately what is going on in a sealed-bid auction. You can tell people that they have to pay their

---

really only cares about how much he expects to pay and how likely he is to win the item. He can pay more to increase his chance of winning, and those with higher values for the object will do so. This insight was one of several contributions to auction theory that led to Myerson's Nobel Prize in 2007. See his seminal paper "Optimal Auction Design," *Mathematics of Operations Research* 6 (1981): 58–73.

* In general, you would guess that the other bidders are equally spaced between you and 0. Thus with one other bidder, that person is halfway to 0. With two other bidders, you expect them to be at 20 and 40. With three other bidders, you expect them to be at 15, 30, and 45. You bid at the expected value of the highest of your rivals. As the number of bidders increases, you can see that bids converge to valuations. As the number of bidders increases, the market approaches perfect competition and all of the surplus goes to the seller.

bid rather than the second highest bid. In response, they'll change what they write down. Instead of bidding their true value, they will shade their bid downward to the point where it equals what they expect the second highest value to be.

To see if you are a believer, try your new intuition out on the world's biggest auction, namely the market for T-bills.

T-BILLS

Each week, the U.S. Treasury holds an auction that determines the interest rate on the national debt, at least the part that comes due that week. Until the early 1990s, the way the auction worked was that the winning bidders paid their bids. After some prodding from Milton Friedman and other economists, the Treasury experimented with uniform pricing in 1992 and made the move permanent in 1998. (The secretary of the treasury at the time was Larry Summers, a distinguished economist.)

We'll explain the difference between two cases through an example. Imagine that the Treasury had $100 million in notes to sell one week. There were ten bids that came in at

Amount Bid at Interest Rate	Cumulative Amount
$10 million at 3.1%	$10 million
$20 million at 3.25%	$30 million
$20 million at 3.33%	$50 million
$15 million at 3.5%	$65 million
$25 million at 3.6%	$90 million
$20 million at 3.72%	$110 million
$25 million at 3.75%	$135 million
$30 million at 3.80%	$165 million
$25 million at 3.82%	$190 million

The Treasury wants to pay the lowest interest rate possible. That means they will start by first accepting the lowest bids. Thus all of the bidders who were willing to take 3.6% or below are winners along with half of the bidders who were willing to take 3.72%.

Under the old rule, the $10 million bid at 3.1% would win and those bidders would get only 3.1% on their Treasury note. The $20 million bid at 3.25% would be awarded notes paying 3.25% and so on all the way up to the $20 million bid at 3.72%. Note that there is more bid at 3.72% than can be fulfilled with the $100 million for sale so that only half that amount will be sold and the other half will walk away empty-handed.*

Under the new rule, all of the bids between 3.1% and 3.6% are winning bids, as are half of those bidding 3.72%. With the uniform price rule, everyone gets the highest rate of any winning bid, in this case 3.72%.

Your first reaction might be to think that the uniform pricing rule is much worse for the government (and better for investors). Instead of paying between 3.1% and 3.72%, the Treasury pays everyone 3.72%.

Based on the numbers used in our example, you'd be correct. The problem with this analysis is that people won't bid the same way in the two auctions. We used the same number only to illustrate the mechanics of the auction. This is the game theory analog of Newton's Third Law of Motion—for every action there is a reaction. If you change the rules of the game, you must expect that players will bid differently.

Let's take a simple example to drive this point home. Imagine that the Treasury had said that instead of getting the interest rate that you bid, you would get 1% less. So a bid of 3.1% would only pay 2.1%. Do you think that would change how much interest they would have to pay?

If we stuck with the same eight bids as above, the answer is yes,

* There was also a neat provision that allowed small bidders to get the average rate of all the winning bidders. If you wanted to bid but didn't want to have to outsmart Goldman Sachs and other savvy investors, you could simply state the amount you wanted without stating a rate. You were guaranteed to win, and your rate was the average of the winning bidders. The large investment banks weren't allowed to bid under this rule, just small investors.

as the 3.1% becomes 2.1% and the 3.25% becomes 2.25%, and so on. But under this new regime, anyone who had previously planned on bidding 3.1% would now bid 4.1%. Everyone would bid 1% higher, and after the Treasury adjustment, things would play out just as before.

Indeed, this takes us to the second part of Newton's Third Law: For every action, there is a reaction, equal and opposite. That latter part may also apply to bidding, at least for the cases we've looked at. The reaction of the bidders offsets the changes in the rules.

After bidders adjust their strategies, the Treasury should expect to pay the same interest rates using a uniform price rule, as when winners get paid their bid. But life is much easier for bidders. A bidder who is willing to accept 3.33% no longer has to strategize about whether to bid 3.6% or 3.72%. If they value the bonds at 3.33%, they can bid 3.33% and know that, if they win, they will get at least 3.33% and most likely something higher. The Treasury doesn't lose any money, and bidders have a much simpler job.*

Many games that might not at first look like an auction turn out to be one. We turn now to look at two battle of wills, the preemption game and the war of attrition. In both contests, the situation is much like an auction.

THE PREEMPTION GAME

On August 3, 1993, Apple Computer launched the Original Newton Message. The Newton was more than a flop. It was an embarrassment. The handwriting recognition software developed by Soviet programmers didn't seem to understand English. In a *Simpsons* episode, the Newton misinterpreted "Beat up Martin" as

---

* The uniform price Treasury auction isn't exactly a Vickrey auction. The complication is that by bidding for more units, the bidder can lower the interest rate it gets on all of its winnings. This leads to some element of strategic bidding. To turn it into a multi-unit Vickrey auction, each bidder would have to receive the highest winning interest rate in the thought experiment where the bidder wasn't in the game.

"Eat up Martha." Doonesbury cartoons lampooned the mistakes made by its handwriting recognition.

The Newton was scrapped five years later, on February 27, 1998. While Apple was busy failing, in March of 1996 Jeff Hawkins introduced the Palm Pilot 1000 handheld organizer, which quickly grew to a billion in annual sales.

The Newton was a great idea, but it was not ready for prime time. That's the paradox. Wait until you're fully ready and miss the opportunity. Jump in too soon and fail. The launch of *USA Today* faced this same issue.

Most countries have long-standing national newspapers. France has *Le Monde* and *Le Figaro*, England has *The Times*, the *Observer*, and the *Guardian*. Japan has the *Asahi Shimbun* and *Yomiuri Shimbun*, China has the *People's Daily*, and Russia has *Pravda*. India has *The Times*, the *Hindu*, *Dainik Jagran*, and some sixty others. Americans were alone in not having a national daily. They had national magazines (*Time*, *Newsweek*) and the weekly *Christian Science Monitor* but no national daily paper. It was only in 1982 that Al Neuharth persuaded Gannett's board to launch *USA Today*.

Creating a national newspaper in the United States was a logistical nightmare. Newspaper distribution is inherently a local business. That meant *USA Today* would have to be printed at plants across the country. With the Internet, that would have been straightforward. But in 1982, the only practical option was satellite transmission. With color pages, *USA Today* was a bleeding-edge technology.

Because we see those blue boxes nearly everywhere now, we

tend to think that *USA Today* must have been a good idea. But just because something is successful today doesn't mean that it was worth the cost. It took Gannett twelve years before they broke even on the paper. Along the way, they lost over a billion dollars. And that was when a billion was real money.

If only Gannett had waited a few more years, the technology would have made their journey much easier. The problem was that the potential market for national papers in the United States was at most one. Neuharth was worried that Knight Ridder would launch first and then the window would be gone for good.

Both Apple and *USA Today* are cases where companies were playing a preemption game. The first person to launch has a chance to own the market, provided they succeed. The question is when to pull the trigger. Pull too early and you'll miss. Wait too long and you'll get beaten.

The way we describe a preemption game suggests a duel and that analogy is apt. If you fire too soon and miss, your rival will be able to advance and hit with certainty. But if you wait too long, you may end up dead without having fired a shot.* We can model the duel as an auction. Think of the time to shoot as the bid. The person who bids lowest gets the first chance to shoot. The only problem with bidding low is that the chance of success also goes down.

It might come as a surprise that both players will want to fire at the same time. That is to be expected when the two players have the same skill. But the result holds even when the two have different abilities.

Imagine that it were otherwise. Say you were planning to wait until time 10 before shooting. Meanwhile, your rival was planning to shoot at 8. That pair of strategies can't be an equilibrium. Your rival should change his strategy. He can now wait until time 9.99 and thereby increase his chance of success without risking being shot first. Whoever plans to go first should wait until the moment before the rival is about to shoot.

---

* Our Yale colleague, Ben Polak, illustrates the preemption game by using a duel with a pair of wet sponges. You can try this at home (or in your class). Start far apart and slowly walk toward each other. When do you throw?

If waiting until time 10 really makes sense, you have to be willing to be shot at and hope your rival misses. That has to be every bit as good as jumping the gun and shooting first. The right time to fire is when your chance of success equals the rival's chance of failure. And since the chance of

TRIP TO THE GYM NO. 8

Imagine that you and your rival both write down the time at which you will shoot. The chance of success at time t is $p(t)$ for you and $q(t)$ for your rival. If the first shot hits, the game is over. If it misses, then the other person waits to the end and hits with certainty. When should you shoot?

failure is 1 minus the chance of success, this implies that you fire the first moment when the two chances of success add up to 1. As you can see, if the two probabilities add up to 1 for you, they also add up to 1 for your rival. Thus the time to shoot is the same for both players. You get to prove this in our trip to the gym.

The way we modeled this game, both sides had correct understanding of the other side's chance of success. This might not always be true. We also assumed that the payoff from trying and failing was the same as the payoff from letting the other side go first and have it win. As they might say, sometimes it is better to have tried and lost than never to have tried at all.

THE WAR OF ATTRITION

The opposite of the preemption game is a war of attrition. Instead of seeing who jumps in first, here the objective is to outlast your rival. Instead of who goes in first, the game is who gives in first. This, too, can be seen as an auction. Think of your bid as the time that you are willing to stay in the game and lose money. It is a bit of a strange auction in that all the participants end up paying their bid. The high bidder still wins. And here it may even make sense to bid more than your value.

In 1986, British Satellite Broadcasting (BSB) won the official license to provide satellite TV to the English market. This had the potential to be one of the most valuable franchises in all of history. For years, English TV viewers were limited in their choices to the two BBC channels and ITV. Channel 4 brought the total to, you guessed it, four. This was a country with 21 million households, high income, and plenty of rain. Moreover, unlike the United States, there was hardly any presence of cable TV.* Thus it was not at all unrealistic to imagine that the satellite TV franchise in the UK could bring in £2 billion revenue annually. Such untapped markets are few and far between.

Everything was looking up for BSB until June 1988, when Rupert Murdoch decided to spoil the fun. Working with an old-fashioned Astra satellite positioned over the Netherlands, Murdoch was able to beam his four channels to England. Now the Brits could finally enjoy *Dallas* (and soon *Baywatch*).

While the market might have seemed large enough for both Murdoch and BSB, the brutal competition between them destroyed all hopes for profit. They got into bidding wars over Hollywood movies and price wars over the cost of ad time. Because their broadcast technologies were incompatible, many people decided to wait and see who would win before investing in a dish.

After a year of competition, the two firms had lost a combined £1.5 billion. This was entirely predictable. Murdoch well understood that BSB wasn't going to fold. And BSB's strategy was to see if they could drive Murdoch into bankruptcy. The reason both firms were willing to suffer such massive losses is that the prize for winning was so large. If either one managed to outlast the other, it would have all of the profits to itself. The fact that you may have already lost £600 million is irrelevant. You've lost that amount whether you continue to play or give up. The only question is whether the additional cost of hanging on is justified by the pot of gold to the winner.

We can model this as an auction in which each side's bid is how long it will stay in the game, measured in terms of financial losses. The company that lasts longest wins. What makes this type of auction especially tricky is that there isn't any one best bidding strategy. If you think the other side is just about to fold, then you should always stay in another period. The reason you might think

---

* Less than 1 percent of homes subscribed to cable, and access to cable was limited by law to regions where on-air reception was unavailable.

they are about to fold is because you think that they think you are going to stay in the game.

As you can see, your bidding strategy all depends on what you think they are doing, which in turn depends on what they think you are doing. Of course you don't actually know what they are doing. You have to decide in your head what they think you are up to. Because there's no consistency check, the two of you can each be overconfident about your ability to outlast the other. This can lead to massive overbidding or massive losses for both players.

Our suggestion is that this is a dangerous game to play. Your best move is to work out a deal with the other player. That's just what Murdoch did. At the eleventh hour, he formed a merger with BSB. The ability to withstand losses determined the split of the joint venture. And the fact that both firms were in danger of going under forced the government's hand in allowing the only two players to merge.

There's a second moral to this game: never bet against Murdoch.

CASE STUDY: SPECTRUM AUCTIONS

The mother of all auctions has been the sale of spectrum for cell phone licenses. Between 1994 and 2005, the Federal Communications Commission raised over $40 billion. In England, an auction for 3G (third-generation) spectrum raised an eye-popping £22.5 billion, making it the biggest single auction of all time.[1]

Unlike the traditional ascending bid auction, some of these auctions were more complicated because they allowed participants to simultaneously bid on several different licenses. In this case, we are going to give you a simplified version of the first U.S. spectrum auction and ask that you develop a bidding strategy. We'll see how you do relative to the actual auction participants.

In our stripped-down auction there will be just two bidders, AT&T and MCI, and just two licenses, NY and LA. Both firms are interested in both licenses, but there is only one of each.

One way to run the auctions would be to sell the two licenses in sequence. First NY and then LA. Or should it be first LA and then NY? There's no obvious answer as to which license should be sold first. Either way causes a problem. Say NY is sold first. AT&T might prefer LA to NY but feel forced to bid on NY knowing that winning LA is far from certain. AT&T would rather end up with something than nothing. But having won NY, it may not then have the budget to bid on LA.

With help from some game theorists, the FCC developed an ingenious solution to this problem: they ran a simultaneous auction. Both NY and LA were up on the auction block at the same time. In effect, participants could call out their bids for either of the two licenses. If AT&T got outbid for LA, it could either raise its offer on LA or move on to bid for NY.

The simultaneous auction was over only when none of the bidders were willing to raise the price on any of the properties up for sale. In practice, the way this worked was that the bidding was divided up into rounds. Each round, players could raise or stay put.

We illustrate how this works using the example below. At the end of round 4, AT&T is the high bidder in NY, and MCI is the high bidder in LA.

	NY	LA
AT&T	6	7
MCI	5	8

In the bidding for round 5, AT&T could bid on LA, and MCI could choose to bid on NY. There's no point in AT&T bidding again on NY, as it is already the high bidder. Ditto for MCI and LA.

Imagine that only AT&T bids. In that case the new result might be:

	NY	LA
AT&T	6	9
MCI	5	8

Now AT&T is the high bidder on both properties. It can't bid. But the auction isn't over yet. The auction only ends when neither party bids in a round. Since AT&T bid in the previous round, there must be at least one more round, and MCI will have a chance to bid. If MCI doesn't bid, the auction is over. Remember that AT&T can't bid. If MCI does bid, say 7 for NY, then the auction continues. In the round that follows, AT&T could bid for NY, and MCI would have another chance to top the bid in LA.

The point of the above example was to make the rules of the auction clear. Now we will ask you to play the auction starting from scratch. To help you out, we'll share our market intelligence with you. The two firms spent millions of dollars preparing for the auction. As part of their preparation, they figured out both their own value for each of the licenses and what they thought their rival's might be. Here are the valuations:

	NY	LA
AT&T	10	9
MCI	9	8

According to the table above, AT&T values both licenses more than MCI. We want you to take this as a given. Furthermore, these valuations are known to both parties. AT&T not only knows its valuation, it knows MCI's numbers and knows that MCI knows AT&T's numbers and that MCI knows AT&T knows MCI's numbers, and so on and so forth. Everyone knows everything. Of course this is an extreme assumption, but the firms did spend a huge amount of money on what is called competitive intelligence, and so the fact that they had good knowledge about the other is pretty accurate.

Now you know the rules of the auction and all the valuations. Let's play. Since we are gentlemen, we'll let you pick which side to take. You picked AT&T? That's the right choice. They have the highest valuations, so you certainly have the advantage in this game. (If you didn't pick AT&T, would you mind picking again?)

It's time to place your bid(s). Please write them down. We've written our bid down and you can trust us to have made our bids without looking at what you've written.

Case Discussion

Before revealing our bid, let's consider some options you may have tried.

Did you bid 10 in NY and 9 for LA? If so, you've certainly won both auctions. But you've made no profit at all. This is one of the more subtle points about bidding in an auction. If you have to pay your bid—as you do in this case—then it makes little sense to bid your value. Think of this as akin to bidding $10 to win a $10 bill. The result is a wash.

The potential confusion here is that it may seem as if there is an extra prize from winning the auction, separate from what you win. Or, if you think of the valuation numbers as maximum bids, but not what you really think the item is worth, then again you might be happy to win at a bid equal to your value.

We don't want you to take either of these approaches. When we say your valuation is 10 for NY, what we mean by that is you are happy to walk away at 10 without whining or winning. At a price of 9.99 you would prefer to win, but only by a tiny amount. At a price of 10.01, you would prefer not to win, although the loss would be small.

Taking this perspective into account, you can see that bidding 10 for NY and 9 for LA is actually a case of a (weakly) dominated strategy. With this strategy, you are guaranteed to end up with zero. This is your payoff whether you win or lose. Any strategy that gives you a chance of doing better than zero while never losing any money will weakly dominate the strategy of bidding 10 and 9 right off the bat.

Perhaps you bid 9 in NY and 8 for LA. If so, you've certainly done better than bidding 10 and 9. Based on our bid, you'll win both auctions. (We won't bid more than our valuations.) So, congratulations.

How did you do? You made a profit of 1 in each city or 2 in total. The key question is whether you can do better.

You obviously can't do better bidding 10 and 9. Nor can you do better repeating your bids of 9 and 8. What other strategies might you consider? Let's assume that you bid 5 and 5. (The way the game will play out for other bids will be quite similar.) Now it's time for us to reveal our bid: we started with 0 (or no bid) in NY and 1 in LA. Given the way the first round of bidding has turned out, you are the high bidder in both cities. Thus you can't bid this round (as there is no point in having you top your own bid). Since we are losing out in both cities, we will bid again.

Think of the situation from our shoes. We can't go back home empty-handed to our CEO and say that we dropped out of the auction when the bids were at 5. We can only go home empty-handed if the prices have escalated to 9 and 8, so that it isn't worth our while to bid anymore. Thus we'll raise our bid in LA to 6. Since we just outbid you, the auction is extended another period. (Remember that the auction is extended another round whenever someone bids.) What will you do?

Imagine that you raise us in LA with a bid of 7. When it comes time for us to bid in the next round, we'll bid in NY this time with an offer of 6. We'd rather win NY at 6 than LA at 8. Of course, you can then outbid us back in NY.

You can see where this is all headed. Depending on who bids when, you will win both licenses at prices of 9 or 10 in NY and 8 or 9 in LA. This is certainly no better than the result when you just started out with a bid of 9 in NY and 8 in LA. It doesn't appear that our experiment has led to any improvement in payoffs. That happens. As you try out different strategies you can't expect them all to work. But was there something else you could have done that would have led to a profit greater than 2?

Let's go back and replay the last auction. What else might you have done after we bid 6 for LA? Recall that at that time, you were the high bidder in NY at a price of 5. Actually, you could have done nothing. You could have stopped bidding. We had no interest in outbidding you in NY. We were plenty happy to win the LA license at a price of 6. The only reason we bid again is that we couldn't go away empty-handed—unless, of course, prices escalated to 9 and 8.

If you had stopped bidding, the auction would have ended then and there. You would only have won just one license, NY, at 5. Since you value that license at 10, this result is worth 5 to you, a big improvement over the gain of 2 you expect with bids of 9 and 8.

Think again from our perspective. We know that we can't beat you in both licenses. You have a higher valuation than we do. We are more than happy to walk away with a single license at any price we can below 9 and 8.

With all this practice, we should give you one last chance to bid and prove you really understand how this game works. Ready? Did you bid 1 in NY and 0 in LA? We hope so—because we bid 0 for NY and 1 for LA. At this point, we each have another chance to bid (as the bids from the previous round mean that the auction gets extended). You can't bid for NY, as you are already the high bidder. What about LA? Do you bid? We certainly hope . . . not. We didn't bid. So if you didn't bid, the auction is over. Remember that the auction ends as soon as there is a round with no bids. If the auction ends at that point, you walk away with just one license, but at the bargain price of 1, and thus you end up making 9.

It may be frustrating to have us win the second license at 1 when you value it well above that level and even more than we do. The following perspective might help soothe your spirits.

Before we walk away with no license, we will bid all the way up to 9 and 8. If you intend to deny us any license, you have to be prepared to bid a total of 17. Right now you have one license at a price of 1. Thus the true cost of winning the second license is 16, which is well in excess of your value.

You have a choice. You can win one license at a price of 1 or two licenses at a combined price of 17. Winning one is the better option. Just because you can beat us in both auctions doesn't mean that you should.

At this point, we'll bet that you still have some questions. For example, how would you know that we would be bidding on LA and leave you the opportunity to bid on NY? In truth, you wouldn't. We were lucky the way things worked out in this case. But, even if we had both bid on NY in the first round, it wouldn't have taken too long to sort things out.

You might also be wondering if this is collusion. Strictly speaking, the answer is no. While it is true that the two firms both end up better off (and the seller is the big loser), observe that neither party needs to make an agreement with the other. Each side is acting in its own best interest. MCI understands all on its own that it can't win both licenses in the auction. This is no surprise, as AT&T has higher values for each item. Thus MCI will be happy to win either license. As for AT&T, it can appreciate that the true cost of the second license is the additional amount it will have to pay on both. Outbidding MCI on LA can raise the price in both LA and NY. The true cost of winning the second license is 16, more than its value.

What we see here is often called tacit cooperation. Each of the two players in the game understands the long-run cost of bidding for two licenses and thus recognizes the advantage of winning one license on the cheap. If you were the seller, you would want to avoid this outcome. One approach is to sell the two licenses in sequence. Now, it wouldn't work for MCI to let AT&T get the NY license for 1. The reason is that AT&T would still have every incentive to go after the LA license in the next auction. The key difference is that MCI can't go back and rebid in the NY auction, so AT&T has nothing to lose when bidding for the LA license.

The larger lesson here is that when two games are combined into one, this creates an opportunity to employ strategies that go across the two games. When Fuji entered the U.S. film market, Kodak had the opportunity to respond in the United States or in Japan. While starting a price war in the United States would have been costly to Kodak, doing so in Japan was costly to Fuji (and not to Kodak, who had little share in Japan). Thus the interaction between multiple games played simultaneously creates opportunities for punishment and cooperation that might otherwise be impossible, at least without explicit collusion.

Moral: If you don't like the game you are playing, look for the larger game.

For more auction case studies, have a look at chapter 14: "The Safer Duel," "The Risk of Winning," and "What Price a Dollar?"

CHAPTER 11

Bargaining

A NEWLY ELECTED trade union leader went to his first tough bargaining session in the company boardroom. Nervous and intimidated by the setting, he blurted out his demand: "We want ten dollars an hour or else."

"Or else what?" challenged the boss.

The union leader replied, "Nine fifty."

Few union leaders are so quick to back down, and bosses need the threat of Chinese competition, not their own power, to secure wage concessions. But the situation poses several important questions about the bargaining process. Will there be an agreement? Will it occur amicably, or only after a strike? Who will concede and when? Who will get how much of the pie that is the object of the haggling?

In chapter 2, we sketched a simple story of the ultimatum game. The example illustrated the strategic principle of looking ahead and reasoning back. Many realities of the bargaining process were sacrificed in order to make that principle stand out. This chapter uses the same principle, but with more attention to issues that arise during bargaining in business, politics, and elsewhere.

We begin by recapitulating the basic idea in the context of

union-management negotiation over wages. To look forward and reason backward, it helps to start at a fixed point in the future, so let us think of an enterprise with a natural conclusion, such as a hotel in a summer resort. The season lasts 101 days. Each day the hotel operates, it makes a profit of $1,000. At the beginning of the season, the employees' union confronts the management over wages. The union presents its demand. The management either accepts this or rejects it and returns the next day with a counteroffer. The hotel can open only after an agreement is reached.

First suppose bargaining has gone on for so long that, even if the next round leads to an agreement, the hotel can open for only the last day of the season. In theory, bargaining will not go on that long, but because of the logic of looking ahead and reasoning back, what actually happens is governed by a thought process that starts at this extreme. Suppose it is the union's turn to present its demand. At this point the management should accept anything as being better than nothing. So the union can get away with the whole $1,000.*

Now look at the day before the last day of the season, when it is the management's turn to make an offer. It knows that the union can always reject this, let the process go on to the last day, and get $1,000. Therefore the management cannot offer any less. And the union cannot do any better than get $1,000 on the last day, so the management need not offer any more on the day before.† Therefore the management's offer at this stage is clear: of the $2,000 profit over the last two days, it asks half. Each side gets $500 per day.

Next let the reasoning move back one more day. By the same logic, the union will offer the management $1,000 and ask for

---

* We could make the more realistic assumption that the management will need some minimal share, such as $100, but that would only complicate the arithmetic and wouldn't change the basic idea of the story. This is the same issue we discussed in the original ultimatum game. You have to give the other side enough so that they won't turn down the offer out of spite.

† Again, there is the issue of a little sweetener, which we ignore for the sake of exposition.

$2,000; this gives the union $667 per day and the management $333. We show the full process in the following table:

Successive rounds of wage bargaining

Days to go	Offer by	Union's share		Management's share	
		Total	Per day	Total	Per day
1	Union	$1,000	$1,000	$0	$0
2	Management	$1,000	$500	$1,000	$500
3	Union	$2,000	$667	$1,000	$333
4	Management	$2,000	$500	$2,000	$500
5	Union	$3,000	$600	$2,000	$400
...					
100	Management	$50,000	$500	$50,000	$500
101	Union	$51,000	$505	$50,000	$495

Each time the union makes an offer, it has an advantage, which stems from its ability to make the last all-or-nothing offer. But the advantage gets smaller as the number of rounds increases. At the start of a season 101 days long, the two sides' positions are almost identical: $505 vs. $495. Almost the same division would emerge if the management were to make the last offer, or indeed if there were no rigid rules like one offer a day, alternating offers, etc.[1]

The appendix to this chapter shows how this framework generalizes to include negotiations in which there is no predetermined last period. Our restrictions to alternating offers and a known finite horizon were simply devices to help us look ahead. They become innocuous when the time between offers is short and the bargaining horizon is long—in these cases, looking ahead and reasoning backward leads to a simple and appealing rule: split the total down the middle.

There is a second prediction of the theory: the agreement will occur on the first day of the negotiation process. Because the two sides look ahead to predict the same outcome, there is no reason why they should fail to agree and jointly lose $1,000 a day. Not all instances of union-management bargaining have such a happy beginning. Breakdowns in negotiations do occur, strikes or lock-outs happen, and settlements favor one side or the other. By refining our example and changing some of the premises, we can explain these facts.

THE HANDICAP SYSTEM IN NEGOTIATIONS

One important element that determines how the pie will be split is each side's cost of waiting. Although both sides may lose an equal amount of profits, one party may have other alternatives that help partially recapture this loss. Suppose that the members of the union can earn $300 a day in outside activities while negotiations with the hotel management go on. Now each time the management's turn comes, it must offer the union not only what the union could get a day later but also at least $300 for the current day. The entries in our table shift in the union's favor; we show this in a new table. Once again the agreement occurs at the season opening and without any strike, but the union does much better.

Successive rounds of wage bargaining
(with outside activities)

Days to go	Offer by	Union's share		Management's share	
		Total	Per day	Total	Per day
1	Union	$1,000	$1,000	$0	$0
2	Management	$1,300	$650	$700	$350
3	Union	$2,300	$767	$700	$233
4	Management	$2,600	$650	$1,400	$350
5	Union	$3,600	$720	$1,400	$280
...					
100	Management	$65,000	$650	$35,000	$350
101	Union	$66,000	$653	$35,000	$347

This result can be seen as a natural modification of the principle of equal division, allowing for the possibility that the parties start the process with different "handicaps," as in golf. The union starts at $300, the sum its members could earn on the outside. This leaves $700 to be negotiated, and the principle is to split it evenly, $350 for each side. The union gets $650 and the management only $350.

In other circumstances the management could have an advantage. For example, it might be able to operate the hotel using scabs while the negotiations with the union go on. But because those workers are less efficient or must be paid more, or because some guests are reluctant to cross the union's picket lines, the management's profit from such operation will be only $500 a day. Suppose the union members have no outside income possibilities. Once again there will be an immediate settlement with the union without an actual strike. But the prospect of the scab operation will give the management an advantage in the negotiation, and it will get $750 a day while the union gets $250.

If the union members have an outside income possibility of $300 *and* the management can operate the hotel with a profit of $500 during negotiations, then only $200 remains free to be bargained over. They split that $200 evenly so that the management gets $600 and the union gets $400. The general idea is that the better a party can do by itself in the absence of an agreement, the larger its share of the bargaining pie will be.

MEASURING THE PIE

The first step in any negotiation is to measure the pie correctly. In the example just above, the two sides are not really negotiating over $1,000. If they reach an agreement, they can split $1,000 per day. But if they don't reach an agreement, then the union has a fallback of $300 and the management has a fallback of $500. Thus an agreement only brings them an additional $200. In this case, the best way to think about the size of the pie is that it is $200. More generally, the size of the pie is measured by how much value is created when the two sides reach an agreement compared to when they don't.

In the lingo of bargaining, the fallback numbers of $300 for the union and $500 for management are called BATNAs, a term coined by Roger Fisher and William Ury. It stands for Best Alternative to a Negotiated Agreement.[2] (You can also think that it stands for Best Alternative to No Agreement.) It is the best you can get if you don't reach an agreement with this party.

Since everyone can get their BATNA without having to negotiate, the whole point of the negotiation is how much value can be created above and beyond the sum of their BATNAs. The best way to think about the pie is how much more value can be created beyond giving everyone his or her BATNA. This idea is both profound and deceptively simple. To see how easy it is to lose sight of BATNAs, consider the following bargaining problem adapted from a real-world case.

Two companies, one in Dallas and one is San Francisco, were using the same New York–based lawyer. As a result of coordinating their schedules, the lawyer was able to fly NY–Houston–SF–NY, a triangle route, rather than make two separate trips.

The one-way airfares were:

NY–Houston	$666
Houston–SF	$909
SF–NY	$1,243
Total:	$2,818

The total cost of the trip was $2,818. Had the lawyer done each of the trips separately, the round-trip fares would have been just double the one-way fares (as there was no time to book the trip in advance).

Our question considers how the two companies might negotiate the division of the airfare. We realize that the stakes are small here, but it is the principle we are looking for. The simplest approach would be to split the airfare in two: $1,409 to each of Houston and

San Francisco.* In response to such a proposal you might well hear from Houston: we have a problem. It would have been cheaper for Houston to have paid for the round trip to Houston all by itself. That fare is only twice $666, or $1,332. Houston would never agree to such a split.

Another approach is to have Houston pay for the NY–Houston leg, to have SF pay for the SF–NY leg, and for the two to split the Houston–SF leg. Under that approach, SF would pay $1,697.50 and Houston would pay $1,120.50.

The two companies could also agree to split the total costs proportionately, using the same ratio as their two round-trip fares. Under this plan, SF would pay $1,835, about twice as much as Houston, who would pay $983.

When faced with such a question, we tend to come up with ad hoc proposals, some of which are more reasonable than others. Our preferred approach is to start with the BATNA perspective and measure the pie. What will happen if the two companies can't agree? The fallback is that the lawyer would make two separate trips. In that case, the cost would be $1,332 to Houston and $2,486 to SF, for a total of $3,818. Recall that the triangle route cost only $2,818. This is the key point: the extra cost of doing the two round-trips over the triangle route is $1,000. That is the pie.

The value of reaching an agreement is that it creates $1,000 in savings that is otherwise lost. Each of the two companies is equally valuable in reaching that agreement. Thus, to the extent that they are equally patient in the negotiations, we would expect them to split this amount evenly. Each party saves $500 over the round-trip fare: Houston pays $832 and SF pays $1,986.

You can see that this is a much lower number for Houston than any of the other approaches. It suggests that the division between two parties should not be based on the mileage or the relative airfares. Although the airfare to Houston is smaller, that doesn't

---

* If you thought the lawyer might just bill the Houston client for $1,332 (the round-trip fare) and the SF client for $2,486 (the round-trip fare) and pocket the difference, perhaps you might have a career at Enron. Whoops, too late.

mean they should end up with less of the savings. Remember, if they don't agree to the deal the whole $1,000 is lost. We would like to think that this is a case where you might have started off with one of the alternative answers, but, having seen how to apply BATNAs and thereby measure the pie correctly, you are persuaded that the new answer is the most equitable outcome. If you started right away with Houston paying $832 and SF paying $1,986, hats off to you. It turns out that this approach to dividing costs can be traced back to the Talmud's principle of the divided cloth.[3]

In the negotiations we've looked at, the BATNAs were fixed. The union was able to get $300 and management, $500. The round-trip airfares for NY–Houston and NY–SF were given exogenously. In other cases, the BATNAs are not fixed. That opens up the strategy of influencing the BATNAs. Generally speaking, you will want to raise your BATNA and lower the BATNA of the other side. Sometimes these two objectives will be in conflict. We now turn to this subject.

THIS WILL HURT YOU MORE THAN IT HURTS ME

When a strategic bargainer observes that a better outside opportunity translates into a better share in a bargain, he will look for strategic moves that improve his outside opportunities. Moreover, he will notice that what matters is his outside opportunity *relative* to that of his rival. He will do better in the bargaining even if he makes a commitment or a threat that lowers both parties' outside opportunities, so long as that of the rival is damaged more severely.

In our example, when the union members could earn $300 a day on the outside while the management could make a profit of $500 a day using scab labor, the result of the bargaining was $400 for the union and $600 for the management. Now suppose the union members give up $100 a day of outside income to intensify their picketing, and this reduces the management's profit by $200 a day. Then the bargaining process gives the union a starting point of $200 ($300 minus $100) and the management $300 ($500 minus

$200). The two starting points add up to $500, and the remaining $500 of daily profit from regular operation of the hotel is split equally between them. Therefore the union gets $450 and the management gets $550. The union's threat of hurting both (but hurting the management more) has earned it an extra $50.

Major League Baseball players employed just such a tactic in their wage negotiations in 1980. They went on strike during the exhibition season, returned to work at the start of the regular season, and threatened to strike again starting on Memorial Day weekend. To see how this "hurt the team owners more," note that during the exhibition season the players got no salaries, while the owners earned revenue from vacationers and locals. During the regular season the players got the same salary each week. For the owners, the gate and television revenues are low initially and rise substantially during and after the Memorial Day weekend. Therefore the loss of the owners *relative* to that of the players was highest during the exhibition season and again starting Memorial Day weekend. It seems the players knew the right strategy.[4]

The owners gave in just before the second half of the threatened strike. But the first half actually occurred. Our theory of looking ahead and reasoning back is clearly incomplete. Why is it that agreements are not always reached before any damage is done— why are there strikes?

BRINKMANSHIP AND STRIKES

Before an old contract expires, the union and the firm begin the negotiations for a new labor contract. But there is no sense of urgency during this period. Work goes on, no output is sacrificed, and there is no apparent advantage to achieving an agreement sooner rather than later. It would seem that each party should wait until the last moment and state its demand just as the old contract is about to expire and a strike looms. That does happen sometimes, but often an agreement is reached much sooner.

In fact, delaying agreement can be costly even during the tranquil phase when the old contract still operates. The process of negotiation has its own risk. There can be misperception of the other side's impatience or outside opportunities, tension, personality clashes, and suspicion that the other side is not bargaining in good faith. The process may break down despite the fact that both parties want it to succeed.

Although both sides may want the agreement to succeed, they may have different ideas about what constitutes success. The two parties do not always look forward and see the same end. They may not have the same information or share the same perspective, so they see things differently. Each side must make a guess about the other's cost of waiting. Since a side with a low waiting cost does better, it is to each side's advantage to claim its cost is low. But these statements will not be taken at face value; they have to be proven. The way to prove one's waiting costs are low is to begin incurring the costs and then show you can hold out longer, or to take a greater risk of incurring the costs—lower costs make higher risks acceptable. It is the lack of a common view about where the negotiations will end that leads to the beginning of a strike.

Think of a strike as an example of signaling. While anyone can say that he or she has a low cost of going on strike or taking on a strike, to actually do so is the best proof possible. As always, actions speak louder than words. And, as always, conveying information by a signal entails a cost, or sacrifice of efficiency. Both the firm and the workers would like to be able to prove their low costs without having to create all the losses associated with a work disruption.

The situation is tailor-made for the exercise of brinkmanship. The union could threaten an immediate breakdown of talks followed by a strike, but strikes are costly to union members as well. While time for continued negotiation remains, such a dire threat lacks credibility. But a smaller threat can remain credible: tempers and tensions are gradually rising, and a breakdown may occur even though the union doesn't really want it to. If this bothers the management more than it bothers the union, it is a good strategy from the union's perspective. The argument works the other way around, too; the strategy of brinkmanship is a weapon for the stronger of the two parties—namely, the one that fears a breakdown less.

Sometimes wage negotiations go on after the old contract has expired but without a strike, and work continues under the terms of the old contract. This might seem to be a better arrangement, because the machinery and the workers are not idle and output is not lost. But one of the parties, usually the union, is seeking a revision of the terms of the contract in its favor, and for it the arrangement is singularly disadvantageous.* Why should the management concede? Why should it not let the negotiations spin on forever while the old contract remains in force de facto?

Again the threat in the situation is the probability that the process may break down and a strike may ensue. The union practices brinkmanship, but now it does so after the old contract has expired. The time for routine negotiations is past. Continued work under an expired contract while negotiations go on is widely regarded as a sign of union weakness. There must be some chance of a strike to motivate the firm to meet the union's demands.

When the strike does happen, what keeps it going? The key to commitment is to reduce the threat in order to make it credible. Brinkmanship carries the strike along on a day-by-day basis. The threat never to return to work would not be credible, especially if the management comes close to meeting the union's demands. But waiting one more day or week is a credible threat. The losses to the workers are smaller than their potential gains. Provided they believe they will win (and soon), it is worth their while to wait. If the workers are correct in their beliefs, management will find it cheaper to give in and in fact should do so immediately. Hence the workers' threat would cost them nothing. The problem is that the firm may not perceive the situation the same way. If it believes the workers are about to concede, then losing just one

---

* One explanation is that the employees are waiting for the right time to strike. UPS employees who go on strike right before Christmas will do much more damage than workers who strike in the dog days of August.

more day's or week's profits is worth getting a more favorable contract. In this way, both sides continue to hold out, and the strike continues.

Earlier, we talked about the risk of brinkmanship as the chance that both sides would fall together down the slippery slope. As the conflict continues, both sides risk a large loss with a small but increasing probability. It is this increasing exposure to risk that induces one side to back down. Brinkmanship in the form of a strike imposes costs differently, but the effect is the same. Instead of a small chance of a large loss, there is a large chance, even certainty, of a small loss when a strike begins. As the strike continues unresolved, the small loss grows, just as the chance of falling off the brink increases. The way to prove determination is to accept more risk or watch strike losses escalate. Only when one side discovers that the other is truly the stronger does it decide to back down. Strength can take many forms. One side may suffer less from waiting, perhaps because it has valuable alternatives; winning may be very important, perhaps because of negotiations with other unions; losing may be very costly, so that the strike losses look smaller.

Brinkmanship applies to the bargaining between nations as well as that between firms. When the United States tries to get its allies to pay a greater share of the defense costs, it suffers from the weakness of negotiating while working under an expired contract. The old arrangement in which the Americans bear the brunt of the burden continues in the meantime, and the U.S. allies are happy to let the negotiations drag on. Can—and should—the United States resort to brinkmanship?

Risk and brinkmanship change the process of bargaining in a fundamental way. In the earlier accounts of sequences of offers, the prospect of what would come later induced the parties to reach an agreement on the first round. An integral aspect of brinkmanship is that sometimes the parties do go over the brink. Breakdowns and strikes can occur. They may be genuinely regretted by both parties but may acquire a momentum of their own and last surprisingly long.

SIMULTANEOUS BARGAINING OVER MANY ISSUES

Our account of bargaining has so far focused on just one dimension, namely the total sum of money and its split between the two sides. In fact, there are many dimensions to bargaining: the union and management care not just about wages but health benefits, pension plans, conditions of work, and so on. The United States and its trading partners care not just about total CO_2 emissions but how they are allocated. In principle, many of these are reducible to equivalent sums of money, but with one important difference: each side may value the items differently.

Such differences open up new possibilities for mutually acceptable bargains. Suppose the company is able to secure group health coverage on better terms than the individual workers would obtain on their own—say, $1,000 per year instead of $2,000 per year for a family of four. The workers would rather have health coverage than an extra $1,500 a year in wages, and the company would rather offer health coverage than an extra $1,500 in wages, too. It would seem that the negotiators should throw all the issues of mutual interest into a common bargaining pot, and exploit the difference in their relative valuations to achieve outcomes that are better for everyone. This works in some instances; for example, broad negotiations toward trade liberalization in the General Agreement on Tariffs and Trade (GATT) and its successor, the World Trade Organization (WTO), have had better success than ones narrowly focused on particular sectors or commodities.

But joining issues together opens up the possibility of using one bargaining game to generate threats in another. For example, the United States may have more success in extracting concessions in negotiations to open up the Japanese market to its exports if it threatened a breakdown of the military relationship, thereby exposing Japan to a risk of Korean or Chinese aggression. The United States has no interest in actually having this happen; it would be merely a threat that would induce Japan to make the economic concession. Therefore, Japan would insist that the economic and military issues be negotiated separately.[5]

THE VIRTUES OF A VIRTUAL STRIKE

Our discussion of negotiation has also left out the effect on all the players who aren't a party to the deal. When UPS workers go on strike, customers end up without packages. When Air France baggage handlers go on strike, holidays are ruined. A strike hurts more than the two parties negotiating. A lack of agreement on global warming and CO_2 emissions could prove devastating to all future generations (who don't get a seat at the table).

But the parties negotiating have to be willing to walk away in order to demonstrate the strength of their BATNA or to hurt the other side more. Even for an ordinary strike, the collateral damage can easily eclipse the size of the dispute. Until President Bush stepped in on October 3, 2002, invoking the Taft-Hartley Act, the ten-day dock-worker lockout disrupted the U.S. economy to the tune of more than $10 billion. The conflict was over $20 million of productivity enhancements. The collateral damage was *500 times* larger than the amounts that the workers and managers were squabbling over.

Is there some way that the two parties can resolve their differences without imposing such large costs on the rest of us? It turns out that for more than fifty years there has been a clever idea to virtually eliminate all of the waste of strikes and lockouts without altering the relative bargaining power of labor and management.[6] Instead of a traditional strike, the idea is to have a *virtual* strike (or virtual lockout), in which the workers keep working as normal and the firm keeps producing as normal. The trick is that during the virtual strike neither side gets paid.

In a regular strike, workers lose their wages and an employer loses its profits. So during a virtual strike, the workers would work for nothing and the employer would give up all of its profits. Profits might be too hard to measure and short-term profits might also understate the true cost to the firm. Instead, we have the firm give up all of its revenue. As to where the money would go, the revenue could go to Uncle Sam or a charity. Or, the product could be free so that the revenues would be given to customers. During a virtual strike, there is no disruption to the rest of economy. The consumer is not left stranded without service. Management and labor feel the pain and thus have an incentive to settle, but the government, charities, or customers get a windfall.

An actual strike (or a lockout that the management initiates to preempt a strike) can permanently destroy consumer demand and risk the future of the whole enterprise. The National Hockey League imposed a lockout in response to a threatened strike during the 2004–5 season. The whole season was lost, there was no Stanley Cup, and it took a long while for attendances to recover after the dispute was finally settled.

The virtual strike is not just a wild idea waiting to be tested. During World War II, the navy used a virtual strike to settle a labor dispute at the Jenkins Company valve plant in Bridgeport, Connecticut. A virtual strike arrangement was also used in a 1960 Miami bus strike. Here, the customers got a free ride, literally.

In 1999, Meridiana Airline's pilots and flight attendants staged Italy's first virtual strike. The employees worked as usual but without being paid, while Meridiana donated the receipts from its flights to charities. The virtual strike worked just as predicted. The flights that were virtually struck were not disrupted. Other Italian transport strikes have followed the Meridiana lead. In 2000, Italy's Transport Union forfeited 100 million lire from a virtual strike carried out by 300 of its pilots. The virtual pilots' strike provided a public relations opportunity, as the strike payments were used to buy a fancy medical device for a children's hospital. Instead of destroying consumer demand, as in the 2004–5 NHL lockout, the virtual strike windfall provides an opportunity to increase the brand's reputation.

Somewhat perversely, the public relations benefit of virtual strikes may make them harder to implement. Indeed, a strike is often designed to inconvenience consumers so that they put pressure on management to settle. Thus asking an employer to forfeit its profits may not replicate the true costs of a traditional strike. It is notable that in all four historical examples, management agreed to forfeit more than its profits—and instead forfeited its entire gross revenue on all sales during the duration of the strike.

Why would workers ever agree to work for nothing? For the same reason that workers are willing to strike now—to impose pain on management and to prove that they have a low cost of waiting. Indeed, during a virtual strike, we might expect to see labor work *harder* because every additional sale represents additional pain to the manufacturer, who has to forfeit the entire revenue on the sale.

Our point is to replicate the costs and benefits of the negotiation to the parties involved while at the same time leaving everyone else unharmed. So long as the two sides have the same BATNAs in the virtual strike as they do in the real one, they have no advantage in employing the real strike over a virtual one. The right time to go virtual is when the two sides are still talking. Rather than wait until the strike is real, labor and management might agree in advance to employ a virtual strike in the event their next contract negotiations fail. The potential gains from eliminating the entire inefficiency of traditional strikes and lockouts justify efforts to experiment with this new vision for managing labor conflict.

CASE STUDY: 'TIS BETTER TO GIVE THAN TO RECEIVE?

Recall our bargaining problem in which a hotel's management and its labor were negotiating over how to divide the season's profits. Now, instead of labor and management alternating offers, imagine that *only* the management gets to make offers, and labor can only accept or reject.

As before, the season lasts 101 days. Each day the hotel operates, it makes a profit of $1,000. Negotiations start at the beginning of the season. Each day, the management presents its offer, which is either accepted or rejected by labor. If accepted, the hotel opens and begins making money, and the remaining profits are split according to the agreement. If rejected, the negotiations continue until either an offer is accepted or the season ends and the entire profits are lost.

The following table illustrates the declining potential profits as the season progresses. If both labor and management's only concern is to maximize its own payoff, what do you expect will happen (and when)? If you were labor, what would you do to improve your position?

Wage bargaining—Management makes all offers

Days to go	Offer by	Total profits to divide	Amount offered to labor
1	Management	$1,000	?
2	Management	$2,000	?
3	Management	$3,000	?
4	Management	$4,000	?
5	Management	$5,000	?
...			
100	Management	$100,000	?
101	Management	$101,000	?

Case Discussion

In this case, we expect the outcome to differ substantially from 50:50. Because management has the sole power to propose, it is in the stronger bargaining position. Management should be able to get close to the entire amount and reach agreement on the first day.

To predict the bargaining outcome, we start at the end and work backward. On the last day there is no value in continuing, so labor should be willing to accept any positive amount, say $1. On the penultimate day, labor recognizes that rejecting today's offer will bring only $1 tomorrow; hence they prefer to accept $2 today. The argument continues right up to the first day of the season. Management proposes to give labor $101, and labor, seeing no better alternative in the future, accepts. This suggests that in the case of making offers, 'tis better to give than to receive.

This analysis clearly exaggerates management's true bargaining power. Postponing agreement, even by one day, costs management $999 and labor only $1. To the extent that labor cares not only

Wage bargaining—Management makes all offers

Days to go	Offer by	Total profits to divide	Amount offered to labor
1	Management	$1,000	$1
2	Management	$2,000	$2
3	Management	$3,000	$3
4	Management	$4,000	$4
5	Management	$5,000	$5
...			
100	Management	$100,000	$100
101	Management	$101,000	$101

about its payments but also how these payments compare to management's, this type of radically unequal division will not be possible. But that does not mean we must return to an even split. Management still has more bargaining power. Its goal should be to find the minimally acceptable amount to give to labor so that labor prefers getting that amount over nothing, even though management may get more. For example, in the last period, labor might be willing to accept $200 while management gets $800 if labor's alternative is zero. If so, management can perpetuate a 4:1 split throughout each of the 101 days and capture 80 percent of the total profit.

The value of this technique for solving bargaining problems is that it suggests some of the different sources of bargaining power. Splitting the difference or even division is a common but not universal solution to a bargaining problem. Look forward and reason backward provides a reason why we might expect to find unequal division. Yet there is reason to be suspicious of the look forward and reason backward conclusion. What if you try it and it doesn't work? Then what should you do?

The possibility that the other side could prove your analysis wrong makes this repeated version of the game different from the one-shot version. In the one-shot version of divide the $100, you can assume that the receiver will find it enough in his interest to accept $20 so that you can get $80. If you end up wrong in this assumption, the game is over and it is too late to change your strategy. Thus the other side doesn't have an opportunity to teach you a lesson with the hope of changing your future strategy. In contrast, when you play 101 iterations of the ultimatum game, the side receiving the offer might have an incentive to play tough at first and thereby establish that he is perhaps irrational (or at least has a strong conviction for the 50:50 norm).*

What should you do if you offer an 80:20 split on day one and the other side says no? This question is easiest to answer in the case where there are only two days total so that the next iteration will be the last. Do you now think that the person is the type who will reject anything other than 50:50? Or do you think that was just a ruse to get you to offer 50:50 in the final round?

If the other party says yes, he will get 200 for both days, for a total of 400. Even a cold, calculating machine would say no to 80:20 if he thought that doing so would get him an even split in the last period, or 500. But if this is just a bluff, you can stick with 80:20 in the final round and be confident that it will be accepted.

The analysis gets more complicated if your initial offer was 67:33 and that gets turned down. Had the receiver said yes, he would have ended up with a total of 333 for two days, or 666. But now that he's said no, the best he can reasonably hope for is a 50:50 split in the final round, or 500. Even if he gets his way, he will end up worse off. At this point, you have some evidence that this isn't a bluff. Now it might well make sense to offer 50:50 in the final round.

In sum, what makes a multiround game different from the one-shot version, even if only one side is making all the offers, is that

---

* In presenting this option, we have subtly changed the game by introducing some uncertainty about the preferences of the other player. Most likely, the person will take any offer that will maximize his payoff. But there is now some small chance that the other player will only accept 50:50 splits, a particular form of fairness norm. Even though it is unlikely that the other player is of this type, many maximizing players would like to convince you that they are a 50:50 type in order to induce you to give them more of the pie.

the receiving side has an opportunity to show you that your theory isn't working as predicted. At that point, do you stick with the theory or change your strategy? The paradox is that the other side will often gain by appearing to be irrational, so you can't simply accept irrationality at face value. But they might be able to do so much damage to themselves (and to you along the way) that a bluff wouldn't help them. In that case, you might very well want to reassess the objectives of the other party.

APPENDIX: RUBINSTEIN BARGAINING

You might think that it is impossible to solve the bargaining problem if there is no end date to the game. But through an ingenious approach developed by Ariel Rubinstein, it is possible to find an answer.[7]

In Rubinstein's bargaining game, the two sides alternate making offers. Each offer is a proposal for how to divide the pie. For simplicity, we assume that the pie is of size 1. A proposal is something like $(X, 1 - X)$. The proposal describes who gets what; thus if $X = 3/4$, that means 3/4 for me, 1/4 for you. As soon as one side accepts the other's proposal, the game is over. Until then, the offers alternate back and forth. Turning down an offer is expensive, as this leads to a delay in reaching an agreement. Any agreement that the parties reach tomorrow would be more valuable if reached today. An immediate settlement is in their joint best interest.

Time is money in many different ways. Most simply, a dollar received earlier is worth more than the same dollar received later, because it can be invested and earn interest or dividends in the meantime. If the rate of return on investments is 10 percent a year, then a dollar received right now is worth $1.10 received a year later. The same idea applies to union and management, but there are some additional features that may add to the impatience factor. Each week the agreement is delayed, there is a risk that old, loyal customers will develop long-term relationships with other suppliers, and the firm will be threatened with permanent closure. The workers and the managers will then have to move to other jobs that don't pay as well, the union leaders' reputation will suffer, and the management's stock options will become worthless. The extent to which an immediate agreement is better than one a week later is the probability that this will come to pass in the course of the week.

Just as with the ultimatum game, the person whose turn it is to make the proposal has an advantage. The size of the advantage depends on the degree of impatience. We measure impatience by how much is left if one does the deal in the next round rather than today. Take the case where there is an offer each week. If a dollar next week is worth 99¢ today, then 99 percent of the value remains (99¢ in the hand is worth a $1 in next week's bush). We represent the cost of waiting by the variable δ. In this example, $\delta = 0.99$. When δ is close to one, such as 0.99, then people are patient; if δ is small, say 1/3, then waiting is costly and the bargainers are impatient. Indeed, with $\delta = 1/3$, two-thirds the value is lost each week.

The degree of impatience will generally depend on how much time elapses between bargaining rounds. If it takes a week to make a counteroffer, then perhaps $\delta = 0.99$. If it only takes a minute, then $\delta = 0.999999$, and almost nothing is lost.

Once we know the degree of impatience, we can find the bargaining division by considering the least one might ever accept and the most one will ever be offered. Is it possible that the least amount you would ever accept is zero? No. Say it were and the other side offers you zero. Then you know that if you were to turn down the zero today and it comes time to make your counteroffer tomorrow, you can offer the other side δ and he will accept. He will accept because he would rather get δ tomorrow than have to wait one more period to get 1. (He would only get 1 in his best-case scenario that you accept 0 in two periods.) So once you know that he will surely take δ tomorrow, that means you can count on $1 - \delta$ tomorrow, and so you should never accept anything less than $\delta(1 - \delta)$ today. Hence neither today nor in two periods should you accept zero.*

---

* Unless, of course, $\delta = 0$, in which case you are completely impatient and future periods have no value.

The argument wasn't fully consistent, in that we found the minimal amount you would accept assuming that you would take zero in two periods. What we really want to find is the minimal amount you would accept where that number holds steady over time. What we are looking for is the number such that when everyone understands this is the least you will ever accept, it leads you to a position where you should accept nothing less.

Here is how we solve that circular reasoning. Assume that the worst (or lowest) division you will ever accept gives you L, where L stands for lowest. To figure out what L must be, let's imagine that you decide to turn down today's offer in order to make a counteroffer. As you contemplate possible counteroffers, you can anticipate that the other side can never hope for more than $1 - L$ when it is their turn again. (They know you won't accept less than L, and so they cannot get more than $1 - L$.) Since that is the best they can do two periods later, they should accept $\delta(1 - L)$ tomorrow.

Thus today when you are contemplating accepting their offer, you can be confident that, were you to reject their offer today and counter with $\delta(1 - L)$ tomorrow, they would accept. Now we are almost done. Once you know that you can always get them to accept $\delta(1 - L)$ tomorrow, that leaves you $1 - \delta(1 - L)$ tomorrow for sure.

Therefore, you should never take anything today less than

$$\delta(1 - \delta(1 - L)).$$

That gives us a minimum value for L:

$$L \geq \delta(1 - \delta(1 - L))$$

or

$$L \geq \frac{\delta(1 - \delta)}{(1 - \delta^2)} = \frac{\delta}{(1 + \delta)}$$

You should never accept anything less than $\delta/(1 + \delta)$, because you can get more by waiting and making a counteroffer that the other side is sure to accept. What is true for you is also true for the other side. By the same logic, the other side will also never accept less than $\delta/(1 + \delta)$. That tells us what the most you can ever hope for is.

Using M for most, let's look for a number that is so large you should never turn it down. Since you know that the other side will never accept less than $\delta/(1 + \delta)$ next period, in the best possible case you can get at most $1 - \delta/(1 + \delta) = 1/(1 + \delta)$ next period. If that is the best you can do next period, then today you should always accept $\delta(1/(1 + \delta)) = \delta/(1 + \delta)$.

So we have

$$L \geq \frac{\delta}{(1 + \delta)}$$

and

$$M \leq \frac{\delta}{(1 + \delta)}.$$

That means that the least you will ever accept is $\delta/(1 + \delta)$ and that you will always accept anything at or above $\delta/(1 + \delta)$. Since these two amounts are exactly the same, that is what you will get. The other side won't offer less, as you will turn it down. They won't offer you more, as you will surely accept $\delta/(1 + \delta)$.

The division makes sense. As the time period between offers and counteroffers shrinks, it is reasonable to say that participants are less impatient; or, mathematically, δ gets close to 1. Look at the extreme case, where $\delta = 1$. Then the proposed division is

$$\frac{\delta}{(1 + \delta)} = \frac{1}{2}.$$

The pie is split evenly between the two sides. If waiting a turn is essentially costless, then the person who goes first doesn't have any advantage, and so the division is 50:50.

At the other extreme, imagine that the pie all disappears if the offer isn't accepted. This is the ultimatum game. If the value of an agreement tomorrow is essentially zero, then $\delta = 0$, and the split is $(0, 1)$, just as in the ultimatum game (with all the caveats, too).

To take an intermediate case, imagine that time is of the essence so that each delay loses half the pie, $\delta = 1/2$. Now the division is

$$\frac{\delta}{(1 + \delta)} = \frac{\frac{1}{2}}{(1 + \frac{1}{2})} = \frac{1}{3}.$$

Think of it this way. The person making me an offer has a claim to all of the pie that will be lost if I say no. That gives him 1/2 right there. Of the half that remains, you can get half of that or 1/4 total, as this amount would be lost if he doesn't accept your offer. Now after two rounds, he will have collected 1/2 and you will have 1/4 and we are back to where we started. Thus in each pair of offers, he can collect twice as much as you, leading to the 2:1 division.

The way we solved the game, the two sides are equally patient. You can use this same approach to find a solution when the two parties have differing costs of waiting. As you might expect, the side that is more patient gets a bigger slice of the pie. Indeed, as the time period between offers gets shorter, the pie is split in the ratio of waiting costs. Thus if one side is twice as impatient as the other, it gets one-third of the pie, or half as much as the other.*

The fact that the greater share in bargaining agreements goes to the more patient side is unfortunate for the United States. Our system of government, and its coverage in the media, fosters impatience. When negotiations with other nations on military and economic matters are making slow progress, interested lobbyists seek support from congressmen, senators, and the media, who pressure the administration for quicker results. Our rival nations in the negotiations know this very well and are able to secure greater concessions from us.

---

* For example, a union and management may assess the risks of delay and their consequences differently. To make things precise, suppose the union regards $1.00 right now as equivalent to $1.01 a week later (its $\delta = 0.99$), and for the management the figure is $1.02 ($\delta = 0.98$). In other words, the union's weekly "interest rate" is 1 percent; the management's, 2 percent. The management is twice as impatient as the union and would therefore end up with half as much.

CHAPTER 12

Voting

People on whom I do not bother to dote
Are people who do not bother to vote

—Ogden Nash,
"Election Day Is a Holiday"

THE FOUNDATION OF a democratic government is that it respects the will of the people as expressed through the ballot box. Unfortunately, these lofty ideals are not so easily implemented. Strategic issues arise in voting, just as in any other multiperson game. Voters will often have an incentive to misrepresent their true preferences. Neither majority rule nor any other voting scheme can solve this problem, for there does not exist any one perfect system for aggregating individuals' preferences into a will of the people.[1]

Actually, simple majority rule works fine in a two-candidate race. If you prefer A over B, then vote for A. There's no need to strategize.* The problems start to arise when there are three or more candidates on the ballot. The voter's problem is whether to vote honestly for the most preferred candidate or to vote strategically for a second or third choice who is a viable candidate.

---

* There is the qualification that you might care about the candidate's margin of victory. You might want your candidate to win, but only with a small margin of victory (in order to temper his megalomania, for example). In that case, you might choose to vote against your preferred candidate, provided you were confident that he would win.

We saw this issue clearly in the presidential election of 2000. The presence of Ralph Nader on the ballot swung the election from Al Gore to George W. Bush. Here we don't mean that the hanging chads or the butterfly ballot turned the election. We mean that if Ralph Nader hadn't run, Al Gore would have won Florida and the election.

Recall that Nader had 97,488 votes in Florida and Bush won with 537 votes. It doesn't take much imagination to see that a large majority of Nader voters would have chosen Gore over Bush.

Nader argues that there were many causes of Gore's defeat. He reminds us that Gore lost his home state of Tennessee, that thousands of Florida voters had been misidentified as ex-felons and removed from the state's rolls, and that 12 percent of Florida's Democrats voted for Bush (or mistakenly for Buchanan). Yes, there were many explanations for Gore's loss. But one of them was Nader.

Our point here is not to bash Nader or any other third-party candidate. Our point is to bash the way we vote. We would like people who genuinely want Ralph Nader to be president to have a way to express that view without having to give up their vote in Bush vs. Gore.*

The challenges of voting in a three-way race hasn't just helped Republicans. Bill Clinton's election in 1992 was much more lopsided as a result of Ross Perot getting 19 percent of the vote. Clinton had 370 electoral votes to George H. W. Bush's 168. It is easy to imagine that several red states (Colorado, Georgia, Kentucky, New Hampshire, Montana) could have gone the other way absent Perot.[2] Unlike in 2000, Clinton would still have won, but the electoral vote could have been much closer.

In the first round of the 2002 French presidential election, the

---

* Actually, there was a solution that we proposed to Ralph Nader and that he rejected. The American voting system is unusual in that people voter for electors in the Electoral College, not the actual candidate. Assuming that Nader preferred Gore over Bush, he could have chosen the same electors as Gore. Thus a vote for Nader would have counted as a vote for Gore (as the electors are the same). That way, voters could express their support for Nader, they could help Nader collect matching funds, all without tilting the election to Bush.

three leading candidates were the incumbent, Jacques Chirac, socialist Lionel Jospin, and the extreme rightist Jean-Marie Le Pen. There were also several candidates of fringe left-wing parties—Maoists, Trotskyites, and the like. It was widely expected that Chirac and Jospin would emerge as the top two vote-getters in the first round and face each other in the runoff election. Therefore many left-wingers indulged themselves by naïvely voting for their most preferred fringe candidates in the first round. They were then stunned when Prime Minister Jospin received fewer votes than Le Pen. In the second round they had to do something unthinkable—vote for the right-winger Chirac, whom they hated, so as to keep out the extremist Le Pen, whom they despised even more.

These cases illustrate where strategy and ethics may collide. Think about when your vote matters. If the election will be won by Bush (or Gore) or Chirac (or Jospin) whether you vote or not, then you might as well vote with your heart. That is because your vote doesn't matter. Your vote really counts when it breaks a tie (or causes a tie). This is what is called being a *pivotal* voter.

If you vote assuming that your vote will count, then a vote for Nader (or a fringe leftist party in France) is a missed opportunity. Even Nader supporters should vote as if they are the one to break the tie between Bush and Gore. This is a bit paradoxical. To the extent that your vote doesn't matter, you can afford to vote with your heart. But, when your vote does matter, then you should be strategic. That's the paradox: it is only okay to speak the truth when it doesn't matter.

You might think that the chance that your vote will ever matter is so small that it can be ignored. In the case of a presidential election, that is pretty much true in a solid blue state like Rhode Island or a solid red state like Texas. But in more balanced states such as New Mexico, Ohio, and Florida, the election result can be close indeed. And while the chance of breaking a tie is still quite small, the effect of such a change is quite large.

The strategic vote problem is an even greater problem for primaries, because there are often four or more candidates. The problem arises both when it comes to voting and when it comes to fundraising. Supporters don't want to waste their vote or campaign contributions on a nonviable candidate. Thus polls and media characterizations that pronounce front-runners have the potential to become self-fulfilling prophecies. The reverse problem can also arise: people expect that someone is a shoe-in and then feel free to vote with their heart for a fringe candidate, only to discover that their second-choice and viable candidate (for example, Jospin) was eliminated.

We are not advocates of strategic voting but the messengers of bad news. We would like nothing more than to propose a voting system that encouraged people to play it straight. Ideally, the voting system could aggregate preferences in a way that expressed the will of the people without leading people to be strategic. Unfortunately, Kenneth Arrow showed that there is no such holy grail. Any way of adding up votes is bound to be flawed.[3] What that means in practical terms is that people will always have an incentive to vote strategically. Thus the election result will be determined by the process just as much as by the voter preferences. That said, you might judge some voting systems to be more flawed than others. We look at some different ways to decide elections below, highlighting the problems and the advantages of each.

NAÏVE VOTING

The most commonly used election procedure is simple majority voting. And yet the results of the majority-rule system can have paradoxical properties, even more peculiar than those demonstrated in the 2000 election. This possibility was first recognized over two hundred years ago by the French Revolution hero Marquis de Condorcet. In his honor, we illustrate his fundamental paradox of majority rule using revolutionary France as the setting.

After the fall of the Bastille, who would be the new populist leader of France? Suppose three candidates, Mr. Robespierre (R), Mr. Danton (D), and Madame Lafarge (L), are competing for the position. The population is divided into three groups, left, middle, and right, with the following preferences:

Left	Middle	Right
40	25	35

Left	Middle	Right
R	D	L
D	L	R
L	R	D

There are 40 voters on the left, 25 in the middle, and 35 on the right. In a vote of Robespierre against Danton, Robespierre wins, 75 to 25. Then in a vote of Robespierre against Lafarge, Lafarge beats Robespierre, 60 to 40. But in a vote of Lafarge against Danton, Danton wins, 65 to 35. There is no overall winner. No one candidate can beat all the others in a head-to-head election. If any candidate were elected, there is another whom a majority would prefer.

This possibility of endless cycles makes it impossible to specify any of the alternatives as representing the will of the people. When Condorcet was faced with this very issue, he proposed that elections that were decided by a larger majority should take precedence over ones that were closer. His reasoning was that there was some true will of the people and that the cycle must therefore reflect a mistake. It was more likely that the small majority was mistaken than the large one.

Based on this logic, the 75 to 25 victory of Robespierre against Danton and 65 to 35 victory of Danton over Lafarge should take priority over the smallest majority, the 60 to 40 victory of Lafarge over Robespierre. In Condorcet's view, Robespierre is clearly preferred over Danton, and Danton is preferred over Lafarge. Thus Robespierre is the best candidate, and the slim majority that favors Lafarge over Robespierre is a mistake. Another way of putting this is that Robespierre should be declared the victor because the maximum vote against Robespierre was 60, while all the other candidates were beaten by an even larger margin.

The irony here is that the French use a different system today, what is often called runoff voting. In their elections, assuming no one gets an absolute majority, the two candidates with the greatest number of votes are selected to go against each other in a runoff election.

Consider what would happen if we used the French system with the three candidates in our example. In the first round, Robespierre would come in first, with 40 votes (as he is the first choice of all 40 voters on the left). Lafarge would come in second, with 35 votes. Danton would come in last, with only 25 votes.

Based on these results, Danton would be eliminated, and the two top vote getters, Robespierre and Lafarge, would meet in a runoff election. In that runoff, we can predict that the Danton supporters would throw their support to Lafarge, who would then win, 60 to 40. Here is more evidence, if it is needed, that the outcome of the election is determined by the rules of voting just as much as by the preferences of the voters.

Of course, we have assumed that the voters are naïve in their decisions. If polls were able to accurately predict voter preferences, then the Robespierre supporters could anticipate that their candidate would lose in a runoff election against Lafarge. That would leave them with their worst possible outcome. As a result, they would have an incentive to vote strategically for Danton, who could then win outright in the first ballot, with 65 percent of the vote.

CONDORCET RULES

Condorcet's insight can offer a solution to the problem of voting in primaries or even the general election when there are three or more candidates. What Condorcet proposes is to have each pair of candidates compete in a pairwise vote. Thus in 2000, there would have been a vote of Bush vs. Gore, Bush vs. Nader, and Gore vs. Nader. The electoral victor would be the candidate who has the smallest maximum vote against him.

Imagine that Gore would have beat Bush, 51–49; Gore would have beat Nader, 80–20; and that Bush would have beat Nader,

70–30. In that case, the maximum vote against Gore was 49, and this is smaller than the maximum against either Bush (51) or Nader (80). Indeed, Gore is what is called a Condorcet winner in that he beats all of the other candidates in head-to-head contests.*

One might think this is interesting in theory but wildly impractical. How could we ask people to vote in three separate elections? And in a primary with six candidates, people would have to vote 15 different times to express their opinions about all two-way races. That seems truly impossible.

Fortunately, there is a simple approach that makes all of this quite practical. All voters have to do is rank their candidates on the ballot. Given that ranking, the computer knows how to vote for any matchup. Thus a voter who ranks the candidates in the order

Gore

Nader

Bush

would vote for Gore over Nader, Nader over Bush, and Gore over Bush. A voter who provides a ranking for the six candidates in a primary has implicitly given a ranking for all possible 15 pairwise choices. If the contest is between her #2 and #5 choices, the vote goes to #2. (If the ranking is incomplete, that's okay, too. A ranked candidate beats all unranked candidates, and the person abstains when the choice is between two unranked ones.)

At Yale School of Management, we implemented the Condorcet voting system to hand out the annual teaching prize. Prior to this, the winner was determined by plurality rule. With some 50 faculty and hence 50 eligible candidates, it was theoretically possible to win the prize with just over 2 percent of the vote (if the votes were nearly evenly split between all of the candidates). More

---

* Since we know that no voting system is perfect, in some cases it will pay to be strategic even when the Condorcet voting system is employed. However, the way in which one is supposed to be strategic is rather complicated, and so we might worry much less about this influencing elections if people can't quite figure out how they should distort their vote for maximum effect.

realistically, there were always a half-dozen strong contenders and another half-dozen with some support. Twenty-five percent was typically enough to win, and so the winner was determined by which candidate's support team managed to focus their vote. Now the students simply rank their professors in order and the computer does all of the voting for them. The winners seem more in line with student demand.

Is it worth the effort to change the way we vote? The next section shows how controlling the agenda can determine the outcome. With the presence of a voting cycle, the outcome is highly sensitive to the voting procedure.

ORDER IN THE COURT

The way the U.S. judicial system works, a defendant is first found to be innocent or guilty. The punishment sentence is determined only after a defendant has been found guilty. It might seem that this is a relatively minor procedural issue. Yet the order of this decision making can mean the difference between life and death, or even between conviction and acquittal. We use the case of a defendant charged with a capital offense to make our point.

There are three alternative procedures to determine the outcome of a criminal court case. Each has its merits, and you might want to choose from among them based on some underlying principles.

1. Status Quo: First determine innocence or guilt; then, if guilty, consider the appropriate punishment.

2. Roman Tradition: After hearing the evidence, start with the most serious punishment and work down the list. First decide if the death penalty should be imposed for this case. If not, decide whether a life sentence is justified. If, after proceeding down the list, no sentence is imposed, the defendant is acquitted.

3. Mandatory Sentencing: First specify the sentence for the crime. Then determine whether the defendant should be convicted.

The only difference between these systems is one of agenda: what gets decided first. To illustrate how important this can be, we consider a case with only three possible outcomes: the death penalty, life imprisonment, and acquittal.[4] This story is based on a true case; it is a modern update of the dilemma faced by Pliny the Younger, a Roman senator under Emperor Trajan around A.D. 100.[5]

The defendant's fate rests in the hands of three deeply divided judges. Their decision is determined by a majority vote. One judge (Judge A) holds that the defendant is guilty and should be given the maximum possible sentence. This judge seeks to impose the death penalty. Life imprisonment is his second choice and acquittal is his worst outcome.

The second judge (Judge B) also believes that the defendant is guilty. However, this judge adamantly opposes the death penalty. His preferred outcome is life imprisonment. The precedent of imposing a death sentence is sufficiently troublesome that he would prefer to see the defendant acquitted rather than executed by the state.

The third judge, Judge C, is alone in holding that the defendant is innocent and thus seeks acquittal. He is on the other side of the fence from the second judge, believing that life in prison is a fate worse than death. (On this the defendant concurs.) Consequently, if acquittal fails, his second best outcome would be to see the defendant sentenced to death. Life in prison would be the worst outcome.

	Judge A's ranking	Judge B's ranking	Judge C's ranking
Best	Death sentence	Life in prison	Acquittal
Middle	Life in prison	Acquittal	Death sentence
Worst	Acquittal	Death sentence	Life in prison

Under the status quo system, the first vote is to determine innocence versus guilt. But these judges are sophisticated decision makers. They look ahead and reason backward. They correctly predict that, if the defendant is found guilty, the vote will be two to one in favor of the death penalty. This effectively means that the original vote is between acquittal and the death penalty. Acquittal wins two to one, as Judge B tips the vote.

It didn't have to turn out that way. The judges might decide to follow the Roman tradition and work their way down the list of charges, starting with the most serious ones. They first decide whether or not to impose the death penalty. If the death penalty is chosen, there are no more decisions to be made. If the death penalty is rejected, the remaining options are life imprisonment and acquittal. By looking forward, the judges recognize that life imprisonment will be the outcome of the second stage. Reasoning backward, the first question reduces to a choice between life in prison and a death sentence. The death sentence wins two to one, with only Judge B dissenting.

A third reasonable alternative is to first determine the appropriate punishment for the crime at hand. Here we are thinking along the lines of a mandatory sentencing code. Once the sentence has been determined, the judges must then decide whether the defendant in the case at hand is guilty of the crime. In this case, if the predetermined sentence is life imprisonment, then the defendant will be found guilty, as Judges A and B vote for conviction. But if the death penalty is to be required, then we see that the defendant will be acquitted, as Judges B and C are unwilling to convict. Thus the choice of sentencing penalty comes down to the choice of life imprisonment versus acquittal. The vote is for life imprisonment, with Judge C casting the lone dissenting vote.

You may find it remarkable and perhaps troubling that any of the three outcomes is possible based solely on the order in which votes are taken. Your choice of judicial system might then depend on the outcome rather than the underlying principles. What this means is that the structure of the game matters. For example, when Congress has to choose between many competing bills, the order in which votes are taken can have a great influence on the final outcome.

THE MEDIAN VOTER

In thinking about voting, we've assumed so far that the candidates simply emerge with a position. The way in which candidates choose their positions is equally strategic. Thus we now turn our attention to the question of how voters try to influence the position of candidates and where the candidates will end up.

One way to help keep your vote from getting lost in the crowd is to make it stand out: take an extreme position away from the crowd. Someone who thinks that the country is too liberal could vote for a moderately conservative candidate. Or she could go all the way to the extreme right and support Rush Limbaugh (should he run). To the extent that candidates compromise by taking central positions, it may be in some voters' interests to appear more extreme than they are. This tactic is effective only up to a point. If you go overboard, you are thought of as a crackpot, and the result is that your opinion is ignored. The trick is to take the most extreme stand consistent with appearing rational.

To make this a little more precise, imagine that we can align all the candidates on a 0 to 100 scale of liberal to conservative. The Green Party is way on the left, around 0, while Rush Limbaugh takes the most conservative stance, somewhere near 100. Voters express their preference by picking some point along the spectrum. Suppose the winner of the election is the candidate whose position is the average of all voters' positions. The way you might think of this happening is that, through negotiations and compromises, the leading candidate's position is chosen to reflect the average position of the electorate. The parallel in bargaining is to settle disputes by offering to "split the difference."

Consider yourself a middle-of-the-roader: if it were in your hands, you would prefer a candidate who stands at the position 50 on our scale. But it may turn out that the country is a bit more conservative than that. Without you, the average is 60. For concreteness, you are one of a hundred voters polled to determine the average position. If you state your actual preference, the candidate will move to $(99 \times 60 + 50)/100 = 59.9$. If, instead, you exaggerate and claim to want 0, the final outcome will be at 59.4. By exaggerating your claim, you are six times as effective in influencing the candidate's position. Here, extremism in the defense of liberalism is no vice.

Of course, you won't be the only one doing this. All those more liberal than 60 will be claiming to be at 0, while those more conservative will be arguing for 100. In the end, everyone will appear to be polarized, although the candidate will still take some central position. The extent of the compromise will depend on the relative numbers pushing in each direction.

The problem with this averaging approach is that it tries to take into account both intensity and direction of preferences. People have an incentive to tell the truth about direction but exaggerate when it comes to intensity. The same problem arises with "split the difference": if that is the rule for settling disputes, everyone will begin with an extreme position.

One solution to this problem is related to Harold Hotelling's observation (discussed in chapter 9) that political parties will converge to the median voter's position. No voter will take an extreme position if the candidate follows the preferences of the median voter—that is, he chooses the platform where there are exactly as many voters who want the candidate to move left as to move right. Unlike the mean, the median position does not depend on the intensity of the voters' preferences, only their preferred direction. To find the median point, a candidate could start at 0 and keep moving to the right as long as a majority supports this change. At the median, the support for any further rightward move is exactly balanced by the equal number of voters who prefer a shift left.

When a candidate adopts the median position, no voter has an incentive to distort her preferences. Why? There are only three cases to consider: (i) a voter to the left of the median, (ii) a voter exactly at the median, and (iii) a voter to the right of the median. In the first case, exaggerating preferences leftward does not alter the median, and therefore the position adopted, at all. The only way that this voter can change the outcome is to support a move rightward. But this is exactly counter to his interest. In the second case, the voter's ideal position is being adopted anyway, and there is nothing to gain by a distortion of preferences. The third case parallels the first. Moving more to the right has no effect on the median, while voting for a move left is counter to the voter's interests.

The way the argument was phrased suggested that the voter knows the median point for the voting population and whether she is to the right or the left of it. Yet the incentive to tell the truth had nothing to do with which of those outcomes occurred. You can think about all three of the above cases as possibilities and then realize that whichever outcome materializes, the voter will want to reveal her position honestly. The advantage of the rule that adopts the median position is that no voter has an incentive to distort her preferences; truthful voting is the dominant strategy for everyone.

The only problem with adopting the median voter's position is its limited applicability. This option is available only when everything can be reduced to a one-dimensional choice, as in liberal versus conservative. But not all issues are so easily classified. Once voters' preferences are more than one-dimensional, there will not be a median, and this neat solution no longer works.

WHY THE CONSTITUTION WORKS

Warning: The material in this section is hard, even for a trip to the gym. We include it because it provides an example of how game theory helps us understand why the U.S. Constitution has proved so durable. The fact that the result is based on research from one of your authors might also play some small part.

We said that things got much more complicated when the candidate positions can no longer be ordered along a single dimension. We turn now to the case where the electorate cares about two issues—say, taxes and social issues.

When everything was one-dimensional, the candidate's position could be represented by a score from 0 to 100, which you can think of as a position on a line. Now the candidate's position on these two issues can be represented as a point in a plane. If there are three issues that matter, then the candidates would have to be located in a three-dimensional space, which is much harder to draw in a two-dimensional book.

We represent a candidate's position on each of the two issues by where he or she is located.

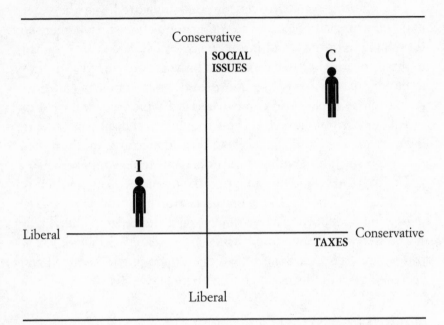

As drawn, the incumbent (I) is middle-of-the-road, slightly liberal on taxes and slightly conservative on social issues. In contrast, the challenger (C) has taken a very conservative position on both taxes and social issues.

Each voter can be thought of as being located at a point in space. The location is the voter's most preferred position. Voters have a simple rule: they vote for the candidate who is closest to their preferred position.

Our next diagram illustrates how the votes will be divided between the two candidates. All those to the left will vote for the incumbent and those to the right will vote for the challenger.

Now that we've explained the rules of the game, where do you imagine the challenger will want to locate? And, if the incumbent

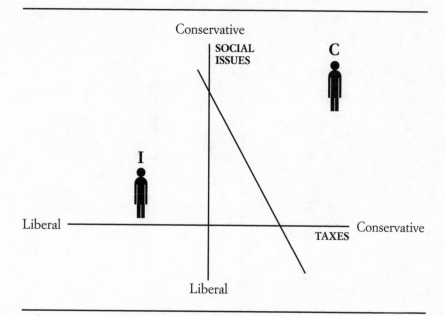

is smart enough to best position herself to fend off the challenger, where will she start out?

Note that as the challenger moves closer to the incumbent, he picks up some votes but loses none. (For example, the move from C to C* expands the group of voters who prefer C; the dividing line is now the dashed line.) That is because anyone who prefers the challenger's position to the incumbent's also prefers something halfway between the two to the incumbent's. Likewise a person who prefers a $1.00 gas tax to no tax is also likely to prefer a 50¢ tax to no tax. What this means is that the challenger has the incentive to locate right next to the incumbent, coming at the incumbent from a direction where there are the most voters. In the picture (on the following page), the challenger would come at the incumbent directly from the northeast.

The challenge for the incumbent is much like the famous cake-cutting problem. In the cake-cutting problem, there are two kids who have to share a cake. The question is to develop a procedure for them to divide it up so as to ensure that each feels he has gotten (at least) half of the cake.

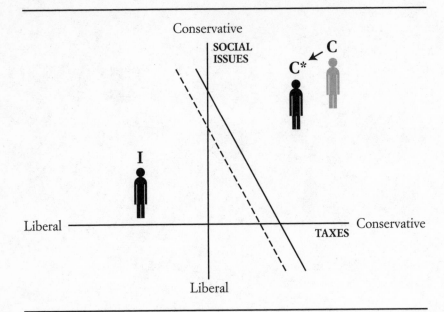

The solution is "I cut, you choose." One kid cuts the cake while the other chooses. This gives the first child an incentive to cut the cake as evenly as possible. Since the second kid has a choice of halves, he will not feel cheated.

This problem is a bit different. Here the challenger gets to cut the cake and to choose. But the incumbent gets to stake out a position that the challenger has to cut through. For example, if all of the voters are uniformly located in a disk, then the incumbent can position herself right in the center. However the challenger tries to position himself in relation to the incumbent, the incumbent can still attract half the voters. In the figure that follows, the dashed line represents a challenger coming from the northwest. The disk is still split in two. The center of the disk is always closest to at least half the points in the disk.

The situation is more complicated if voters are uniformly located in a triangle. (For simplicity, we leave off the issue axes.) Now where should the incumbent position herself, and what is the greatest number of votes that she can assure herself of?

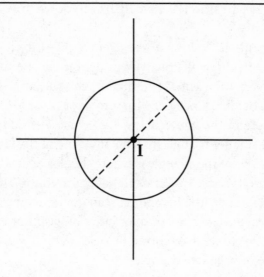

In the picture below, the incumbent has positioned herself poorly. If the challenger were to approach from either the left or the right, the incumbent can still attract support from half the voters. But were the challenger to approach from below, she can garner well more than half the votes. The incumbent would have done

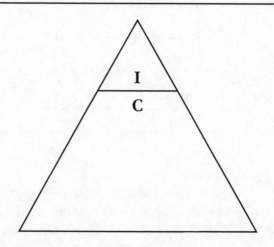

better to have positioned herself lower down to preempt this attack.

It turns out that locating at the average point in the set, what is known as the center of gravity, will guarantee the incumbent at least 4/9 of the total vote. The incumbent will attract 2/3 of the vote in each of two dimensions for a total of $(2/3) \times (2/3) = 4/9$.

You can see in the figure below that we've divided the triangle up into nine smaller triangles, each a mini-me of the large one. The center of gravity of the triangle is where the three lines intersect. (It is also the preferred position of the average voter.) By locating at the center of gravity, the incumbent can guarantee herself support from voters in at least four of the nine triangles. For example, the challenger can attack from straight below and capture all of the voters in the bottom five triangles.

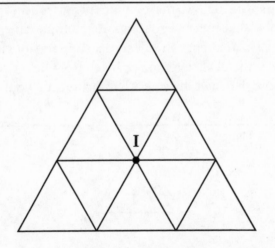

If we extend this to a triangle in three dimensions, then the incumbent still does best by locating at the center of gravity but now can be guaranteed only $(3/4) \times (3/4) \times (3/4) = 27/64$ of the vote.

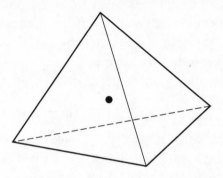

A rather surprising finding is that the triangle (and its multidimensional analogs) turns out to be the worst possible result for the incumbent across all convex sets in any number of dimensions. (A set is convex if for any two points in the set, the line connecting them is also in the set. Thus a disk and a triangle are convex, while the letter T is not.)

Now for the real surprise. Across all convex sets, the incumbent, by locating at the center of gravity, can guarantee herself at least $1/e = 1/2.71828 \approx 36$ percent of the vote. The result even holds when voters are normally distributed (like a bell curve) rather than uniform. What that means is that if a 64 percent majority is required to dislodge the status quo, then it is possible to find a stable outcome by picking the point that is the average of all the voters' preferences. No matter the challenger's position, the incumbent is able to attract at least 36 percent of the vote and thus remain in place.[6] All that is required is that the distribution of voter preferences isn't too extreme. It is okay for some people to take extreme positions so long as there are relatively more people in the middle, as arises with the normal distribution.

The incumbent could be a policy or a precedent, not just a politician. This might then explain the stability of the U.S. Constitution. If all it took was a simple majority (50 percent) to amend the Constitution, it could spin around in cycles. But as it requires more than a 64 percent majority, namely a two-thirds vote, then there is some position that is undefeatable against all comers. That doesn't mean any status quo would be undefeated by all other alternatives. It means that there is some status quo, namely the average position in the voter population, which can't be beaten according to a 67–33 vote.

We want a majority rule that is small enough to allow for flexibility or change when preferences change but not so small as to create instability. A simple majority rule is the most flexible, but is too flexible. It has the potential for cycles and instability. At the other extreme, a 100 percent or unanimity rule would eliminate cycles but lock in the status quo. The goal is to pick the smallest majority size that ensures a stable outcome. It looks like two-thirds majority rule is just on the right side of 64 percent to do the trick. The U.S. Constitution got it right.

We appreciate that this all went a bit quickly. The results here are based on the research of Andrew Caplin done jointly with Barry Nalebuff.[7]

ALL-TIME GREATS

Back to Earth. After the White House, election to Cooperstown may be the next most coveted national honor. Membership in the Baseball Hall of Fame is determined by an election. There is a group of eligible candidates—a player with ten years of experience becomes eligible five years after retirement.[*] The electors are the members of the Baseball Writers' Association. Each voter may vote for up to ten candidates. All candidates capturing votes from more than 75 percent of the total number of ballots returned are elected.

As you would now expect, a problem with this system is that the electors don't have the right incentives to vote for their true prefer-

---

[*] However, if the player has been on the ballot for fifteen years and failed to get elected, then eligibility is lost. For otherwise ineligible players, there is an alternative route to election. An Old-Timers' Committee considers special cases and sometimes elects one or two candidates a year.

ences. The rule that limits each voter to ten choices forces the voters to consider electability as well as merit. (You might think that ten votes are enough, but remember there are around thirty candidates on the ballot.) Some sportswriters may believe a candidate is deserving, but they don't want to throw away the vote if the player is unlikely to make the cutoff. This same issue arose for voting in presidential primaries, and it appears in any election in which each voter is given a fixed number of votes to distribute among the candidates.

Two experts in game theory propose an alternative way to run elections. Steven Brams and Peter Fishburn, one a political scientist and the other an economist, argue that "approval voting" allows voters to express their true preferences without concern for electability.[8] Under approval voting, each voter may vote for as many candidates as he wishes. Voting for one person does not exclude voting for any number of others. Thus there is no harm in voting for a candidate who has no hope of winning. Of course if people can vote for as many candidates as they wish, who gets elected? Like the Cooperstown rule, the electoral rule could specify in advance a percentage of the vote needed to win. Or it could prespecify the number of winning candidates, and then the positions are filled by those who gather the greatest number of votes.

Approval voting has begun to catch on and is used by many professional societies. How would it work for the Baseball Hall of Fame? Would Congress do better if it used approval voting when deciding which expenditure projects should be included in the annual budget? We look at the strategic issues associated with approval voting when a cutoff percentage determines the winners.

Imagine that election to the different sports halls of fame was decided by approval voting, in which all candidates capturing above a fixed percentage of the votes are elected. At first glance, the voters have no incentive to misstate their preferences. The candidates are not in competition with one another but only with an absolute standard of quality implicit in the rule that specifies the required percentage of approval. If I think Mark McGwire should be in the Baseball Hall of Fame, I can only reduce his chances by withholding my approval, and if I think he doesn't belong there, I can only make his admission more likely by voting contrary to my view.

However, candidates may compete against one another in the voters' minds, even though nothing in the rules mandates it. This will usually happen because voters have preferences concerning the size or the structure of the membership. Suppose Mark McGwire and Sammy Sosa come up for election to the Baseball Hall of Fame.* I think McGwire is the better hitter, although I will admit that Sosa also meets the standard for a Hall of Fame berth. However, I think it most important that two sluggers not be elected in the same year. My guess is that the rest of the electorate regards Sosa more highly and he would get in no matter how I vote but that McGwire's case will be a close call, and my approval is likely to tip him over. Voting truthfully means naming McGwire, which is likely to lead to the outcome in which both are admitted. Therefore I have the incentive to misstate my preference and vote for Sosa.

If this seems a bit convoluted, it is. That's the type of reasoning that would be required for people to act strategically under approval voting. It is possible, though unlikely. A similar problem arises if two players complement each other, rather than compete with each other, in the voters' minds.

I may think neither Geoff Boycott nor Sunil Gavaskar belongs in the Cricket Hall of Fame, but it would be a gross injustice to have one and not the other. If in my judgment the rest of the electorate would choose Boycott even if I don't vote for him, while my vote may be crucial in deciding Gavaskar's selection, then I have an incentive to misstate my preference and vote for Gavaskar.

In contrast, a quota rule explicitly places candidates in competition with one another. Suppose the Baseball Hall of Fame limits

* In 2007, there were 32 candidates on the ballot and 545 electors. At least 75 percent of the votes, or 409 votes, were required for election. Mark McGwire got 128. Cal Ripken Jr. established a record by being named on 537 ballots, breaking the previous mark of 491 by Nolan Ryan in 1999. Ripken's 98.53 percent is the third highest in history behind Tom Seaver (98.83 percent in 1992) and Nolan Ryan (98.79 percent in 1999). Sammy Sosa won't be eligible till 2010 (at the earliest).

admission to only two new people each year. Let each voter be given two votes; he can divide them between two candidates or give both to the same candidate. The candidates' votes are totaled, and the top two are admitted. Now suppose there are three candidates—Joe DiMaggio, Marv Throneberry, and Bob Uecker.* Everyone rates DiMaggio at the top, but the electors are split equally between the other two. I know that DiMaggio is sure to get in, so as a Marv Throneberry fan I give my two votes to him to increase his chances over Bob Uecker. Of course everyone else is equally subtle. The result: Throneberry and Uecker are elected and DiMaggio gets no votes.

Government expenditure projects naturally compete with one another so long as the total budget is limited or congressmen and senators have strong preferences over the size of the budget. We will leave you to think which, if any, is the DiMaggio project, and which ones are the Throneberrys and Ueckers of federal spending.

LOVE A LOATH'D ENEMY

The incentive to distort one's preferences is a common problem. One instance occurs when you can move first and use this opportunity to influence others.[9] Take, for example, the case of charitable contributions by foundations. Suppose there are two foundations, each with a budget of $250,000. They are presented with three grant applications: one from an organization helping the homeless, one from the University of Michigan, and one from Yale. Both foundations agree that a grant of $200,000 to the homeless is the top priority. Of the two other applications, the first foundation would like to see more money go to Michigan, while the second would prefer to fund Yale. Suppose the second steals a march and sends a check for its total budget, $250,000, to Yale. The first is then left with no

* Marv Throneberry played first base for the '62 Mets, possibly the worst team in the history of baseball. His performance was instrumental to the team's reputation. Bob Uecker is much better known for his performance in Miller Lite commercials than for his play on the baseball field.

alternative but to provide $200,000 to the homeless, leaving only $50,000 for Michigan. If the two foundations had split the grant to the homeless, then Michigan would have received $150,000, as would Yale. Thus the second foundation has engineered a transfer of $100,000 from Michigan to Yale through the homeless.

In a sense, the foundation has distorted its preferences—it has not given anything to its top charity priority. But the strategic commitment does serve its true interests. In fact, this type of funding game is quite common.* By acting first, small foundations exercise more influence over which secondary priorities get funded. Large foundations and especially the federal government are then left to fund the most pressing needs.

This strategic rearranging of priorities has a direct parallel with voting. Before the 1974 Budget Act, Congress employed many of the same tricks. Unimportant expenditures were voted on and approved first. Later on, when the crunch appeared, the remaining expenditures were too important to be denied. To solve this problem, Congress now votes first on budget totals and then works within them.

When you can rely on others to save you later, you have an incentive to distort your priorities by exaggerating your claim and taking advantage of the others' preferences. You might be willing to gain at the expense of putting something you want at risk, if you can count on someone else bearing the cost of the rescue.

CASE STUDY: THE TIE OF POWER

Recent presidential elections have emphasized the importance of the selection of the vice president. This person will be just a

---

* A similar example is the strategic interaction played between the Marshall and Rhodes scholarships. The Marshall Scholarship moves second (via a waitlist) and thereby has the maximum influence over who is given a scholarship to study in England. If someone has the potential to win both a Marshall and a Rhodes, the Marshall Scholarship allows the person to study as a Rhodes Scholar. That brings the person to England at no cost to the Marshall Scholarship and thus allows them to select one more person.

heartbeat away from the presidency. But most candidates for president spurn the suggestion of the second spot on the ticket, and most vice presidents do not seem to enjoy the experience.*

Only one clause of the Constitution specifies any actual activity for the vice president. Article I, Section 3.4, says: "The Vice-President of the United States shall be President of the Senate, but shall have no vote, unless they be equally divided." The presiding is "ceremony, idle ceremony," and most of the time the vice president delegates this responsibility to a rotation of junior senators chosen by the senate majority leader. Is the tie-breaking vote important, or is it just more ceremony?

Case Discussion

At first glance, both logic and evidence seem to support the ceremonial viewpoint. The vice president's vote just does not seem important. The chance of a tie vote is small. The most favorable circumstances for a tie arise when each senator is just as likely to vote one way as the other, and an even number of senators vote. The result will be roughly one tie vote in twelve.†

The most active tie-breaking vice president was our first, John Adams. He cast 29 tie-breaking votes during his eight years. This is not surprising, since his Senate consisted of only 20 members, and a tie was almost three times more likely than it is today, with our 100-member Senate. In fact, over the first 218 years, there have been only 243 occasions for the vice president to vote. Richard Nixon, under Eisenhower, tied for the most active twentieth-century vice president, casting a total of 8 tie-breaking votes—out

---

* No doubt they console themselves by thinking of the even worse plight of Britain's Prince Charles. John Nance Garner, FDR's first VP, expressed this succinctly: "The vice-presidency ain't worth a pitcher of warm spit."

† The biggest chance that a fixed group of 50 senators votes aye and the remaining 50 vote nay is $(1/2)^{50} \times (1/2)^{50}$. Multiplying this by the number of ways of finding 50 supporters out of the total 100, we get approximately 1/12. Of course senators' votes are far from random. Only when the two parties are roughly equal or when there is an especially divisive issue that splits some of the party lines does the vice president's vote get counted.

of 1,229 decisions reached by the Senate during the period 1953–1961.* This fall in tie-breaking votes also reflects the fact that the two-party system is much more entrenched, so that fewer issues are likely to cross party lines.

But this ceremonial picture of the vice president's vote is misleading. More important than how often the vice president votes is the impact of the vote. Measured correctly, the vice president's vote is roughly equal in importance to that of any senator. One reason the vice president's vote matters is that it tends to decide only the most important and divisive issues. For example, George H. W. Bush's vote, as vice president, saved the MX missile program— and with it helped hasten the collapse of the Soviet Union. This suggests that we should look more closely at just when it is that a vote matters.

A vote can have one of two effects. It can be instrumental in determining the outcome, or it can be a "voice" that influences the margin of victory or defeat without altering the outcome. In a decision-making body like the Senate, the first aspect is the more important one.

To demonstrate the importance of the vice president's current position, imagine that the vice president is given a regular vote as president of the Senate. When does this have any additional impact? For important issues, all 100 senators will try to be present.† *The only time the outcome hinges on the vice president's 101st vote is when the Senate is split 50:50, just the same as when the vice president has only a tie-breaking vote.*

The best example of this is the 107th Congress during the first George W. Bush administration. The Senate was evenly split, 50:50. Thus Vice President Cheney's tie-breaking vote gave the Republicans control of the Senate. All 50 Republicans senators

* Nixon is tied with Thomas R. Marshall (under Woodrow Wilson) and Alben Barkley (under Harry Truman).

† Or senators on opposite sides of the issue will try to pair off their absences. If the 100 senators are split 51:49 or more lopsidedly, then the outcome is the same no matter which way the vice president votes.

were pivotal. If any one had been replaced, control would have shifted to the Democrats.

We recognize that our account of a vice president's voting power leaves out aspects of reality. Some of these imply less power for the vice president; others, more. Much of a senator's power comes from the work in committees, in which the vice president does not partake. On the other hand, the vice president has the ear and the veto power of the president on his side.

Our illustration of the vice president's vote leads to an important moral of wider applicability: anyone's vote affects the outcome only when it creates or breaks a tie. Think how important your own vote is in different contexts. How influential can you be in a presidential election? Your town's mayoral election? Your club's secretarial election?

"The Shark Repellent That Backfired" in chapter 14 offers another case study on voting.

CHAPTER 13

Incentives

WHY DID THE socialist economic systems fail so miserably? The best-laid Five-Year Plans of Stalin and his successors "gang agley" because the workers and the managers lacked adequate incentives. Most importantly, the system offered no reward for doing a good job rather than a merely adequate one. People had no reason to show initiative or innovation, and every reason to cut corners wherever they could—fulfilling quantity quotas and slacking on quality, for example. In the old Soviet economy, this was captured by the quip: They only pretend to pay us, so we only pretend to work.

A market economy has a better natural incentive mechanism, namely the profit motive. A company that succeeds in cutting costs or introducing a new product makes a greater profit; one that lags behind stands to lose money. But even this does not work perfectly. Each employee or manager in a company is not fully exposed to the chill wind of competition in the market, and the top management of the firm has to devise its own internal carrots and sticks to obtain the desired standards of performance from those below. When two firms join forces for a particular project, they have the added problem of designing a contract that will share the incentives between them in the right way.

We develop the components required for a smart incentive scheme through a series of examples.

INCENTIVES FOR EFFORT

Of all the steps involved in writing a book, surely the most tedious for an author is the correction of the printer's proofs. For readers unfamiliar with the process, let us briefly explain what that entails. The printer sets the type from the final manuscript. These days this is done electronically and is therefore relatively error-free, but strange errors—missing words and lines, chunks of text shifted to wrong places, bad line and page breaks—may still creep in. Moreover, this is the author's last chance to correct any slips of writing or even of thinking. Therefore the author has to read the printer's copy in parallel with his own manuscript, and locate and mark all errors for the typesetter to correct.

The author is reading the same text for the umpteenth time, so it is not surprising that his eyes glaze over and he misses several errors. Therefore it is better to hire someone, typically a student in our case, to do the proofreading. A good student can not only catch the typographical errors but also spot and alert the author to more substantive errors of writing and thinking.

But hiring a student to read proofs brings its own problems. The authors have a natural incentive to make the book as error-free as possible; the student is less motivated. Therefore it becomes necessary to give the student the correct incentives, usually by paying him in relation to how well he does the job.

The author wants the student to catch all the typesetting errors there may be. But the only way the author can tell whether the student has done a perfect job is to do a perfect check himself, which defeats the whole purpose of hiring the student. The student's effort is unobservable—he takes away the materials and comes back a week or so later to give the author a list of the errors he has found. What is worse, even the outcome cannot be observed immediately. The author will find out about any error that the student failed to catch only when some other reader (such as yourself) finds and reports the errors, which may not happen for several months or even years.

So the student has the temptation to shirk—just hold on to the materials for a few days and then say that there were no errors. Thus it won't do to offer the student a fixed flat sum for the job. But if offered a piece rate (so much per error he finds), he may worry that the typesetter has done a perfect job, in which case he will have to spend a week or more on the work and get no money at the end of it. He will be reluctant to take the job on these terms.

We have a problem of information asymmetry, but it is different from the ones we considered in chapter 8. The author is the informationally disadvantaged party: he cannot observe the student's effort. This is not something innate to the student; it is the student's deliberate choice. Therefore the problem is not one of adverse selection.* Rather, it is similar to the problem that an insured homeowner may be less careful about locking all doors and windows. Of course insurance companies regard such behavior as almost immoral, and they have coined the term *moral hazard* to describe it. Economists and game theorists take a more relaxed attitude. They think it perfectly natural that people would respond in their own best interests to the incentives they face. If they can get away with shirking on the job, they will do so. What else should one expect from rational players? The onus is on the other player to design the incentive scheme better.

Although moral hazard and adverse selection are different issues, there are some similarities in the methods for coping with them. Just as screening mechanisms have to consider the restrictions of incentive compatibility and participation, so do incentive payment schemes that cope with moral hazard.

A fixed sum does not handle the incentive aspect well, whereas a pure piece-rate payment does not handle the participation aspect

* Adverse selection may also exist; a student who is willing to work for the wages the professor offers may be of too low a quality to command better offers elsewhere. But professors have other ways of finding out the quality of a student: how well the student performed in the professor's courses, recommendations from colleagues, and so on.

well. Therefore the compensation scheme has to be a compromise between the two extremes—a flat sum plus a bonus per error the student discovers. This should give him enough assurance of the total compensation to make the job sufficiently attractive and also enough incentive to attempt a thorough reading.

One of us (Dixit) recently hired a student to read the proofs of a 600-page book. He offered a fixed payment of $600 (a dollar per page), plus the outcome-based incentive of $1 per error found. (The student found 274.) The work took the student about 70 hours, so the average per hour, $12.49, was quite a decent wage by undergraduate student standards. We don't claim that the scheme was fully optimal or the best deal Avinash could have gotten. And the outcome of the job was quite good but not perfect: about 30 errors the student missed have come to light since.* But the example illustrates the general idea of a mixture and how it works in practice.†

We see the same principle applied in many jobs and contracts. How do you pay a software designer or a copywriter? It is hard to monitor their hours. Is the time spent playing foosball, web surfing, or doodling part of the creative process or just slacking off? Even more important, it is even harder to measure how hard they have worked. The answer is to base part of their compensation on the success of the project and the success of the company, and that can be done using the company's stock or stock options. The principle is to combine a basic salary and an incentive bonus tied to the outcome. The same applies with even greater force to the compensation of higher-level management. Of course, like everything else,

---

* You think that is too many errors in all? Try writing a long and complicated book yourself.

† Perhaps it would have been better to pay $2 per error but then dock the student $10 for each one missed. Since the missed errors are only discovered over time, that would require holding part of the payment in escrow, which might be more complicated than it is worth. When is the escrow released? Is there a maximum amount docked? Simplicity is a third constraint on incentive schemes. The people being incentivized need to understand how the system works.

these schemes can be manipulated, but the general principle behind their use as incentive payment mechanisms remains valid.

Many extensions and applications of this principle have been studied by game theorists, economists, business analysts, psychologists, and others. We will give you a brief taste of some of this work and point you to references where you can pursue the subject as far as you wish.[1]

HOW TO WRITE AN INCENTIVE CONTRACT

The main problem of moral hazard is the unobservability of a worker's action or effort. Therefore payments cannot be based on effort, even though more or better effort is what you as an employer want to achieve. Payments must be based on some observable metric, such as the outcome or your profit. If there were a perfect and certain one-to-one relationship between such observable outcomes and the unobservable underlying action, perfect control of effort would be possible. But in practice outcome depends on other chance factors, not just the effort.

For example, the profits of an insurance company depend on its sales force and claim agents, the pricing, and Mother Nature. In a season with many hurricanes, profits will be depressed no matter how hard people work. In fact, they will often have to work harder due to the increased number of claims.

The observable outcome is only an imperfect indicator of the unobservable effort. The two are related, and therefore incentive payments based on the outcome are still useful in influencing effort, but they don't work to perfection. Rewarding an employee for a good outcome in part rewards him for good luck, and penalizing him for a poor outcome in part penalizes him for bad luck. If the chance element is too large, the reward is only poorly related to effort, and therefore the effect of outcome-based incentives on effort is weak. Realizing this, you would not offer such incentives to any great extent. Conversely, if the chance element is small, then stronger and sharper incentives can be used. This contrast will appear repeatedly in what follows.

Nonlinear Incentive Schemes

A special feature of many incentive schemes, for example payment per error found by the proofreader, or a fixed percentage of the sale paid to a salesperson, or payment in stock that constitutes a given fraction of the firm's profit, is their linearity: the incremental payment is strictly proportional to the improvement in the outcome. Other commonly used payment schemes are distinctly nonlinear. The most obvious is a bonus that is paid if the outcome exceeds a specified threshold or an assigned quota. What are the relative merits of a quota and bonus scheme, as compared to a linear or proportional compensation scheme?

Take the context of a salesman, and consider a pure quota bonus scheme: the salesman is paid a low fixed sum if he fails to meet the quota during the year and a higher fixed sum if he does. First suppose the quota is set at such a level that by exerting a hard level of effort he stands a good chance of meeting it, but the chance drops substantially if he slacks off even a little. Then the bonus provides a powerful incentive: the salesman stands to gain a lot of income or lose a lot, depending on whether he decides to work hard or shirk.

But suppose the quota is set at such a demanding level that the salesman has little chance of meeting it even with superhuman effort. Then he will see no point in exerting himself for the sake of that unlikely bonus. And circumstances may change during the year, turning what you thought was a well-set quota into something too demanding and therefore ineffective.

For example, suppose that the quota for the full year is not absurdly high, but the salesman has some bad luck in the first half of the year, making it unlikely that he would meet the year's quota in the six months that remain. That would cause him to give up and take things easy for the rest of the year, which is surely not what the employer wants. Conversely, if the salesman gets lucky and meets the quota in June, again he can relax for the rest of the year: there is no added reward to any further effort in that year. In fact, the salesman could conspire with some customers to

TRIP TO THE GYM NO. 9

The typical real estate agent commission is 6%, which is a linear incentive. How much of an incentive does your agent have to get a higher price? What would an extra $20,000 in the price bring in? Hint: The answer is not $1,200. How would you design a better incentive scheme? What might be some of the issues with your alternative scheme?

postpone their orders to the next year to give him a good start toward next year's quota. Again that is hardly in your interests.

This illustrates in an extreme form the drawbacks of many nonlinear payment schemes. They have to be designed just right, otherwise their incentives may be ineffective. And they are prone to manipulation. Linear schemes may not give extra incentives at just the right points, but they are much more robust to changing circumstances and misuse.

In practice, combinations of linear and nonlinear schemes are often used. For example, salespeople usually get a percentage commission in addition to a bonus for achieving a given quota. And there may be larger bonuses for achieving further thresholds, such as 150 percent or 200 percent of the base quota. Such mixtures can achieve some of the useful purposes of the quota without risking the larger drawbacks.

Carrots versus Sticks

An incentive payment scheme has two key aspects: the average payment to the worker, which must be enough to fulfill the participation constraint, and the spread of payments in good versus bad outcomes, which is what provides the incentive to exert more or better effort. The bigger the spread, the more powerful the incentive.

Even holding the spread constant, the incentive scheme can be designed as either a carrot or a stick. Imagine that the spread is 50 (and the average payment is 100). Under a carrot reward scheme, the employee would get, for example, 99 most of the time and 149 in the event of exceptional performance. The required output for the exceptional performance reward is set so high that the chance of hitting this target is only 2%, assuming the worker puts forth the desired amount of effort. Conversely, under a stick reward scheme, the employee would get 101 almost all the time, but a punishment of 51 in the event of exceptionally poor performance. Here, the bar for poor performance is set so low that the chance of hitting it is only 2%, assuming the desired level of effort is provided. Although the schemes feel quite different, both the spread and the average payments are the same.

The average is determined by the participation constraint, and therefore in turn by how much the worker could have earned in opportunities other than this employment. The employer wants to keep the payment to the worker low so as to increase his own profit. He may deliberately seek agents with poor alternatives, but workers who come for such low average payments may have low skills; an adverse selection problem may raise its ugly head.

In some cases the employer may have strategies that deliberately reduce the worker's alternatives. That is exactly what Stalin attempted. The Soviet state did not pay its workers very well even when they performed well, but if they didn't, they went to Siberia. They could not escape from the country; they had no outside opportunities.

This could have been a good incentive scheme, in the sense that it provided powerful incentives and was cheap to operate. But it failed because the punishment scheme wasn't closely tied to effort. People found that they might be denounced and punished almost as easily if they worked hard or shirked, so they did not have the incentive to work hard after all. Luckily, private employers and even the state in modern democracies lack these arbitrary powers to reduce the workers' alternative opportunities.

But think of the compensation schemes of CEOs in this light. It seems that CEOs gain huge sums as incentive rewards if their companies do well on their watch, but they get only slightly less huge sums if the companies just do okay, and a "golden parachute" if the company actually fails on their watch. The average of these large sums, calculated using the probabilities of the possible outcomes, must be greatly above what is truly needed to induce these people to take the jobs. In the theory's jargon, their participation constraints appear to be grossly overfulfilled.

The reason for this is competition for the CEO candidates. Compared to the alternative of driving a cab or retiring early to play golf, the pay is well more than what is needed to keep the person working. But if another company is willing to guarantee $10 million no matter what happens, then the participation constraint with your company is not compared with driving a taxi or playing golf but with taking a $10 million CEO package elsewhere. In Europe, where CEO pay is generally much lower, companies are still able to hire and motivate CEOs. The lower pay still beats playing golf, and because many candidates are unwilling to move their families to the United States, the relevant participation constraint is with the other European companies.

THE MANY DIMENSIONS OF INCENTIVE PAYMENT SCHEMES

We have so far emphasized a single task, such as reading the proofs of a book, or selling a product. In reality, each context where incentive schemes are used has multiple dimensions—many tasks, many workers—even many workers engaged simultaneously in the same or similar tasks—and many years before the outcome is fully known. Incentive schemes must take into account the interactions among all these dimensions. This makes for some difficult analysis, but some simple ideas emerge. Let us look at some of them.

Career Concerns

When the job is expected to last for several years, the worker may be motivated in the early years not by immediate cash (or stock) payments, but by the prospect of future salary increases and promotions—that is, by incentives that extend over the full career. Such considerations are stronger for people who have longer futures with the company ahead of them. They are less useful for workers near retirement, and not very useful for youngsters who are just starting out in the labor market and expect to move and change jobs a few times before settling on a career. Promotion incentives are most useful for younger employees at lower and middle levels. To use an example from our experience, assistant professors are motivated to do research by the prospects of tenure and promotion much more than they are by immediate salary increases within the rank.

In the example of the student reading the proofs of the professor's book, a longer-term interaction may exist because the professor is supervising the student's research, or the student will need letters of recommendation from the professor in the future for jobs where similar skills are relevant. These career concerns make immediate cash incentive less important. The student will do a good job for the sake of the implicit future rewards—closer attention to his research and better letters of recommendation. These aspects don't even have to be spelled out explicitly by the professor; everyone in the environment understands the larger game being played.

Repeated Relationships

Another aspect of ongoing employment is that the same worker performs similar actions repeatedly. Each time, there is an element of chance that makes the outcome an inaccurate indicator of the effort, so incentives cannot be perfect. But if luck at different times is independent, then by the law of large numbers the average output is a more accurate measure of average effort. This makes more powerful incentives possible. The idea is that the employer may believe the employee's hard luck story once; the claim of persistent bad luck is less plausible.

Efficiency Wages

You are considering hiring someone for a job in your firm. This job requires careful effort, and good work would be worth $60,000 a year to you. The worker would rather take things easy; he suffers some mental or even physical cost in exerting effort. The worker's subjective valuation of this cost is $8,000 per year.

You need to pay enough to entice the worker to your company and pay in a way that induces careful effort. There are dead-end jobs requiring no special effort that pay $40,000. You need to beat that.

In terms of motivating effort, you cannot observe directly whether the worker is exerting careful effort or not. If he does not, there is some chance that things will go wrong in a way you will notice. Suppose the probability of this is 25 percent. What incentive payment will induce the worker to put in the requisite effort?

You can offer a contract of the following kind: "I will pay you some amount above your other opportunities so long as no shirking on your part comes to light. But if that ever happens, I will fire you, and spread the word of your misbehavior among all the other employers, with the result that you will never earn anything more than the basic $40,000 again."

How high does the salary have to be so that the risk of losing it will deter the worker from cheating? Clearly, you will have to pay more than $48,000. Otherwise, the worker will only take your job if he plans to shirk. The question is how much more. We will call the extra amount X, for a total salary of $48,000 + X. That means the worker is X better off at your job compared to the alternative.

Suppose the worker does cheat one year. In that year he will not have to incur the subjective cost of effort, so he will have gained the equivalent of $8,000. But he will run a 25 percent risk of being found out and losing $X this year and every year after that. Is the one-time gain of $8,000 worth the prospective loss of 0.25X every year thereafter?

That depends on how money at different times is compared—that is, on the interest rate. Suppose the interest rate is 10%. Then getting an extra X annually is like owning a bond with a face value of $10X (which at 10% pays X annually). The immediate gain of the equivalent of $8,000 should be compared with the 25 percent chance of losing $10X. If $8,000 < 0.25 × 10X, the worker will calculate that he should not shirk. This means $X > $8,000/2.5 = $3,200.

If you offer the worker an annual salary of $48,000 + $3,200 = $51,200 so long as no shirking comes to light, he will not in fact shirk. It isn't worth risking the extra $3,200 forever in order to slack off and get a quick $8,000 this year. And since good effort is worth $60,000 per year to you, it is in your interest to offer this higher wage.

The purpose of the wage is to get the worker to put in the requisite effort and work more efficiently, and so it is called an *efficiency wage.* The excess above the basic wage elsewhere, which is $11,200 in our example, is called the *efficiency premium.*

The principle behind the efficiency wage enters your daily life at many points. If you go to the same car mechanic regularly, you would do well to pay a little more than the lowest rate for the job. The prospect of the steady extra profit will deter the mechanic from cheating you.* You are paying him a premium, not for efficiency in this case but for honesty.

Multiple Tasks

Employees usually perform multiple tasks. To take an example close to home, professors teach and carry out research. In such cases, the incentives for the different tasks can interact. The overall effect depends on whether the tasks are substitutes (in the sense that when the worker devotes more effort to one task, the net productivity of effort on the other task suffers) or complements (in the sense that more effort to one task raises the net productivity of the effort devoted to the other task). Think of a farmhand working in the cornfields and in the dairy. The more he works on corn, the more tired he will get, and each hour he spends in the dairy will be less productive. Conversely, think of a farm worker who looks after beehives and apple orchards. The more effort he puts into beekeeping, the more productive his effort in growing apples.

When tasks are substitutes, giving strong incentives to each kind of effort hurts the outcome of the other. Therefore both incentives have to be kept weaker than you would think if you considered each task in isolation. But when tasks are mutual complements, incentives to induce better effort on each task help the outcome of the other. Then the owner can make both incentives strong, to take advantage of these synergies without having to worry about any dysfunctional interactions.

---

* To fill out the analogy, imagine that the mechanic can "invent" a problem that brings him $1,000 of extra profits, which at a 10% interest rate is like an extra $100 per year. However, there is a 25 percent chance that you will catch him, in which case you will never return to this garage. If your future business will bring in more than $400 of profit every year, then he would rather play it straight than risk losing your future business and the profits that go with it.

This has ramifications for organization design. Suppose you want to get many tasks performed. You should try, to the extent possible, to assign them to the employees in such a way that each performs a set of tasks that complement one another. In the same way, a big enterprise should be structured into divisions, each of which is responsible for a subset of mutually complementary tasks, and substitute tasks should be assigned to different divisions. That way you can take advantage of strong incentives for each employee, and within each division.

The consequences of failing to follow this rule can be seen by everyone who has ever flown to or through London's Heathrow Airport. The function of an airport is to receive departing passengers at the curbside and deliver them to their airplanes, and receive arriving passengers from their airplanes and deliver them to their ground transport. All the activities that occur in each of these processes—check-in, security, shopping, and so on—are complements. Conversely, multiple airports serving a city are mutual substitutes (albeit not perfect ones—they differ in their locations relative to the city, the ground transport facilities to which they are connected, and so on). The principle of grouping together complementary activities and separating substitute activities says that all the functions within one airport should be under the control of one management, and then different airports should compete with each other for the airlines' and the passengers' business.

The UK government has done exactly the opposite. All three airports serving London—Heathrow, Gatwick, and Stanstead—are owned and managed by one company, the British Airports Authority (BAA). But within each airport, different functions are controlled by different bodies—the BAA owns and leases the shopping areas, the police are in charge of security but the BAA provides the physical setup of the security checks, a regulatory agency sets landing fees, and so on. No wonder incentives are dysfunctional. The BAA profits from the leases on shops and so provides too little space for security checks, the regulator sets landing fees too low with the aim of benefiting consumers, but this causes too many airlines to choose Heathrow, which is nearer central

London, and so on. Both authors have suffered from this, along with millions of other "users" of these airports.

Turning to an application even closer to the authors' experience, are teaching and research substitutes or complements? If they are substitutes, the two should be performed in separate institutions, as is done in France, where universities do mostly teaching and the research is done in specialized institutes. If they are complements, the optimal arrangement is to combine research and teaching within one institution, as is the case in major U.S. universities. The comparative success of these two organizational forms is evidence in favor of the complements case.

Are the multiple tasks performed within the new Department of Homeland Security substitutes or complements? In other words, is that department a good way to organize these activities? We do not know the answer, but it is surely a question deserving careful attention from the nation's top policymakers.

Competition between Workers

In many firms and other organizations, there are many people simultaneously performing similar or even identical tasks. Different shifts work the same assembly line, and investment fund managers deal with the same overall market conditions. The outcome of each task is a mixture of the person's effort, skill, and an element of chance. Because the tasks are similar and performed under similar conditions at the same time, the chance part is highly correlated across the workers: if one has good luck, it is likely that all the others have good luck, too. Then comparisons of the outcomes produced by different workers yield a good indication of the relative efforts (and skill) exerted by them. In other words, the employer says to a worker who pleads bad luck to explain his poor outcome: "Then how come the others performed so much better?" In this situation, incentive schemes can be designed based on relative performance. Investment fund managers are ranked based on how well they do relative to their peers. In other cases, incentives are provided by a competition, with prizes for the top performers.

Consider the professor who employs a student to correct the

proofs of his book. He can hire two students (who do not know each other) and split the work between them, but with some overlap of pages to be corrected by both of them. A student who finds only a few errors in the overlap segment will be proved to have been slacking if the other finds many more. So the payment can be based on the "relative performance" on the overlap, to sharpen the incentives. Of course the professor should not tell each student who the other is (else they can collude), nor should he tell either what pages overlap between them (else they will be careful with that part and careless with the rest).

Indeed, the inefficiency caused by the overlap can be more than offset by the improved incentives. This is one of the advantages of dual sourcing. Each supplier helps create the baseline by which to judge the other.

For this book, Barry distributed copies to students enrolled in Yale's undergraduate game theory class. The reward was $2/typo, but you had to be the first to find it. This obviously led to massive duplication of effort, but in this case, the students were reading the book as part of the course. While many students did well, the big money winner was Barry's assistant, Catherine Pichotta. The reason she did best wasn't just that she found the greatest number of typos. The reason was that, unlike the Yale students, she thought ahead and started at the back of the book.

Motivated Workers

We have assumed that workers don't care about doing the job well for its own sake or about the employer's success other than through how it directly affects their pay and careers. Many organizations attract workers who care about the job itself and about the success of the organization. This is especially true in nonprofits, health care, education, and some public sector agencies. It is also true of tasks that require innovation or creativity. More generally, people are intrinsically motivated when performing activities that improve their self-image and give them a sense of autonomy.

Going back to the example of the student reading proofs, a student willing to do an academic-related job on campus for relatively low payment instead of more lucrative outside jobs, for example as a software consultant for local businesses, may be more genuinely interested in the subject of the book. Such a student has intrinsic motivation to do a good job of proofreading. And such a student is also more likely to want to become an academic, and therefore will be more aware of, and more strongly motivated by, the "career concerns" mentioned above.

Intrinsically rewarding tasks and do-good organizations need fewer or weaker material incentives. In fact, psychologists have found that the "extrinsic" monetary incentives can diminish the "intrinsic" incentives of workers in such settings. They come to feel that they are doing it just for the money, rather than for any warm glow that comes from helping others or from the achievement itself. And the existence of material penalties such as lower pay or dismissal for failure may undermine the enjoyment of doing a challenging or worthwhile task.

Uri Gneezy and Aldo Rustichini conducted an experiment where subjects were given fifty questions from an IQ test.[2] One group was asked to do the best they could. Another was given 3¢ per correct answer. A third group was rewarded with 30¢ per correct answer and a fourth was paid 90¢ per correct answer. As you might have predicted, the two groups being paid 30¢ and 90¢ both outperformed the ones with no bonus—on average, they got 34 questions right compared to 28. The surprise was that the group with only 3¢ payment did the worst of all, getting only 23 right on average. Once money enters the picture, it becomes the main motivation, and 3¢ just wasn't enough. It may also have conveyed that the task wasn't that important. Thus Gneezy and Rustichini conclude that you should offer significant financial rewards or none at all. Paying just a little might lead to the worst of all outcomes.

Hierarchical Organizations

Most organizations of any size have multiple tiers—companies have a hierarchy of shareholders, board of directors, top management, middle management, supervisors, and line workers. Each is the boss of those lower down in the hierarchy, and in charge of providing them with the appropriate incentives. In these situations, the boss of each level must be aware of the danger of strategic behavior among those lower down the hierarchy. For example, suppose the incentive scheme for a worker depends on the quality of the work as certified by the immediate supervisor. Then the supervisor passes shoddy work in order to meet a target to get his own bonus. The supervisor can't punish the worker without also hurting himself in the process. When the upper-level boss designs incentive schemes to mitigate such practices, the effect is generally to weaken incentives at those levels, because that reduces the potential benefit from deception and fraud.

Multiple Owners

In some organizations the control structure is not a pyramid. In places the pyramid gets inverted: one worker is responsible to several bosses. This happens even in private companies but is much more common in the public sector. Most public sector agencies have to answer to the executive, the legislature, the courts, the media, various lobbies, and so on.

The interests of these multiple owners are often imperfectly aligned or even totally opposed. Then each owner can try to undermine the incentive schemes of the others by placing offsetting features in his own scheme. For example, a regulatory agency may be in the executive branch while the Congress controls its annual budget; the Congress can threaten budget cuts when the agency responds more to the wishes of the executive. When the different bosses offset each other's incentives in this way, the effect is weakness of incentives in the aggregate.

Imagine that one parent gives a reward for good grades and the other a reward for success on the athletic field. Instead of working synergistically, each reward is likely to offset the other. The reason is that as the kid spends more time studying, this will take some time away from athletics and thus reduce the chance of getting the sports award. The expected gain from an extra hour hitting the books won't be, say, $1, but $1 minus the likely reduction in the sports prize. The two rewards might not totally offset each other, as the kid could spend more time studying and practicing with less time for sleeping and eating.

In fact, mathematical models show that the overall strength of incentives in such situations is inversely proportional to the number of different bosses. That may explain why it is so difficult to get anything done in international bodies like the United Nations and the World Trade Organization—all sovereign nations are separate bosses.

In the extreme situation where the owners' interests are totally opposed, aggregate incentives may be totally without power. That is like the admonition in the Bible: "No man can serve two masters . . . God and mammon."[3] The idea is that interests of God and mammon are totally opposed; when the two are joint bosses, the incentives provided by one exactly cancel those provided by the other.

HOW TO REWARD WORK EFFORT

We have illustrated the key elements of a well-designed incentive scheme. Now we want to flesh out some of these principles through more developed examples.

Imagine you are the owner of a high-tech company trying to develop and market a new computer chess game, Wizard 1.0. If you succeed, you will make a profit of $200,000 from the sales. If you fail, you make nothing. Success or failure hinges on what your expert player-programmer does. She can either put her heart and soul into the work or just give it a routine shot. With high-quality effort, the chances of success are 80 percent, but for routine effort, the figure drops to 60 percent. Chess programmers can be hired for $50,000, but they like to daydream and will give only their routine effort for this sum. For high-quality effort, you have to pay $70,000. What should you do?

As shown in the following table, a routine effort will get you $200,000 with a 60 percent chance, which comes out to $120,000 on average. Subtracting the $50,000 salary leaves an average profit of $70,000. The corresponding calculation if you hire a high-effort expert is 80 percent of $200,000 minus $70,000—that is, $90,000. Clearly you do better to hire a high-effort expert at the higher salary.

	Chance of success	Average revenue	Salary payments	Average profit
Low-quality effort	60%	$120,000	$50,000	$70,000
High-quality effort	80%	$160,000	$70,000	$90,000

But there is a problem. You can't tell by looking at the expert's work whether she is exerting routine effort or quality effort. The creative process is mysterious. The drawings on your programmer's pad may be the key to a great graphics display that will ensure the success of Wizard 1.0 or just doodles of pawns and bishops to accompany her daydreaming. Knowing that you can't tell the difference between routine effort and quality effort, what is to prevent the expert from accepting the salary of $70,000 appropriate for high effort but exerting routine effort just the same? Even if the project fails, that can always be blamed on chance. After all, even with genuine quality effort, the project can fail 20 percent of the time.

When you can't observe the quality of effort, we know that you have to base your reward scheme on something you can observe. In the present instance the only thing that can be observed is the ultimate outcome, namely success or failure of the programming effort. This does have a link to effort, albeit an imperfect one—higher quality of effort means a greater chance of success. This link can be exploited to generate an incentive scheme.

What you do is offer the expert a remuneration that depends on the outcome: a larger sum upon success and a smaller sum in the event of failure. The difference, or the bonus for success, should be just enough to make it in the employee's own interest to provide high-quality effort. In this case, the bonus must be big enough so that the expert expects a high effort will raise her earnings by $20,000, from $50,000 to $70,000. Hence the bonus for success has to be at least $100,000: a 20 percent increase (from 60 to 80 percent) in the chance of getting a $100,000 bonus provides the necessary $20,000 expected payment for motivating high-quality effort.

We now know the bonus, but we don't know the base rate, the amount paid in the event of a failure. That needs a little calculation. Since even low effort has a 60 percent chance of success, the $100,000 bonus provides an expected $60,000 payment for low effort. This is $10,000 more than the market requires.

Thus the base pay is –$10,000. You should pay the employee $90,000 for success, and she should pay you a fine of $10,000 in the event of failure. Thus, with this incentive scheme, the programmer's incremental reward for success is $100,000, the minimum necessary for inducing quality effort. The average payment to her is $70,000 (an 80 percent chance of $90,000 and a 20 percent chance of –$10,000).

This pay scheme leaves you, the owner, an average profit of $90,000 (an 80 percent chance of $200,000, minus the average salary of $70,000). Another way of saying this is that on average your revenue is $160,000 and your average cost is what the worker expects to earn, namely $70,000. This is exactly what you could have gotten if you could observe quality of effort by direct supervision. The incentive scheme has done a perfect job; the unobservability of effort hasn't made any difference.

In essence, this incentive scheme sells 50 percent of the firm to the programmer in exchange for $10,000 and her effort.* Her net payments are then either $90,000 or –$10,000, and with so much riding on the outcome of the project it becomes in her interest to supply high-quality effort in order to increase the chance of success (and her profit share of $100,000). The only difference between this contract and the fine/bonus scheme is in the name. While the name may matter, we see there is more than one way to achieve the same effect.

But these solutions may not be possible, either because assessing a fine on an employee may not be legal or because the worker does not have sufficient capital to pay the $10,000 for her 50 percent

---

* Recall that a successful project is worth $200,000. Since the employee is paid a bonus of $100,000 for success, it is just as if the employee owns half the business.

stake. What do you do then? The answer is to go as close to the fine solution or equity-sharing as you can. Since the minimum effective bonus is $100,000, the worker gets $100,000 in the event of success and nothing upon failure. Now the employee's average receipt is $80,000, and your profit falls to $80,000 (since your average revenue remains $160,000). With equity-sharing, the worker has only her labor and no capital to invest in the project. But she still has to be given a 50 percent share to motivate her to supply high-quality effort. So the best you can do is sell her 50 percent of the company for her labor alone. The inability to enforce fines or get workers to invest their own capital means that the outcome is less good from your point of view—in this case, by $10,000. Now the unobservability of effort makes a difference.

Another difficulty with the bonus scheme or equity sharing is the problem of risk. The worker's incentives arise from her taking a $100,000 gamble. But this rather large risk may lead the employee to value her compensation at less than its average of $70,000. In this case, the worker has to be compensated both for supplying high-quality effort and for bearing risk. The bigger the risk, the bigger the compensation. This extra compensation is another cost of a firm's inability to monitor its workers' efforts. Often the best solution is a compromise; risk is reduced by giving the worker less than ideal incentives and consequently, this motivates less than an ideal amount of effort.

In other instances you may have other indicators of the quality of effort, and you can and should use them when designing your incentive scheme. Perhaps the most interesting and common situation is one in which there are several such projects. Even though success is only an inexact statistical indicator of the quality of effort, it can be made more precise if there are more observations. There are two ways in which this can be done. If the same expert works for you on many projects, then you can keep a record of her string of successes and failures. You can be more confident in attributing repeated failure to poor effort rather than to chance. The greater accuracy of your inference allows you to design a better incentive scheme. The second possibility is that you have several experts working on related projects, and there is some correlation in the success or failure of the projects. If one expert fails while others around her are succeeding, then you can be more confident that she is a shirker and not just unlucky. Therefore rewards based on relative performance—in other words, prizes—will generate suitable incentives.

CASE STUDY: TREAT THEM LIKE ROYALTY

The typical way that authors get paid for writing a book is via a royalty arrangement. For every book sold, the author gets a certain percentage, something like 15% of the list price on hardcover sales and 10% for paperback. The author might also get an advance against future royalties. This advance is usually paid in parts; one part upon signing of the contract, another upon delivery (and acceptance) of the manuscript, and the rest upon publication. How does this payment system create the right incentives, and where might it create a wedge between the interests of the publishing house and those of the author? Is there a better way to pay authors?

Case Discussion

The only good author is a dead author.   *—Patrick O'Connor*

An editor is one who separates the wheat from the chaff and prints the chaff.   *—Adlai Stevenson*

As these quotes suggest, there are many possible sources of tension in the relationship between authors and publishers. The contract helps resolve some of the problems and creates others. Holding back some of the advance gives the author an incentive to complete the book on time. The advance also transfers risk from the author to the publisher, who might be in a better position to spread the risk over a large number of projects. The size of the advance is also a credible signal that the publisher is truly excited about the prospects for the book. Any publisher can say that they love the book proposal, but actually offering a large advance will

TRIP TO THE GYM NO. 10

How big is the wedge between publishers and authors? Try to estimate how much more the publisher would like to charge compared to the author.

be much more costly if you don't believe that the book will sell a large number of copies.

One place where authors and publishers disagree is over the list price of the book. You might at first think that since authors are getting a percentage of the list price, they would want the price to be high. But what authors are really getting is a percentage of total revenue, say 15% in the case of the hardcover sales. Thus what authors really care about is total revenue. They would like the publishing house to pick a list price that maximizes total revenue.

The publisher, on the other hand, seeks to maximize its profits. Profits are revenue net of cost. What that means is that the publisher always wants to charge a higher price than would maximize revenue. If the publisher were to start at the revenue-maximizing price and go up a little bit, that would keep revenue almost constant but would reduce sales and thus cut costs. In our case, we anticipated this issue in advance and negotiated the list price as part of the contract. You are welcome. And thanks for reading the book.

There are two more cases on incentives in the next chapter: "Bay Bridge" and "But One Life to Lay Down for Your Country."

CHAPTER 14

Case Studies

THE OTHER PERSON'S ENVELOPE IS ALWAYS GREENER

The inevitable truth about gambling is that one person's gain must be another person's loss. Thus it is especially important to evaluate a gamble from the other side's perspective before accepting. If they are willing to gamble, they expect to win, which means they expect you to lose. Someone must be wrong, but who? This case study looks at a bet that seems to profit both sides. That can't be right, but where's the flaw?

There are two envelopes, each containing an amount of money; the amount of money is either $5, $10, $20, $40, $80, or $160, and everybody knows this. Furthermore, we are told that one envelope contains exactly twice as much money as the other. The two envelopes are shuffled, and we give one envelope to Ali and one to Baba. After both the envelopes are opened (but the amounts inside are kept private), Ali and Baba are given the opportunity to switch. If both parties want to switch, we let them.

Suppose Baba opens his envelope and sees $20. He reasons as follows: Ali is equally likely to have $10 or $40. Thus my expected reward if I switch envelopes is $(10 + 40)/2 = $25 > $20. For gambles this small, the risk is unimportant, so it is in my interest to switch. By a similar argument, Ali will want to switch whether she sees $10 (since she figures that he will get either $5 or $20, which has an average of $12.50) or $40 (since she figures to get either $20 or $80, which has an average of $50).

Something is wrong here. Both parties can't be better off by switching envelopes, since the amount of money to go around is not getting any bigger by switching. What is the mistaken reasoning? Should Ali and/or Baba offer to switch?

Case Discussion

A switch should never occur if Ali and Baba are both rational and assume that the other is too. The flaw in the reasoning is the assumption that the other side's willingness to switch envelopes does not reveal any information. We solve the problem by looking deeper into what each side thinks about the other's thought process. First we take Ali's perspective about what Baba thinks. Then we use this from Baba's perspective to imagine what Ali might be thinking about him. Finally, we go back to Ali and consider what she should think about how Baba thinks Ali thinks about Baba. Actually, this all sounds much more complicated than it is. Using the example, the steps are easier to follow.

Suppose that Ali opens her envelope and sees $160. In that case, she knows that she has the greater amount and hence is unwilling to participate in a trade. Since Ali won't trade when she has $160, Baba should refuse to switch envelopes when he has $80, for the only time Ali might trade with him occurs when Ali has $40, in which case Baba prefers to keep his original $80. But if Baba won't switch when he has $80, then Ali shouldn't want to trade envelopes when she has $40, since a trade will result only when Baba has $20. Now we have arrived at the case in hand. If Ali doesn't want to switch envelopes when she has $40, then there is no gain from trade when Baba finds $20 in his envelope; he doesn't want to trade his $20 for $10. The only person who is willing to trade is someone who finds $5 in the envelope, but of course the other side doesn't want to trade with him.

HERE'S MUD IN YOUR EYE

One of our colleagues decided to go to a Jackson Browne concert at Saratoga Springs. He was one of the first to arrive and scouted the area for the best place to sit. It had rained recently and the area in front of the stage was all muddy. Our colleague settled on the front row closest to the stage yet still behind the muddied area. Where did he go wrong?

Case Discussion

No, the mistake wasn't in picking Jackson Browne. His 1972 hit song "Doctor My Eyes" is still a classic. The mistake was in not looking ahead. As the crowd arrived, the lawn filled up until there was nowhere behind him left to sit. At that point, latecomers ventured into the muddied region. Of course nobody wanted to sit down there. So they stood. Our colleague's view was completely blocked and his blanket equally darkened by the masses of muddied feet.

Here's a case where look forward and reason backward would have made all the difference. The trick is to not choose the best place to sit independently of what others are doing. You have to anticipate where the late arrivals are going to go, and based on this prediction, choose what you anticipate will be the best seat. As the Great Gretzky said in another context, you have to skate to where the puck will be, not where it is.

RED I WIN, BLACK YOU LOSE

While we might never get the chance to skipper in an America's Cup race, one of us found himself with a very similar problem. At the end of his academic studies, Barry celebrated at one of Cambridge University's May Balls (the English equivalent of a college prom). Part of the festivities included a casino. Everyone was given £20 worth of chips, and the person who had amassed the greatest fortune by evening's end would win a free ticket to next year's ball. When it came time for the last spin of the roulette wheel, by a happy coincidence, Barry led with £700 worth of chips, and the next closest was a young Englishwoman with £300. The rest of the group had been effectively cleaned out. Just before the last bets were to be placed, the woman offered to split next year's ball ticket, but Barry refused. With his substantial lead, there was little reason to settle for half.

To better understand the next strategic move, we take a brief detour to the rules of roulette. The betting in roulette is based on where a ball will land when the spinning wheel stops. There are typically numbers 0 through 36 on the wheel. When the ball lands on 0, the house wins. The safest bet in roulette is to bet on even or odd (denoted by black or red). These bets pay even money—a one-dollar bet returns two dollars—while the chance of winning is only 18/37. Even betting her entire stake would not lead to victory at these odds; therefore, the woman was forced to take one of the more risky gambles. She bet her entire stake on the chance that the ball would land on a multiple of three. This bet pays two to one (so her £300 bet would return £900 if she won) but has only a 12/37 chance of winning. She placed her bet on the table.

At that point it could not be withdrawn. What should Barry have done?

Case Discussion

Barry should have copied the woman's bet and placed £300 on the chance that the ball would land on a multiple of three. This would have guaranteed that he stayed ahead of her by £400 and won the ticket: either they both would lose the bet and Barry would win £400 to £0, or they both would win the bet and Barry would end up ahead £1,300 to £900. The woman had no other choice. If she did not bet, she would have lost anyway; whatever she bet on, Barry could have followed her and stayed ahead.*

---

* Actually, this is what Barry wished he had done. It was 3:00 in the morning and much too much champagne had been drunk for him to have been thinking this clearly. He bet £200 on the even numbers, figuring that he would end up in second place only in the event that he lost and she won, the odds of which were

Her only hope was that Barry would bet first. If Barry had been first to place £200 on black, what should she have done? She should have bet her £300 on red. Betting her stake on black would do her no good, since she would win only when Barry won (and she would place second with £600, compared with Barry's £900). Winning when Barry lost would be her only chance to take the lead, and that dictated a bet on red. The strategic moral is the opposite to that of our tales of Martin Luther and Charles de Gaulle. In this tale of roulette, the person who moved first was at a disadvantage. The woman, by betting first, allowed Barry to choose a strategy that would guarantee victory. If Barry had bet first, the woman could have chosen a response that offered an even chance of winning. The general point is that in games it is not always an advantage to seize the initiative and move first. This reveals your hand, and the other players can use this to their advantage and your cost. Second movers may be in the stronger strategic position.

THE SHARK REPELLENT THAT BACKFIRED

Corporations have adopted many new and innovative ways, often called shark repellent, to prevent outside investors from taking over their company. Without commenting on the efficiency or even morality of these ploys, we present a new and as yet untested variety of shark repellent and ask you to consider how to overcome it.

The target company is Piper's Pickled Peppers. Although now publicly held, the old family ties remain, as the five-member board of directors is completely controlled by five of the founder's grandchildren. The founder recognized the possibility of conflict between his grandchildren as well as the threat of outsiders. To guard against both family squabbles and outsider attacks, he first required that the board of director elections be staggered. This means that even someone who owns 100 percent of the shares cannot replace the entire board—rather, only the members whose

---

approximately 5:1 in his favor. Of course 5:1 events sometimes happen, and this was one of those cases. She won.

terms are expiring. Each of the five members had a staggered five-year term. An outsider could hope to get at most one seat a year. Taken at face value, it appeared that it would take someone three years to get a majority and control of the company.

The founder was worried that his idea of staggered terms would be subject to change if a hostile party wrested control of the shares. A second provision was therefore added. The procedure for board election could be changed *only* by the board itself. Any board member could make a proposal without the need for a seconder. But there was a major catch. The proposer would be required to vote for his own proposal. The voting would then proceed in clockwise order around the boardroom table. To pass, a proposal needed at least 50 percent of the total board (absences were counted as votes against). Given that there were only five members, that meant at least 3 out of 5. Here's the rub. Any person who made a proposal to change either the membership of the board or the rules governing how membership was determined would be deprived of his position on the board and his stock holdings *if his proposal failed*. The holdings would be distributed evenly among the remaining members of the board. In addition, any board member who voted for a proposal that failed would also lose his seat on the board and his holdings.

For a while this provision proved successful in fending off hostile bidders. But then Sea Shells by the Sea Shore Ltd. bought 51 percent of the shares in a hostile takeover attempt. Sea Shells voted itself one seat on the board at the annual election. But it did not appear that loss of control was imminent, as Sea Shells was one lone voice against four.

At their first board meeting, Sea Shells proposed a radical restructuring of the board membership. This was the first such proposal that the board had ever voted on. Not only did the Sea Shells proposal pass; amazingly, it passed unanimously! As a result, Sea Shells got to replace the entire board immediately. The old directors were given a lead parachute (which is still better than nothing) and then were shown the door.

How did Sea Shells do it? Hint: It was pretty devious. Backward reasoning is the key. First work on a scheme to get the resolution to pass, and then you can worry about unanimity. To ensure that the Sea Shells proposal passes, start at the end and make sure that the final two voters have an incentive to vote for the proposal. This will be enough to pass the resolution, since Sea Shells starts the process with a first yes vote.

Case Discussion

Many proposals do the trick. Here's one of them. Sea Shells's restructuring proposal has the following three cases:

1. If the proposal passes unanimously, then Sea Shells chooses an entirely new board. Each board member replaced is given a small compensation.
2. If the proposal passes 4 to 1, then the person voting against is removed from the board, and no compensation is made.
3. If the proposal passes with a vote of 3 to 2, then Sea Shells transfers the entirety of its 51 percent share of Piper's Pickled Peppers to the other two yes voters in equal proportion. The two no voters are removed from the board with no compensation.

At this point, backward reasoning finishes the story. Imagine that the vote comes down to the wire: the last voter is faced with a 2–2 count. If he votes yes, it passes and he gets 25.5 percent of the company's stock. If it fails, Sea Shells's assets (and the other yes-voter's shares) are distributed evenly among the three remaining members, so he gets $(51 + 12.25)/3 = 21.1$ percent of the company's stock. He'll say yes.

Everyone can thereby use backward reasoning to predict that if it comes down to a 2–2 tie-breaking vote, Sea Shells will win when the final vote is cast. Now look at the fourth voter's dilemma. When it is his turn to vote, the other votes are:

i. 1 yes (Sea Shells)

ii. 2 yes

or

iii. 3 yes.

If there are three yes votes, the proposal has already passed. The fourth voter would prefer to get something over nothing and therefore votes yes. If there are two yes votes, he can predict that the final voter will vote yes even if he votes no. The fourth voter cannot stop the proposal from passing. Hence, again it is better to be on the winning side, so he will vote yes. Finally, if he sees only one yes vote, then he would be willing to bring the vote to a 2–2 tie. He can safely predict that the final voter will vote yes, and the two of them will make out very nicely indeed.

The first two Piper's board members are now in a true pickle. They can predict that even if they both vote no, the last two will go against them and the proposal will pass. Given that they can't stop it from passing, it is better to go along and get something.

This case demonstrates the power of backward reasoning. Of course it helps to be devious too.

TOUGH GUY, TENDER OFFER

When Robert Campeau made his first bid for Federated Stores (and its crown jewel, Bloomingdales), he used the strategy of a *two-tiered* tender offer. A two-tiered bid typically offers a high price for the first shares tendered and a lower price to the shares tendered later. To keep numbers simple, we look at a case in which the pre-takeover price is $100 per share. The first tier of the bid offers a higher price, $105 per share to the first shareholders until half of the total shares are tendered. The next 50 percent of the shares tendered fall into the second tier; the price paid for these shares is only $90 per share. For fairness, shares are not placed in the different tiers based on the order in which they are tendered. Rather, everyone gets a blended price: all the shares tendered are placed on a prorated basis into the two tiers. Those who don't tender find all of their shares end up in the second tier if the bid succeeds.[1]

We can express the average payment for shares by a simple algebraic expression: if fewer than 50 percent tender, everyone gets $105 per share; if an amount $X\% \geq 50\%$ of the company's total stock gets tendered, then the average price paid per share is

$$\$150\left(\frac{50}{X}\right) + \$90\left(\frac{X+50}{X}\right) = \$90 + \$15\left(\frac{50}{X}\right).$$

One thing to notice about the way the two-tiered offer is made is that it is unconditional; even if the raider does not get control, the tendered shares are still purchased at the first-tier price. The second feature to note about the way this two-tiered offer works is that if *everyone* tenders, then the average price per share is only $97.50. This is less than the price before the offer. It's also worse than what they expect should the takeover fail; if the raider is defeated, shareholders expect the price to return to the $100 level. Hence they hope that the offer is defeated or that another raider comes along.

In fact, another raider did come along, namely Macy's. Imagine that Macy's makes a conditional tender offer: it offers $102 per share *provided* it gets a majority of the shares. To whom do you tender, and which (if either) offer do you expect to succeed?

Case Discussion

Tendering to the two-tiered offer is a dominant strategy. To verify this, we consider all the possible cases. There are three possibilities to check.

The two-tiered offer attracts less than 50 percent of the total shares and fails.

The two-tiered offer attracts some amount above 50 percent and succeeds.

The two-tiered offer attracts exactly 50 percent. If you tender, the offer will succeed, and without you it fails.

In the first case, the two-tiered offer fails, so that the post-tender price is either $100 if both offers fail or $102 if the competing offer succeeds. But if you tender you get $105 per share, which is bigger than either alternative. In the second case, if you don't tender you get only $90 per share. Tendering gives you at worst $97.50. So again it is better to tender. In the third case, while other people are worse off if the offer succeeds, you are privately better off. The reason is that since there are exactly 50 percent tendered, you will be getting $105 per share. This is worthwhile. Thus you are willing to push the offer over.

Because tendering is a dominant strategy, we expect everyone to tender. When everyone tenders, the average blended price per share may be below the pre-bid price and even below the expected future price should the offer fail. Hence the two-tiered bid enables a raider to pay less than the company is worth. The fact that shareholders have a dominant strategy does not mean that they end up ahead. The raider uses the low price of the second tier to gain an unfair advantage. Usually the manipulative nature of the second tier is less stark than in our example because the coercion is partially hidden by the takeover premium. If the company is really worth $110 after the takeover, then the raider can still gain an unfair advantage by using a second tier below $110 but above $100. Lawyers view the two-tiered bid as coercive and have successfully used this as an argument to fight the raider in court. In the battle for Bloomingdales, Robert Campeau eventually won, but with a modified offer that did not include any tiered structure.

We also see that a conditional bid is not an effective counter-strategy against an unconditional two-tiered bid. In our example, the bid by Macy's would be much more effective if its offer of $102 per share were made unconditionally. An unconditional bid by Macy's destroys the equilibrium in which the two-tiered bid succeeds. The reason is that if people thought that the two-tiered bid were certain to succeed, they would expect a blended price of $97.50, which is less than they would receive by tendering to Macy's. Hence it cannot be that shareholders expect the two-tiered bid to succeed and still tender to it.*

In late 1989, Campeau's operations unraveled because of excessive debt. Federated Stores filed for reorganization under Chapter 11 of

---

* Unfortunately, it is not an equilibrium for Macy's bid to succeed either, for in that case, the two-tiered bid would attract less than 50 percent of the shares and so the price per share offered would be above the bid by Macy's. Alas, this is one of those cases with no equilibrium. Finding a solution requires the use of randomized strategies, as discussed in chapter 5.

the bankruptcy law. When we say Campeau's strategy was successful, we merely mean that it achieved the aim of winning the takeover battle. Success in running the company was a different game.

THE SAFER DUEL

As pistols become more accurate, does that change the deadliness of a duel?

Case Discussion

At first glance, the answer would seem to be obvious: yes. But recall that the players will adapt their strategies to the new situation. Indeed, the answer is easier to see if we flip the question: suppose we try to make dueling safer by reducing the accuracy of the pistols. The new outcome is that the adversaries will come closer to one another before firing.

Recall our discussion of the duel on page 324. Each player waits to shoot until the point where his probability of hitting the other side is just equal to the other side's chance of missing. Note that the accuracy of the pistols doesn't enter into the equation. All that matters is the ultimate chance of success.

To illustrate this point with some numbers, suppose that the adversaries are equally good shots. Then the optimal strategy is for the two to keep on approaching each other until the moment that the probability of hitting reaches 1/2. At that point one duelist takes a shot. (It doesn't matter which person shoots, as the chance of success is a half for the shooter and a half for the person who is being shot at.) The probability each player will survive is the same (1/2) irrespective of the accuracy of the pistols. A change in the rules need not affect the outcome; all the players will adjust their strategies to offset it.

THE THREE-WAY DUEL

Three antagonists, Larry, Moe, and Curly, are engaged in a three-way duel. There are two rounds. In the first round, each player is given one shot: first Larry, then Moe, and then Curly.

After the first round, any survivors are given a second shot, again beginning with Larry, then Moe, and then Curly. For each duelist, the best outcome is to be the sole survivor. Next best is to be one of two survivors. In third place is the outcome in which no one gets killed. Dead last is that you get killed.

Larry is a poor shot, with only a 30 percent chance of hitting a person at whom he aims. Moe is a much better shot, achieving 80 percent accuracy. Curly is a perfect shot—he never misses. What is Larry's optimal strategy in the first round? Who has the greatest chance of survival in this problem?

Case Discussion

Although backward reasoning is the safe way to solve this problem, we can jump ahead a little by using some forward-looking arguments. We start by examining each of Larry's options in turn. What happens if Larry shoots at Moe? What happens if Larry shoots at Curly?

If Larry shoots at Moe and hits, then he signs his own death warrant. It becomes Curly's turn to shoot, and he never misses. Curly will not pass at the chance to shoot Larry, as this leads to his best outcome. Larry shooting at Moe does not seem to be a very attractive option.

If Larry shoots at Curly and hits, then it is Moe's turn. Moe will shoot at Larry. (Think about how we know this to be true.) Hence, if Larry hits Curly, his chance of survival is less than 20 percent, the chance that Moe misses.

So far, neither of these options looks to be very attractive. In fact, Larry's best strategy is to fire up in the air! In this case, Moe will shoot at Curly, and if he misses, Curly will shoot and kill Moe. Then it becomes the second round and it is Larry's turn to shoot again. Since only one other person remains, he has at least a 30 percent chance of survival, since that is the probability that he kills his one remaining opponent.

The moral here is that small fish may do better by passing on their first chance to become stars. We see this every four years in presidential campaigns. When there is a large number of contenders, the leader of the pack often gets derailed by the cumulative attacks of all the medium-sized fish. It can be advantageous to wait, and step into the limelight only after the others have knocked each other and themselves out of the running.

Your chances of survival depend on not only your own ability but also whom you threaten. A weak player who threatens no one may end up surviving if the stronger players kill each other off. Curly, although he is the most accurate, has the lowest chance of survival—only 14 percent. So much for survival of the fittest! Moe has a 56 percent chance of winning. Larry's best strategy turns his 30 percent accuracy into a 41.2 percent chance of winning.[2]

THE RISK OF WINNING

One of the more unusual features of a Vickrey sealed-bid auction is that the winning bidder does not know how much she will have to pay until the auction is over and she has won. Remember, in a Vickrey auction the winning bidder pays only the second highest bid. In contrast, there is no uncertainty in the more standard sealed-bid auction, in which the winner pays her bid. Since everyone knows her own bid, no one has any doubts as to how much she will have to pay if she wins.

The presence of uncertainty suggests that we might want to consider the effect of risk on the participants' bidding strategies. The typical response to uncertainty is negative: the bidders are worse off in a Vickrey auction because they do not know how much they will have to pay if they have submitted the winning bid. Is it reasonable that a bidder will respond to this uncertainty or risk by lowering her bid below the true valuation?

Case Discussion

It is true that the bidders dislike the uncertainty associated with how much they might have to pay if they win. Each is in fact worse off. Yet, in spite of the risk, participants should still bid their true valuations. The reason is that a truthful bid is a dominant strategy. As long as the selling price is below the valuation, the bidder wants to buy the good. The only way to ensure that you win whenever the price is below your value is to bid the true value.

In a Vickrey auction, bidding the true valuation doesn't make you pay more—except when someone else would have outbid you, in which case you would have wanted to raise your bid until the selling price exceeded your valuation. The risk associated with a Vickrey auction is limited; the winner is never forced to pay an amount greater than her bid. While there is uncertainty about what the winner will pay, this uncertainty is only over the degree of good news. Even though the good news might be variable, the best strategy is to win the auction whenever it's profitable. That means bidding your true value. You never miss a profitable opportunity, and whenever you win you pay less than your true value.

BUT ONE LIFE TO LAY DOWN FOR YOUR COUNTRY

How can the commanders of an army motivate its soldiers to risk their lives for their country? Most armies would be finished if each soldier on the battlefield started to make a rational calculation of the costs and the benefits of risking his own life. What are the various devices that can motivate and incentivize soldiers to risk their lives?

Case Discussion

First look at some devices that transform the soldiers' self-regarding rationality. The process begins in boot camp. Basic training in the armed forces everywhere is a traumatic experience. The new recruit is maltreated, humiliated, and put under such immense physical and mental strain that the few weeks quite alter his personality. An important habit acquired in this process is an automatic, unquestioning obedience. There is no reason why socks should be folded, or beds made, in a particular way, except that the officer has so ordered. The idea is that the same obedience will occur when the order is of greater importance. Trained not to question orders, the soldier becomes a fighting machine; commitment is automatic.

Many armies got their soldiers drunk before battle. This may have reduced their fighting efficiency, but it also reduced their capacity for rational calculation of self-preservation.

The seeming irrationality of each soldier turns into strategic rationality. Shakespeare knew this perfectly well; in *Henry V,* the night before the battle of Agincourt (fought on St. Crispin's day, October 25, 1415), King Henry prays (emphasis added):

> O God of battles! steel my soldiers' hearts;
> Possess them not with fear; *take from them now*
> *The sense of reckoning,* if th'opposed numbers
> Pluck their hearts from them

Just before the battle, Henry does something that may at first seem to defeat his purpose. Instead of enforcing any compulsion to fight, he declares:

> . . . he which hath no stomach to this fight,
> Let him depart; his passport shall be made,
> And crowns for convoy put into his purse:
> We would not die in that man's company
> That fears his fellowship to die with us.

The catch is that anyone who wants to take up this offer has to do so in full view of all of his companions. Of course everyone is too ashamed to do so. And the action (actually, inaction) of publicly declining the offer changes soldiers' preferences, even personalities, irrevocably. By their act of rejecting the offer, the soldiers have psychologically burned their ships home. They have established an implicit contract with each other not to flinch from death if the time comes.*

---

* Others have used the same ploy. Roald Amundsen started his journey of exploration to the South Pole using a trick; those who signed up did so in the belief that they were going on a long but much less risky voyage to the Arctic. He revealed his true objective only at the last possible point of return, and offered a passage-paid return to Norway to anyone who did not want to continue. No one took him up on this, even though later there was much muttering: "Why did you say yes? If only you had answered no, I would have done the same" (Roland

Next consider incentives to act. These can be material: in the old days, victorious soldiers had the opportunity to loot from the property and even the bodies of the enemy. Generous death benefits can be promised for next-of-kin if the worst happens. But the incentives to fight and risk lives are mostly nonmaterial: medals, honor, and glory come to the brave whether they live or die in battle; the lucky survivors can boast of their exploits for years to come. Here is Shakespeare's King Henry V again:

> He that shall live this day, and see old age,
> Will yearly on the vigil feast his neighbours,
> . . . he'll remember with advantages
> What feats he did that day . . .
> And Crispin Crispian shall ne'er go by,
> From this day to the ending of the world,
> But we in it shall be remember'd;
> We few, we happy few, we band of brothers;
> For he to-day that sheds his blood with me
> Shall be my brother; . . .
> And gentlemen in England now a-bed
> Shall think themselves accursed they were not here,
> And hold their manhoods cheap whiles any speaks
> That fought with us upon Saint Crispin's day.

Being the king's brother; others holding their manhoods cheap when you speak: what powerful incentives! But think a moment. What does it really mean to be the king's brother? Suppose you live and return to England with the victorious army. Is the king going to say: "Ah, my brother! Come and live with me at the palace." No. You will return to the same old life of poverty that you had before. In concrete terms, the incentive is empty. It is like the "cheap talk" we mentioned in connection with credibility (see footnote, p. 203). But it works. The science of game theory cannot fully explain why. Henry's speech is the art of strategy at its best.

Huntford, *The Last Place on Earth* [New York: Modern Library, 1999], 289). Like Henry V, Amundsen was victorious and became the first man to stand on the geographic South Pole.

There is a related subtext. The night before the battle, Henry goes wandering in disguise among his troops to find out what they are really thinking and feeling. He discovers one disconcerting fact: they are afraid of being killed or captured, and they believe that he does not face the same risk. Even if the enemy gets to him, they will not kill him. It will be more profitable to hold him for ransom and this will then be paid. Henry must dispel this fear if he is to command the soldiers' loyalty and solidarity. It would not do in his speech the following morning to say: "Hey, guys; I hear some of you think that I am not risking my life with you. Let me assure you most earnestly that I am." That would be worse than useless; it would have the effect of reinforcing the soldier's worst suspicions, rather like Richard Nixon's declaration "I am not a crook" during the Watergate crisis. No; in his speeches Henry simply takes it for granted that he is risking his life and turns the question around: "Are you risking your life *with me*?" That is how we should interpret the phrases "we would not die in that man's company" and "he that sheds his blood with me." Once again, it is a beautiful example of the art of strategy.

Of course this is not actual history but Shakespeare's fictionalization of it. However, we think that artists often have more perceptive insights about human emotions, reasoning, and motivation than do psychologists, let alone economists. Therefore we should be willing to learn lessons on the art of strategy from them.

WINNING WITHOUT KNOWING HOW

Chapter 2 introduced games in which players move in sequence and which always end after a finite number of moves. In theory, we could examine every possible sequence of moves and thereby discover the best strategy. This is relatively easy for tic-tac-toe and impossible (at present) for chess. In the game below, the best strategy is unknown. Yet, even without knowing what it is, the very fact that it exists is enough to show that it must lead to a win for the first player.

ZECK is a dot game for two players. The object is to force your opponent to take the last dot. The game starts with dots arranged in any rectangular shape, for example 7 × 4:

. . . .

. . . .

. . . .

. . . .

Each turn, a player removes a dot and with it *all* remaining dots to the northeast. If the first player chooses the fourth dot in the second row, this leaves his opponent with

. . .

. . .

. . . .

. . . .

Each period, at least one dot must be removed. The person who is forced to take the last dot loses.

For any shaped rectangle with more than one dot, the first player must have a winning strategy. Yet this strategy is not currently known. Of course we can look at all the possibilities and then figure it out for any particular game, such as the 7 × 4 above—but we don't know the best strategy for all possible configurations of dots. How can we show who has the winning strategy without knowing what it is?

Case Discussion

If the second player has a winning strategy, that means that for *any* opening move of the first player, the second has a response that puts him in a winning position. In particular, this means that the second player must have a winning response even if the first player just takes the upper-right-hand dot.

But no matter how the second player responds, the board will be left in a configuration that the first player could have created in his first move. If this is truly a winning position, the first player should have and could have opened the game this way. There is nothing the second player can do to the first that the first player can't do unto him beforehand.

A BURQA FOR PRICES

Hertz and Avis advertise that you can rent a car for $19.95/day. But that car rental price typically leaves out the inflated cost of filling up the tank at the return, often twice the price at the pump. Ads for hotel room rates don't mention the $2/minute charge for long-distance calls. When choosing between HP and Lexmark printers, who has the cheaper cost per page? It is hard to tell when the toner cartridges don't let you know how many pages you'll get. Cell phone companies offer plans with a fixed number of minutes per month. Minutes you don't use are lost, and if you go over, there is a steep charge.* The ad promising 800 minutes for $40/month will almost always cost more than 5¢/minute. As a result, it becomes difficult, if not impossible, to understand or compare the real cost. Why does this practice persist?

Case Discussion

Consider what would happen if one car rental company decided to advertise its all-in price. This maverick would have to set a higher daily rental price in order to make up for the lost revenue from overcharging for gas. (That would still be a good idea: wouldn't you rather pay an extra $2/day and then not have to worry about finding a place to fill up as you dash back to the airport? This might save you from missing the flight or even save your marriage.) The problem is that the company who plays it straight puts itself at a disadvantage compared to its rivals. The one honest firm would seem to be charging the *highest* price when customers do a comparison on Expedia. There isn't an asterisk that says, "We don't rip you off on gas like everyone else does."

The problem is that we are stuck in a bad equilibrium, much like the one involving the QWERTY keyboard. Customers assume that the prices will include lots of hidden extras. Unless a firm can cut through the clutter and convince customers that they aren't playing the same game, the honest firm will just seem to be too expensive. Worse still, since customers don't know the true cost at the rival firms, they don't know how much they should pay. Imagine that a cell phone company offered a single flat price per minute. Does 8¢/minute beat $40 for 800 minutes (with a 35¢ per minute surcharge for going over)? Who knows?

The bottom line is companies go on advertising just one component of the total price. The parts they don't mention are then priced at exorbitant levels. But that doesn't mean that firms end up making more money. Because each company can anticipate making high profits on the back end, they are willing to go to extraordinary

---

* AT&T (Cingular) is the exception to this practice.

lengths to attract or steal customers. Thus laser printers are practically given away, as are most cell phones. The firms compete away all of their future profits in the battle to attract customers. The end result is too much switching and the loss of customer loyalty.

If society wants to improve matters for consumers, one way would be to legislate a change in the convention: require that hotels, car rental companies, and cell phone providers advertise the all-in price paid by the average customer. Comparison shopping sites now do this for books sold online, where the all-in price comparison includes the cost of shipping and handling.[3]

KING SOLOMON'S DILEMMA REDUX

King Solomon wanted to find a way to obtain some information: who was the real mother? The two women who possessed the information had conflicting incentives about revealing it. Mere words would not suffice; strategic players would willingly manipulate answers in their own interests. What is needed is some way to make the players put their money, or, more generally, something they value, where their mouths are. How could a game theory king have persuaded the two women to tell the truth?

Case Discussion

Of several devices that work even when both women play strategically, here is the simplest.[4] Call the two women Anna and Bess. Solomon sets up the following game:

Move 1: Solomon decides on a fine or punishment.

Move 2: Anna is asked to either give up her claim, in which case Bess gets the child and the game ends, or to assert her claim, in which case we go on to . . .

Move 3: Bess can either accept Anna's claim, in which case Anna gets the child and the game ends, or challenge Anna's claim. In the latter case, Bess must put in a bid B of her own choosing for the child, and Anna must pay the fine F to Solomon. We go on to . . .

Move 4: Anna can either match Bess's bid, in which case Anna gets the child and pays B to Solomon, while Bess pays the fine F to Solomon; or Anna does not match, in which case Bess gets the child and pays her bid B to Solomon.

Here is the game in tree form:

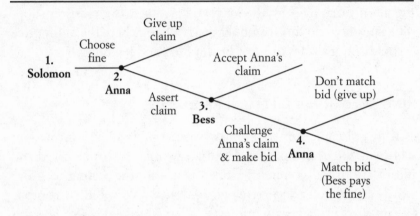

As long as the true mother values the child more than the false claimant, in the subgame perfect equilibrium the true mother gets the child. Solomon does not have to know these values. No fines or bids are actually paid; their sole purpose is to avoid any false claims by either woman.

The reasoning is simple. First suppose Anna is the true mother. Bess knows in move 3 that, unless she bids more than the child is worth to her, Anna will match her bid in move 4, and she (Bess) will end up paying the fine and not getting the child. So Bess will not bid. Knowing this, Anna in move 2 will claim the child and get it. Next suppose Bess is the true mother. Then Anna knows in move 2 that Bess in move 3 will choose a bid that is not worth Anna's while to match in move 4, so she (Anna) is simply going to end up paying the fine F and not getting the child. So in move 2 Anna does best for herself by renouncing her claim.

At this point you are no doubt criticizing us for reducing everything to the sordid world of money. We respond by pointing out that in the actual play that results in the equilibrium of this game, the bids are not actually paid, and neither is the fine. Their only purpose is as a threat; they make it costly for either woman to lie. In this respect, they are similar to the threat of cutting the child in two and, we would argue, a lot less gruesome.

One potential difficulty remains. For the device to work, it must be the case that the true mother is able to bid at least as much as the false claimant. Presumably she loves and values the child at least as much in a subjective sense, but what if she does not have as much money to back up her value? In the original story, the two women came from the same household (actually the book says that they were both prostitutes), so Solomon could reasonably regard their abilities to pay as approximately equal. Even otherwise, the difficulty can be resolved. The bids and fines need not be monetary sums at all. Solomon can specify them in some other "currency" that the two women should be expected to possess in nearly equal amounts, for example having to perform a certain number of days of community service.

BAY BRIDGE

The morning traffic from Oakland to San Francisco across the Bay Bridge gets backed up from 7:30 A.M. to 11:00 A.M. Until the jam clears at 11:00, each additional car that enters the traffic makes all those who come later wait just a little longer. The right way to measure this cost is to sum up the additional waiting times across everyone who is delayed. What is the total waiting-time cost imposed by one additional car that crosses the bridge at 9:00 A.M.?

You may be thinking you don't know enough information. A remarkable feature of this problem is that the externality can be calculated based on the little you've been told. You don't need to know how long it takes the cars to cross the toll plaza, nor the distribution of cars that arrive after 9:00. The answer is the same whether the length of the traffic jam stays constant or varies widely until it clears.

Case Discussion

The trick is to see that all that matters is the sum of the waiting time. We are not concerned with who waits. (In other circumstances, we might want to weigh the waiting times by the monetary value of time for those caught in the jam.) The simplest way to figure out the total extra waiting time is to shuffle around who waits, putting all the burden on one person. Imagine that the extra driver, instead of crossing the bridge at 9:00 A.M., pulls his car over to the side and lets all the other drivers pass. If he passes up his turn in this way, the other drivers are no longer delayed by the extra car. Of course, he has to wait two hours before the traffic clears. But these two hours exactly equal the total waiting time imposed on all the other drivers if he were to cross the bridge rather than wait on the sidelines. The reason is straightforward. The total waiting time is the time it takes for everyone to cross the bridge. Any solution that involves everyone crossing the bridge gives the same total waiting time, just distributed differently. Looking at the solution in which the extra car does all the extra waiting is the easiest way to add up the new total waiting time.

WHAT PRICE A DOLLAR?

Professor Martin Shubik of Yale University designed the following game of entrapment. An auctioneer invites bids for a dollar. Bidding proceeds in steps of five cents. The highest bidder gets the dollar, but *both* the highest *and* the second highest bidders pay their bids to the auctioneer.[5]

Professors have made tidy profits—enough for a lunch or two at the faculty club—from unsuspecting undergraduates playing this game in classroom experiments. Suppose the current highest bid is 60 cents and you are second with 55. The leader stands to make 40 cents, but you stand to lose your 55. By raising to 65, you can put the boot on the other foot. The logic is no different when the leading bid is $3.60 and yours is $3.55. If you do not raise the bidding still further, the "winner" loses $3.60, but you lose $3.55.

How would you play this game?

Case Discussion

This is an example of the slippery slope. Once you start sliding, it is hard to recover. It is better not to take the first step unless you know where you are going.

The game has one equilibrium, in which the first bid is a dollar and there are no further bids. But what happens if the bidding starts at less than a dollar? The escalation has no natural limit other than the amount of money in your wallet: the bidding must stop when you run out of money. That is all we need to apply Rule 1: Look forward and reason backward.

Imagine that Eli and John are the two students in Shubik's auction of a dollar. Each has $2.50 in his wallet, and each knows the other's cash supply.[6] To keep things simple, bidding takes place in dime units.

To start at the end, if Eli ever bids $2.50, he'll win the dollar (and be down $1.50). If he bids $2.40, then John must bid $2.50 in order to win. Since it is not worth spending a dollar to win a dollar, an Eli bid of $2.40 will win if John's current bid is at $1.50 or less.

The same argument works if Eli bids $2.30. John can't bid $2.40 and expect to win, because Eli would counter with $2.50. To beat $2.30, John needs to go all the way up to $2.50. Hence a $2.30 bid beats $1.50 and below. So does a $2.20 bid, a $2.10 bid, all the way down to a $1.60 bid. If Eli bids $1.60, John should predict that Eli won't give up until the bidding reaches $2.50. Eli's $1.60 is already lost, but it is worth his while to spend another 90 cents to capture the dollar.

The first person to bid $1.60 wins, because that establishes a credible commitment to go up to $2.50. In our mind, we should think of $1.60 as the same sort of winning bid as $2.50. In order to beat $1.50, it suffices to bid $1.60, and nothing less will do. That means $1.50 will beat all bids at 60 cents and below. Even a bid of

70 cents will beat all bids at 60 cents and below. Why? Once someone bids 70 cents, it is worthwhile for them to go up to $1.60 and be guaranteed victory. With this commitment, no one with a bid of 60 cents or less finds it worthwhile to challenge.

We expect that either John or Eli will bid 70 cents and the bidding will end. Although the numbers will change, the conclusion does not depend on there being just two bidders. Given that budgets differ, backward reasoning can still find the answer. But it is critical that everyone know everyone else's budget. When budgets are unknown, as one would expect, an equilibrium will exist only in mixed strategies.

Of course there is a much simpler and more profitable solution for the students: collusion. If the bidders agree among themselves, a designated person will bid a dime, no one else will bid at all, and the class will share the profit of 90 cents.

You may take this story as proof of the folly of Yale undergraduates. But was the escalation of the superpowers' nuclear arms arsenals all that different? Both incurred costs in the trillions of dollars in quest of the "dollar" of victory. Collusion, which in this case means peaceful coexistence, is a much more profitable solution.

THE KING LEAR PROBLEM

> Tell me, my daughters
> Since now we will divest us both of rule,
> Interest of territory, cares of state,
> Which of you shall we say doth love us most?
> That we our largest bounty may extend
> Where nature doth with merit challenge.
> —Shakespeare, *King Lear*

King Lear was worried about how his children would treat him in his old age. Much to his regret, he discovered that children do not always deliver what they promise. In addition to love and respect, children are also motivated by the possibility of an inheritance. Here we look at how a strategic use of inheritance can manipulate children to visit their parents.

Imagine that parents want each of their children to visit once and phone twice a week. To give their children the right incentives, they threaten to disinherit any child who fails to meet this quota. The estate will be evenly divided among all the children who meet this quota. (In addition to motivating visits, this scheme has the advantage of avoiding the incentive for children to suffocate their parents with attention.)

The children recognize that their parents are unwilling to disinherit all of them. As a result, they get together and agree to cut back the number of visits, potentially down to zero.

The parents call you in and ask for some help in revising their will. Where there is a will, there is a way to make it work. But how? You are not allowed to disinherit all of the children.

Case Discussion

As before, any child who fails to meet the quota is disinherited. The problem is what to do if all of them are below the quota. In that case, give *all* of the estate to the child who visits the most. This will make the children's reduced visiting cartel impossible to maintain. We have put the children into a multiperson dilemma. The smallest amount of cheating brings a massive reward. A child who makes just one more phone call increases his or her inheritance from an equal share to 100 percent. The only escape is to go along with the parents' wishes. (Obviously, this strategy fails with only children. There is no good solution for couples with an only child. Sorry.)

UNITED STATES V. ALCOA

An established firm in an industry stands to gain by keeping out new competition. Then it can raise prices to monopoly levels. Since monopoly is socially harmful, the antitrust authorities try to detect and prosecute firms that employ strategies to deter rivals from entering the business.

In 1945, the Aluminum Corporation of America (Alcoa) was convicted of such a practice. An appellate panel of circuit court judges found that Alcoa had consistently installed more refining capacity than was justified by demand. In his opinion, Judge Learned Hand said:

> It was not inevitable that it [Alcoa] should always anticipate increases in the demand for ingot and be prepared to supply them. Nothing compelled it to keep doubling and redoubling its capacity before others entered the field. It insists that it never excluded competitors; but we can think of no more effective exclusion than progressively to embrace each new opportunity as it opened and to face every newcomer with new capacity already geared into a great organization.

This case has been debated at length by scholars of antitrust law and economics.[7] Here we ask you to consider the conceptual basis of the case. How could the construction of excess capacity deter new competitors?

Case Discussion

An established firm wants to convince potential new competitors that the business would not be profitable for them. This basically means that if they entered, the price would be too low to cover their costs. Of course the established firm could simply put out the word that it would fight an unrelenting price war against any newcomers. But why would the newcomers believe such a verbal threat? After all, a price war is costly to the established firm too.

Installing capacity in excess of the needs of current production gives credibility to the established firm's threat. When such capacity is in place, output can be expanded more quickly and at less extra cost. It remains only to staff the equipment and get the materials; the capital costs have already been incurred and are bygones. A price war can be fought more easily, more cheaply, and therefore more credibly.

ARMS ACROSS THE OCEAN

In the United States many homeowners own guns for self-defense. In Britain almost no one owns a gun. Cultural differences provide one explanation. The possibility of strategic moves provides another.

In both countries, a majority of homeowners prefer to live in an unarmed society. But they are willing to buy a gun if they have reason to fear that criminals will be armed.* Many criminals prefer to carry a gun as one of the tools of their trade.

The table below suggests a possible ranking of outcomes. Rather than assign specific monetary payoffs to each possibility, the outcomes are ranked 1, 2, 3, and 4 from best to worst for each side.

		Criminals	
		No guns	Guns
Homeowners No guns	1	2	1 / 4
Guns	2	4	3 / 3

If there were no strategic moves, we would analyze this as a game with simultaneous moves and use the techniques from chapter 3. We first look for dominant strategies. Since the criminals' grade in column 2 is always higher than that in a corresponding row in column 1, criminals have a dominant strategy: they prefer to carry guns whether or not homeowners are armed. Homeowners do not have a dominant strategy; they prefer to respond in kind. If criminals are unarmed, a gun is not needed for self-defense.

What is the predicted outcome when we model the game in this way? Following Rule 2, we predict that the side with a dominant strategy uses it; the other side chooses its best response to the dominant strategy of its opponent. Since Guns is the dominant

* The empirical evidence suggests that allowing the public to carry a concealed gun does not reduce the probability of crime, but it does not increase it, either. See Ian Ayres and John Donohue, "Shooting Down the 'More Guns, Less Crime' Hypothesis," *Stanford Law Review* 55 (2003): 1193–1312.

strategy for criminals, this is their predicted course of action. Homeowners choose their best response to Guns; they too will own a gun. The resulting equilibrium is ranked (3, 3), the third-best outcome for both parties.

In spite of their conflicting interests, the two sides can agree on one thing. They both prefer the outcome in which neither side carries guns (1, 2) to the case in which both sides are armed (3, 3). What strategic move makes this possible, and how could it be credible?

Case Discussion

Imagine for a moment that criminals are able to preempt the simultaneity and make a strategic move. They would commit not to carry guns. In this now sequential game, homeowners do not have to predict what criminals will do. They would see that the criminals' move has been made, and they are not carrying guns. Homeowners then choose their best response to the criminals' commitment; they too go unarmed. This outcome is ranked (1, 2), an improvement for *both* sides.

It is not surprising that criminals do better by making a commit-ment.* But homeowners are better off, too. The reason for the mutual gain is that both sides place a greater weight on the others' move than their own. Homeowners can reverse the criminals' move by allowing them to make an unconditional move.†

In reality, homeowners do not constitute one united player, and neither do criminals. Even though criminals as a class may gain by

---

* Could the criminals have done even better? No. Their best outcome is the homeowners' worst. Since homeowners can *guarantee* themselves 3 or better by owning guns, no strategic move by criminals can leave homeowners at 4. Hence a commitment to go unarmed is the best strategic move for criminals. What about a commitment by the criminals to carry arms? This is their dominant strat-egy. Homeowners would anticipate this move anyway. It has no strategic value. As with warnings and assurances, a commitment to a dominant strategy could be called a "declaration": it is informational rather than strategic.

† What happens if homeowners preempt and let the criminals respond? Homeowners can predict that for any unconditional choice of action on their part, criminals will respond by going armed. Hence homeowners will want to go armed, and the result is no better than with simultaneous moves.

taking the initiative and giving up guns, any one member of the group can get an additional advantage by cheating. This prisoners' dilemma would destroy the credibility of the criminals' initiative. They need some way to bond themselves together in a joint commitment.

If the country has a history of strict gun control laws, guns will be unavailable. Homeowners can be confident that criminals will be unarmed. Britain's strict control of guns allows criminals to commit to work unarmed. This commitment is credible, as they have no alternative. In the United States, the greater prevalence of guns denies criminals an ability to commit to work unarmed. As a result, many homeowners are armed for self-defense. Both sides are worse off.

Clearly this argument oversimplifies reality; one of its implications is that criminals should support gun control legislation. Even in Britain, this commitment is difficult to maintain. The political strife over Northern Ireland had the indirect effect of increasing the availability of guns to the criminal population. As a consequence, any commitment from criminals not to carry guns has begun to break down.

In looking back, note that something unusual happened in the transition from a simultaneous-move to a sequential-move game. Criminals chose to forego what was their dominant strategy. In the simultaneous-move game it was dominant for them to carry guns. In the sequential-move game, they chose not to. The reason is that in a sequential-move game, their course of action affects the homeowners' choice. Because of this interaction, they can no longer take the homeowners' response as beyond their influence. They move first, so their action affects the homeowners' choice. Carrying a gun is no longer a dominant strategy in the sequential representation of the game.

FOOLING ALL THE PEOPLE SOME OF THE TIME: THE LAS VEGAS SLOTS

Any gambling guide should tell you that slot machines are your worst bet. The odds are way against you. To counter this perception and encourage slot machine play, some Las Vegas casinos have begun to advertise the payback ratio for their machines—the fraction of each dollar bet returned in prize money. Going one step further, some casinos guarantee that they have machines that are set to a payback ratio greater than 1! These machines actually put the odds in your favor. If you could only find those machines and play them, you would expect to make money. The trick, of course, is that they don't tell you which machines are which. When they advertise that the average payback is 90 percent and that some machines are set at 120 percent, that also means that other machines must be set somewhere below 90 percent. To make it harder for you, there is no guarantee that machines are set the same way each day—today's favorable machines could be tomorrow's losers. How might you go about guessing which machines are which?

Case Discussion

Since this is our final case, we can admit that we do not have the answer—and if we did, we probably wouldn't share it. Nonetheless, strategic thinking can help you make a more educated guess. The trick is to put yourself into the casino owners' shoes. They make money only when people play the disadvantageous machines at least as much as the favorable or loose machines as they are known.

Is it really possible that the casinos could "hide" the machines that are offering the favorable odds? If people play the machines that pay out the most, won't they find the best ones? Not necessarily, and especially not necessarily in time! The payoff of the machine is in large part determined by the chance of a jackpot prize. Look at a slot machine that takes a quarter a pull. A jackpot prize of $10,000 with a 1 in 40,000 chance would give a payoff ratio of 1. If the casino raised the chance to 1 in 30,000, then the payoff ratio would be very favorable at 1.33. But people watching others play the machine would almost always see a person dropping quarter after quarter with no success. A natural conclusion would be that this is one of the least favorable machines. Eventually, when the machine pays its jackpot prize, it could be retooled and set at a lower rate.

In contrast, the least favorable machines could be set to pay back a small prize with a high frequency, and basically eliminate the hope of the big jackpot. Look at a machine set with a payback of 80 percent. If it provided a $1 prize on roughly every fifth draw, then this machine would make a lot of noise, attracting attention and possibly more gamblers' money. Are these the machines they put at the end of the aisles or near the buffet?

Perhaps the experienced slot players have figured all this out. But if so, you can bet that the casinos are just doing the reverse. Whatever happens, the casinos can find out at the end of the day which machines were played the most. They can make sure that the payoff patterns that attract the most play are actually the ones with the lower payoff ratio. For while the difference between a payoff ratio of 1.20 and 0.80 may seem large—and determines the difference between making money and losing money—it can be extremely hard to distinguish based on the number of pulls any one slot player can afford to make. The casinos can design the payoffs to make these inferences harder and even go the wrong way most of the time.

The strategic insight is to recognize that unlike the United Way, Las Vegas casinos are not in the business to give out money. In their search for the favorable machines, the majority of the players can't be right. For if the majority of the people were able to figure it out, the casino would discontinue their offer rather than lose money. So, don't wait in line. You can bet that the most heavily played machines are not the ones with the highest payback.

FURTHER READING

PIONEERING BOOKS are often enjoyable to read. In this spirit, we recommend John von Neumann and Oscar Morgenstern's *Theory of Games and Economic Behavior* (Princeton, NJ: Princeton University Press, 1947), even though the mathematics may be hard to follow in places. Thomas Schelling's *The Strategy of Conflict* (Cambridge, MA: Harvard University Press, 1960) is more than just a pioneering book; it continues to provide instruction and insight.

For an entertaining exposition of zero-sum games, J. D. Williams's *The Compleat Strategyst*, rev. ed. (New York: McGraw-Hill, 1966) still cannot be beat. The most thorough and highly mathematical treatment of pre-Schelling game theory is in Duncan Luce and Howard Raiffa, *Games and Decisions* (New York: Wiley, 1957). Among general expositions of game theory, Morton Davis, *Game Theory: A Nontechnical Introduction*, 2nd ed. (New York: Basic Books, 1983), is probably the easiest to read.

In terms of biographies, surely the most famous book on game theory is Sylvia Nasar, *A Beautiful Mind: The Life of Mathematical Genius and Nobel Laureate John Nash* (New York: Touchstone, 2001). The book is even better than the movie. William Poundstone's *Prisoner's Dilemma* (New York: Anchor, 1993) goes beyond a description of the eponymous game to offer a first-rate biography of John von Neumann, the polymath who invented the modern computer along with game theory.

In terms of textbooks, we are naturally partial to two of our own. Avinash Dixit and Susan Skeath, *Games of Strategy*, 2nd ed. (New York: W. W. Norton & Company, 2004), is designed for undergraduates. Barry Nalebuff and Adam Brandenburger's *Co-opetition* (New York: Doubleday, 1996) offers an application of game theory for MBAs and managers more broadly.

Other excellent textbooks include Robert Gibbons, *Game Theory for Applied Economists* (Princeton, NJ: Princeton University Press, 1992); John McMillan, *Games, Strategies, and Managers: How Managers Can Use Game Theory to Make Better Business Decisions* (New York: Oxford University Press, 1996); Eric Rasmusen, *Games and Information* (London: Basil Blackwell, 1989); Roger B. Myerson, *Game Theory: Analysis of Conflict* (Cambridge, MA: Harvard University Press, 1997); Martin J. Osborne and Ariel Rubinstein, *A Course in Game Theory* (Cambridge, MA: MIT Press, 1994); and Martin J. Osborne, *An Introduction to Game Theory* (New York: Oxford University Press, 2003). We always look forward to Ken Binmore's books. *Playing for Real: A Text on Game Theory* (New York: Oxford University Press, 2007) is the much-anticipated revision of his *Fun and Games* (Lexington, MA: D. C. Heath, 1992). (Warning: The title is a bit misleading. The book is actually quite challenging, both conceptually and mathematically. But it is very rewarding for the well prepared.) Binmore's latest offering is *Game Theory: A Very Short Introduction* (New York: Oxford University Press, 2008).

The following books are much more advanced and are used largely in graduate courses. They are strictly for the very ambitious: David Kreps, *A Course in Microeconomic Theory* (Princeton, NJ: Princeton University Press, 1990) and Drew Fudenberg and Jean Tirole, *Game Theory* (Cambridge, MA: MIT Press, 1991).

One of our sins of omission is a discussion of "cooperative games." Here players choose and implement their actions jointly and produce equilibria like the Core or the Shapley Value. This was done because we think cooperation should emerge as the equilibrium outcome of a noncooperative game in which actions are chosen separately. That is, individuals' incentive to cheat on any agreement should be recognized and made a part of their strategy choice. Interested readers can find treatments of cooperative games in the books by Davis and by Luce and Raiffa mentioned above and more extensively in Martin Shubik's *Game Theory in the Social Sciences* (Cambridge, MA: MIT Press, 1982).

There are several terrific books applying game theory to specific contexts. One of the most powerful applications is to auction design. Here there is no better source than Paul Klemperer's *Auctions: Theory and Practice*, The Toulouse Lectures in Economics (Princeton, NJ: Princeton University Press, 2004). Professor Klemperer was behind the design of many of the spectrum auctions, including the UK auction, which helped bring in some £34 billion and nearly bankrupted the telecom industry in the process. For game theory applied to law, see Douglas Baird, Robert Gertner, and Randal Picker, *Game Theory and the Law* (Cambridge, MA: Harvard University Press, 1998). One of their many contributions is the idea of an information escrow, which turns out to be a particularly useful tool in negotiation.* In the field of politics, noteworthy books include Steven Brams, *Game Theory and Politics* (New York: Free Press, 1979), and his more recent *Mathematics and Democracy: Designing Better Voting and Fair-Division Procedures* (Princeton, NJ: Princeton University Press, 2007); William Riker, *The Art of Political Manipulation* (New Haven, CT: Yale University Press, 1986); and the more technical approach of Peter Ordeshook's *Game Theory and Political Theory* (New York: Cambridge University Press, 1986). For applications to business, Michael Porter's *Competitive Strategy* (New York: Free Press,

---

* In an information escrow, each side makes an offer and then a neutral party evaluates whether the offers cross or not. In the legal environment, the DA offers a plea bargain, say three years. The defendant offers to accept anything under five. Since the defendant is willing to accept the DA's offer, a deal is done. But if the offers don't overlap—if, say, the DA asks for six years—then neither party learns what the other has put on the table.

1982); R. Preston McAfee's *Competitive Solutions: The Strategist's Toolkit* (Princeton, NJ: Princeton University Press, 2005); and Howard Raiffa's *The Art and Science of Negotiation* (Cambridge, MA: Harvard University Press, 1982) are excellent resources.

On the web, www.gametheory.net has the best collection of links to books, movies, and reading lists on game theory and its application.

WORKOUTS

TRIP TO THE GYM NO. 1

You win by leaving the other side with 1, which it is forced to take. That means that starting a round with 2, 3, or 4 is a winning position. Hence a person stuck with 5 loses, as whatever he does leaves the other side with 2, 3, or 4. Taken to the next round of thinking, a person stuck with 9 flags loses. Carrying on the same reasoning, the player starting with 21 has a losing hand (assuming the rival player uses the correct strategy and always takes the total down in groups of four).

Another way to see this is to note that the person who gets to take the next to last flag is the winner, as that leaves the other side with just one, which they are forced to take. Taking the penultimate flag is just like taking the last flag in a game with one fewer flag. In the case with 21 flags, you act as if there are only 20 and try to take the last one out of the twenty. Unfortunately, this is a losing position, at least if the other side understands the game. Incidentally, this shows that the first mover in a game need not always have the advantage, as we pointed out in the footnote on page 45.

TRIP TO THE GYM NO. 2

If you want to calculate the numbers in the tables for yourself, the exact formula for the sales of RE is: quantity sold by RE = 2800 − 100 × RE's price + 80 × BB's price.

The formula for the sales of BB is the mirror image of this. To calculate each store's profits, recall that both have a cost of $20, so that

RE's profit = (RE's price − 20) × quantity sold by RE.

The formula for BB's profit is similar.

Alternately, these formulas can be embedded into an Excel spreadsheet. In the first column (column A), enter the prices of RE for which you want to do the calculations in rows 2, 3, ... With five prices in our range, these are rows 2–6. In the top row (row 1), enter the corresponding prices for BB in columns B, C, . . . , in this case, columns B–F. In cell B2, enter the formula: =MAX(2800−100*$A2+80*B$1,0).

Note carefully the dollar signs; in Excel notation they ensure the appropriate "absolute" and "relative" cell references when the formula is copied and pasted to the other cells with the different price combinations. The formula also ensures that if prices charged by the two firms are too different, the sales of the firm with the higher price do not go negative. This is the table of the quantities sold by RE.

To calculate RE's profits from these quantities, write down in a blank cell somewhere else in the spreadsheet (we used cell J2) RE's cost, namely 20. On the same spreadsheet, directly below the table of quantities, say, in rows 8–12 (leaving row 7 blank for clarity), copy the prices of RE in column A. In cell B8, enter the formula: =B2*($A8−$J$2).

This yields RE's profit when it charges its first price in the set we are considering (42), and BB charges the first of its prices (42). Copy and paste this formula into the other cells to get the full table of profits for RE.

The formulas for the quantities and profits of BB can be entered in rows 14–18 and 20–24. The formula for its quantities is =MAX(2800−100* B$1+80*$A14,0). And, entering BB's cost in a spare cell, J3, the formula for its profits is =B14*(B$1−$J$3).

When all is said and done, you should have ended up with a table that looks a lot like the one on the following page. Of course, if you want to experiment with these equations with different quantities sold or different costs, you should change the numbers accordingly.

	A	B	C	D	E	F	G	H	I	J
1		42	41	40	39	38			Costs	
2	42	1,960	1,880	1,800	1,720	1,640			RE	20
3	41	2,060	1,980	1,900	1,820	1,740	RE's		BB	20
4	40	2,160	2,080	2,000	1,920	1,840	quantities			
5	39	2,260	2,180	2,100	2,020	1,940				
6	38	2,360	2,280	2,200	2,120	2,040				
7										
8	42	43,120	41,360	39,600	37,840	36,080				
9	41	41,260	41,580	39,900	38,220	36,540	RE's			
10	40	43,200	41,600	40,000	38,400	36,800	profits			
11	39	42,940	41,420	39,900	38,380	36,860				
12	38	42,480	41,040	39,600	38,160	36,720				
13										
14	42	1,960	2,060	2,160	2,260	2,360				
15	41	1,880	1,980	2,080	2,180	2,280	BB's			
16	40	1,800	1,900	2,000	2,100	2,200	quantities			
17	39	1,720	1,820	1,920	2,020	2,120				
18	38	1,640	1,740	1,840	1,940	2,040				
19										
20	42	43,120	43,260	43,200	42,940	42,480				
21	41	41,360	41,580	41,600	41,420	41,040	BB's			
22	40	39,600	39,900	40,000	39,900	39,600	profits			
23	39	37,840	38,220	38,400	38,380	38,160				
24	38	36,080	36,540	36,800	36,860	36,720				

TRIP TO THE GYM NO. 3

The Excel spreadsheet is easily modified by changing RE's cost figure in cell J2 from 20 to 11.60:

	A	B	C	D	E	F	G	H	I	J
1		40	39	38	37	36			Costs	
2	36	2,300	2,220	2,140	2,060	1,980			RE	11.60
3	41	2,400	2,320	2,240	2,160	2,080	RE's		BB	20
4	40	2,500	2,420	2,340	2,260	2,180	quantities			
5	39	2,600	2,520	2,440	2,360	2,280				
6	38	2,700	2,620	2,540	2,460	2,380				
7										
8	37	58,420	56,388	54,356	52,324	50,292				
9	36	58,560	56,608	54,656	52,704	50,752	RE's			
10	35	58,500	56,628	54,756	52,884	51,012	profits			
11	34	58,240	56,448	54,656	52,864	51,072				
12	33	57,780	56,068	54,356	52,644	50,932				
13										
14	37	1,760	1,860	1,960	2,060	2,160				
15	36	1,680	1,780	1,880	1,980	2,080	BB's			
16	35	1,600	1,700	1,800	1,900	2,000	quantities			
17	34	1,520	1,620	1,720	1,820	1,920				
18	33	1,440	1,540	1,640	1,740	1,840				
19										
20	37	35,200	35,340	35,280	35,020	34,560				
21	37	33,600	33,820	33,840	33,660	33,280	BB's			
22	35	32,000	32,300	32,400	32,300	32,000	profits			
23	34	30,400	30,780	30,960	30,940	30,720				
24	33	28,800	29,260	29,520	29,580	29,440				

The profit numbers are then entered into the payoff table for the game:

		B. B. Lean's price				
		40	39	38	37	36
	38	35,200 / 58,420	**35,340** / 56,388	35,280 / 54,356	35,020 / 52,324	34,560 / 50,292
Rainbow's End's price	37	33,600 / **58,560**	33,820 / 56,608	**33,840** / 54,656	33,660 / 52,704	33,280 / 50,752
	36	32,000 / 58,500	32,300 / **56,628**	**32,400** / **54,756**	32,300 / **52,884**	32,000 / 51,012
	35	30,400 / 58,240	30,780 / 56,448	**30,960** / 54,656	30,940 / 52,864	30,720 / **51,072**
	34	28,800 / 57,780	29,260 / 56,068	29,520 / 54,356	**29,580** / 52,644	29,440 / 50,932

Observe that we had to use a range of lower prices to locate best responses. In the new Nash equilibrium, BB charges $38 and RE charges $35. RE benefits twice over, once from its lower cost and further as its price cut shifts some customers to it from BB. As a result, BB's profit goes down by a lot (from $40,000 to $32,400) while RE's profit goes up by a lot (from $40,000 to $54,756). Even though RE's cost advantage is only 42 percent ($11.60 is 58 percent of $20), its profit advantage is 69 percent ($54,756 is 1.69 times $32,400). Now you see why businesses are so keen to eke out seemingly small cost advantages, and why firms frequently move to lower-cost locations and countries.

TRIP TO THE GYM NO. 4

Without any U.S. strategic move, the game tree is

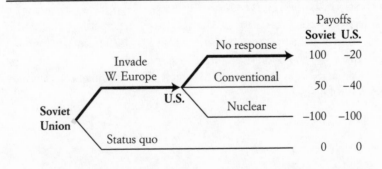

If the Soviets invaded Western Europe, the United States would suffer some loss of prestige if it made no response and accepted the fait accompli. But it would suffer a military defeat, severe casualties, and perhaps an even bigger loss of prestige if it tried to respond with conventional arms, since the Soviet army was much bigger and they cared much less about casualties. And the United States would suffer far more if it responded with nuclear weapons, because the Soviets would then counterattack the United States itself with their own nukes. Therefore the least-bad response for the United States after the fact would be to abandon Western Europe to its fate. If you think this an unlikely scenario, the European members of NATO thought it all too likely and wanted the United States to commit credibly to a response. The U.S. threat "we will respond with nuclear weapons if you attack Western Europe" removes the first two branches from the node where the United States chooses its action and converts the game into the following:

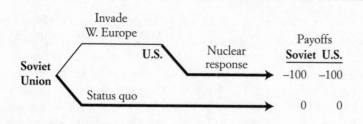

Now the Soviets face a nuclear response with a payoff of −100 if they invade; therefore, they accept the status quo that gives them their less-bad payoff of 0. We discuss in chapter 6 and in chapter 7 how the U.S. threat can be made credible.

TRIP TO THE GYM NO. 5

The first-class fare, $215, is comfortably below the business travelers' willingness to pay for this class, namely $300. So their participation constraint is fulfilled. Tourists get zero consumer surplus ($140 − 140) from an economy-class purchase, but they would get negative surplus ($175 − 215 = −40) from a first-class seat. So they do not want to switch; their incentive compatibility condition is fulfilled.

TRIP TO THE GYM NO. 6

In a Vickrey auction, you would not be willing to pay anything at all to learn the bids of the other players. Remember that bidding your true value is a dominant strategy in a Vickrey auction. Thus you would bid the same amount no matter what you learn the others are doing.

There is one caveat, however. We are assuming that your value in the auction is private to you and not influenced by what the others think it is worth. In the case of a common value Vickrey auction, you might want to change your bid based on what others are doing, but only because it changes what you think the item is worth.

TRIP TO THE GYM NO. 7

To show how to bid in a sealed-bid auction, we transform a Vickrey auction into a sealed-bid auction. We do this in the simple case where there are two bidders, each with a valuation between 0 and 100, where all numbers are equally likely.

Let's start with the Vickrey auction. Your value is 60 and so you bid 60. If we told you that you've won the auction, you'd be pleased but you wouldn't know how much you'll have to pay. All you know is that it is something below 60. All possibilities below 60 are equally likely, so on average you'll pay \$30. If we now offered you the choice of paying \$30 or paying what the second highest bid turned out to be, you'd be indifferent. You expect to pay \$30 either way. Similarly, if your value were \$80, you'd be happy to pay \$40 when told that you have a winning bid in a Vickrey auction. More generally, if your value is \$X, then you expect to pay \$X/2, as the second highest bid, when you win in a Vickrey auction. You'd be just as happy if you had to pay \$X/2 outright when your bid of \$X wins.

Let's take that step. Instead of paying the second highest bid, we'll change the rules so that when you bid \$X, you only have to pay \$X/2 when you win. Since this has the same outcome on average as a Vickrey auction, your optimal bid shouldn't change. Now we let everyone else follow the same rule. Their bids shouldn't change, either.

At this point, we have something very similar to a sealed-bid auction. Everyone is writing down a number, and the high number wins. The only difference is that instead of paying your number, you only have to pay half your number. It's like having to pay in U.S. dollars instead of in British pounds.

Bidders aren't fooled by this game. If saying \$80 means you have to pay

$40, then a bid of "$80" really means $40. If we changed the rules once more so that you have to pay your bid, rather than half your bid, then everyone would just cut their bids in half. In that case, if you are willing to pay $40, you say $40 rather than $80. With this final step, we have arrived at a sealed-bid auction. You'll note that an equilibrium strategy is for both players to bid half their value.

If you want to double-check that this is an equilibrium, you can assume that the other player is bidding half his value and imagine how you would respond. If you bid X, you will win if the other bidder has a value below 2X (and thus bids below X). The chance of this is 2X/100. Thus your payoff from a bid of X when your true value is V is:

$$(\text{Probability X wins}) \times (V - X) = \left(\frac{2X}{100}\right)(V - X).$$

This is maximized at $X = V/2$. If the other player is bidding half his value, then you want to bid half yours. And if you are bidding half your value, then the other player will want to do likewise. Thus we have a Nash equilibrium. As you can see, it is easier to check that something is an equilibrium than to find the equilibrium in the first place.

TRIP TO THE GYM NO. 8

Say you knew your rival would act at $t = 10$. You could either act at 9.99 or wait and let your rival take her chance. If you shoot at $t = 9.99$, your chance of winning is just about $p(10)$. If you wait, you will win if your rival fails. The chance of that is $1 - q(10)$. Hence you should preempt if $p(10) > 1 - q(10)$.

Of course, your rival is doing the same calculation. If she thinks you are going to preempt at $t = 9.99$, she would prefer to move first at $t = 9.98$ if $q(9.98) > 1 - p(9.98)$.

You can see that the condition that determines the time that neither side wants to preempt is:

$$p(t) \leq 1 - q(t) \text{ and } q(t) \leq 1 - p(t).$$

These are one and the same condition:

$$p(t) + q(t) \leq 1.$$

Thus both sides are willing to wait until $p(t) + q(t) = 1$ and then they both shoot.

TRIP TO THE GYM NO. 9

If your house sells for $250,000, the commission will be $15,000, generally split evenly between your agent and the buyer's agent. The problem is that this payment structure provides weak incentives. When your agent works hard and brings in an extra $20,000, that translates to just $600 more in commission after the split. Worse still, the agent typically has to share this commission with the agency, leaving only $300. This is hardly worth the extra effort, so agents have an incentive to do the deal quickly rather than get the best price.

Why not offer a nonlinear scheme: pay 2.5% for the first $200,000 and then 20% on everything above that amount? If the sale price is $250,000, the commission would be the same, $15,000. But if your agent truly succeeds and brings in $270,000, that would boost the commission by $2,000, even after the split.

The problem, of course, is where to set the commission rate threshold. If you think your place can fetch $300,000, then you would want the threshold to be near $250,000. In contrast, the agent will be more conservative and argue that $250,000 is the market price, which leads to the higher commission kicking in at $200,000. This creates a serious conflict between you and your agent right at the start of the relationship.

TRIP TO THE GYM NO. 10

To see how big this effect might be, we dive a bit deeper into the economics. Typically, the publisher gets 50% of the list price as its wholesale price. The cost to print and ship an average hardcover book is around $3. Then at a price of p, which leads to sales of q(p), the publisher makes

$$(0.5p - 0.15p - 3) \times q(p) = 0.35 \times (p - 8.6) \times q(p)$$

Since the publisher gets only half the list price and has to pay the author 15%, the publisher ends up with only about 35% of the list price but has to pay the entire printing cost. As a result, it is as if the effective printing cost is $8.60, almost three times as high.

We can pick a simple case where the demand is linear, say $q(p) = 40 - p$, and demand is measured in thousands. To maximize revenue, the author would pick a list price of $20. In contrast, the publisher would pick a list price of $24.30 in order to maximize profits.

NOTES

Chapter 1

1. Their research is reported in "The Hot Hand in Basketball: On the Misperception of Random Sequences," *Cognitive Psychology* 17 (1985): 295–314.

2. *New York Times*, September 22, 1983.

3. These quotes are from Martin Luther's speech at the Diet of Worms on April 18, 1521, as described in Roland Bainton, *Here I Stand: A Life of Martin Luther* (New York: Abingdon-Cokesbury, 1950).

4. Don Cook, *Charles de Gaulle: A Biography* (New York: Putnam, 1982).

5. David Schoenbrun, *The Three Lives of Charles de Gaulle* (New York: Athenaeum, 1966).

6. See Thomas Schelling, *Arms and Influence* (New Haven, CT: Yale University Press, 1966), 45; and Xenophon, *The Persian Expedition* (London: Penguin, 1949), 136–37, 236.

7. The show, *Life: The Game*, aired on March 16, 2006. A DVD is available for purchase at www.abcnewsstore.com as "PRIMETIME: Game Theory: 3/16/06." A sequel, where this threat was contrasted with positive reinforcement, aired on December 20, 2006, and is available as "PRIMETIME: Basic Instincts – Part 3 – Game Theory: 12/20/06."

8. Warren Buffett, "The Billionaire's Buyout Plan," *New York Times*, September 10, 2000.

9. Truman Capote, *In Cold Blood* (New York: Vintage International, 1994), 226–28.

10. Our quotes are from the *New York Times* coverage of the story, May 29, 2005.

11. One online option is Perry Friedman's AI algorithm at http://chappie.stan ford.edu/cgi-bin/roshambot. It placed sixteenth in the second international RoShamBo programming competition; www.cs.ualberta.ca/~darse/rsbpc .html. For readers looking to brush up their skills, we recommend Douglas Walker and Graham Walker's *The Official Rock Paper Scissors Strategy Guide* (New York: Simon & Schuster, 2004) and a visit to www.worldrps.com.

12. Kevin Conley, "The Players," *The New Yorker*, July 11, 2005, 55.

Chapter 2

1. Louis Untermeyer, ed., *Robert Frost's Poems* (New York: Washington Square Press, 1971).

2. In many states, governors do have the power of line-item veto. Do they have significantly lower budget expenditures and deficits than states without line-item vetoes? A statistical analysis by Professor Douglas Holtz-Eakin of Syracuse University (who went on to be the director of the Congressional Budget Office) showed that they do not ("The Line Item Veto and Public Sector Budgets," *Journal of Public Economics* 36 (1988): 269–92).

3. A good free and open source package of this kind is Gambit. It can be downloaded from http://gambit.sourceforge.net.

4. For a description and brief video of the actual game, go to www.cbs .com/primetime/survivor5/.

5. This is a particularly simple example of a class of games called Nim-type games. To be specific, it is called a subtraction game with one heap. Harvard mathematician Charles Bouton was the first to discuss Nim-type games. His pioneering article is, "Nim, a game with a complete mathematical theory," *Annals of Mathematics* 3, no. 2 (1902): 35–39, in which he proved a general rule for solving them. Almost a century's worth of the research that followed was surveyed by Richard K. Guy, "Impartial Games," in Richard J. Nowakowski, ed., *Games of No Chance* (Cambridge: Cambridge University Press, 1996), 61–78. There is also a Wikipedia article on Nim-type games, http://en.wikipedia.org/wiki/Nim, that gives further details and references.

6. These experiments are too plentiful to cite in full. An excellent survey and discussion can be found in Colin Camerer, *Behavioral Game Theory: Experiments in Strategic Interaction* (Princeton, NJ: Princeton University Press, 2003), 48–83, 467. Camerer also discusses experiments and findings on other related games, most notably the "trust game," which is like the Charlie-Fredo game (see his pp. 83–90). Once again, actual behavior differs from what would be predicted by backward reasoning that assumes purely selfish preferences; considerable trusting behavior and its reciprocation are found.

7. See Jason Dana, Daylian M. Cain, and Robyn M. Dawes, "What You Don't Know Won't Hurt Me: Costly (but Quiet) Exit in Dictator Games," *Organizational Behavior and Human Decision Processes* 100 (2006): 193–201.

8. Alan G. Sanfey, James K. Rilling, Jessica A. Aronson, Leigh E. Nystrom, and Jonathan D. Cohen, "The Neural Basis of Economic Decision Making in the Ultimatum Game," *Science* 300 (June 2003): 1755–57.

9. Camerer, *Behavioral Game Theory*, 68–74.

10. Ibid., 24. Emphasis in the original.

11. Ibid., 101–10, for an exposition and discussion of some such theories.

12. Burnham is the co-author of *Mean Genes* (Cambridge, MA: Perseus, 2000) and the author of *Mean Markets and Lizard Brains: How to Profit from the New Science of Irrationality* (Hoboken, NJ: Wiley, 2005). His paper on this experiment is "High-Testosterone Men Reject Low Ultimatum Game Offers," *Proceedings of the Royal Society B* 274 (2007): 2327–30.

13. For a detailed expert discussion of chess from the game-theoretic perspective, read Herbert A. Simon and Jonathan Schaeffer, "The Game of Chess," in *The Handbook of Game Theory*, Vol. 1, ed. Robert J. Aumann and Sergiu Hart (Amsterdam: North-Holland, 1992). Chess-playing computers have improved greatly since the article was written, but its general analysis retains its validity. Simon won the Nobel Prize in Economics in 1978 for his pioneering research into the decision making process within economic organizations.

Chapter 3

1. From "Brief History of the Groundfishing Industry of New England," on the U.S. government web site www.nefsc.noaa.gov/history/stories/ground fish/grndfsh1.html.

2. Joseph Heller, *Catch-22* (New York: Simon & Schuster, 1955), 455 in Dell paperback edition published in 1961.

3. University of California biologist Garrett Harding brought this class of problems to wide attention in his influential article "The Tragedy of the Commons," *Science* 162 (December 13, 1968): 1243–48.

4. "The Work of John Nash in Game Theory," Nobel Seminar, December 8, 1994. On the web site at http://nobelprize.org/nobel_prizes/economics/lau reates/1994/nash-lecture.pdf.

5. William Poundstone, *Prisoner's Dilemma* (New York: Doubleday, 1992), 8–9; Sylvia Nasar, *A Beautiful Mind* (New York: Simon & Schuster, 1998), 118–19.

6. James Andreoni and Hal Varian have developed an experimental game called Zenda based on this idea. See their "Preplay Communication in the Prisoners' Dilemma," *Proceedings of the National Academy of Sciences* 96, no. 19 (September 14, 1999): 10933–38. We have tried the game in classrooms and found it to be successful in developing cooperation. Its implementation in a more realistic setting is harder.

7. This research comes from their working paper "Identifying Moral Hazard: A Natural Experiment in Major League Baseball," available at http://dd rinen.sewanee.edu/Plunk/dhpaper.pdf.

8. At the time, Schilling was pitching for the National League's Arizona Diamondbacks and Cy Young winner Randy Johnson was his teammate. Quoted in Ken Rosenthal, "Mets Get Shot with Mighty Clemens at the Bat," *Sporting News*, June 13, 2002.

9. The results are due to M. Keith Chen and Marc Hauser, "Modeling Reciprocation and Cooperation in Primates: Evidence for a Punishing Strategy," *Journal of Theoretical Biology* 235, no. 1 (May 2005): 5–12. You can see a video of the experiment at www.som.yale.edu/faculty/keith.chen/datafilm .htm.

10. See Camerer, *Behavioral Game Theory*, 46–48.

11. See Felix Oberholzer-Gee, Joel Waldfogel, and Matthew W. White, "Social Learning and Coordination in High-Stakes Games: Evidence from Friend or Foe," NBER Working Paper No. W9805, June 2003. Available at SSRN: http://ssrn.com/abstract=420319. Also see John A List, "Friend or Foe? A Natural Experiment of the Prisoner's Dilemma," *Review of Economics and Statistics* 88, no. 3 (2006): 463–71.

12. For a detailed account of this experiment, again see Poundstone, *Prisoner's Dilemma*, 8–9; and Sylvia Nasar, *A Beautiful Mind*, 118–19.

13. Jerry E. Bishop, "All for One, One for All? Don't Bet On It," *Wall Street Journal*, December 4, 1986.

14. Reported by Thomas Hayden, "Why We Need Nosy Parkers," *U.S. News and World Report*, June 13, 2005. Details can be found in D. J. de Quervain, U. Fischbacher, V. Treyer, M. Schellhammer, U. Schnyder, and E. Fehr, "The Neural Basis of Altruistic Punishment," *Science* 305, no. 5688 (August 27, 2004): 1254–58.

15. Cornell University economist Robert Frank, in *Passions Within Reason* (New York: W. W. Norton, 1988), argues that emotions, such as guilt and love, evolved and social values, such as trust and honesty, were developed and sustained to counter individuals' short-run temptations to cheat and to secure the long-run advantages of cooperation. And Robert Wright, in *Nonzero* (New York: Pantheon, 2000), develops the idea that the mechanisms that achieve mutually beneficial outcomes in non-zero-sum games explain much of human cultural and social evolution.

16. Eldar Shafir and Amos Tversky, "Thinking through Uncertainty: Nonconsequential Reasoning and Choice," *Cognitive Psychology* 24 (1992): 449–74.

17. *The Wealth of Nations*, vol. 1, book 1, chapter 10 (1776).

18. Kurt Eichenwald gives a brilliant and entertaining account of this case in *The Informant* (New York: Broadway Books, 2000). The "philosophy" quote is on p. 51.

19. David Kreps, *Microeconomics for Managers* (New York: W. W. Norton, 2004), 530–31, gives an account of the turbine industry.

20. See Paul Klemperer, "What Really Matters in Auction Design," *Journal of Economic Perspectives* 16 (Winter 2002): 169–89, for examples and analysis of collusion in auctions.

21. Kreps, *Microeconomics for Managers*, 543.

22. "Picture a pasture open to all. It is to be expected that each herdsman will try to keep as many cattle as possible on this commons. . . . Therein is the tragedy. Each man is locked into a system that compels him to increase his herd without limit, in a world that is limited. Ruin is the destination toward which all men rush, each pursuing his own best interest in a society that believes in the freedom of the commons" (Harding, "The Tragedy of the Commons," 1243–48).

23. Elinor Ostrom, *Governing the Commons* (Cambridge: Cambridge University Press, 1990), and "Coping with the Tragedy of the Commons," *Annual Review of Political Science* 2 (June 1999): 493–535.

24. The literature is huge. Two good popular expositions are Matt Ridley, *The Origins of Virtue* (New York: Viking Penguin, 1997); and Lee Dugatkin, *Cheating Monkeys and Citizen Bees* (Cambridge, MA: Harvard University Press, 1999).

25. Dugatkin, *Cheating Monkeys*, 97–99.

26. Jonathan Weiner, *Beak of the Finch*, 289–90.

Chapter 4

1. See chapter 1, note 7.

2. Keynes's oft-quoted text remains remarkably current: "Professional investment may be likened to those newspaper competitions in which competitors have to pick out the six prettiest faces from one hundred photographs, the prize being awarded to the competitor whose choice most nearly corresponds to the average preference of the competitors as a whole; so that each competitor has to pick, not those faces which he himself finds prettiest, but those which he thinks likeliest to catch the fancy of the other competitors, all of whom are looking at the problem from the same point of view. It is not a case of choosing those which, to the best of one's judgment, are really the prettiest, nor even those which average opinion genuinely thinks the prettiest. We have reached the third degree where we devote our intelligences to anticipating what average opinion expects the average opinion to be." See *The General Theory of Employment, Interest, and Money*, vol. 7, of *The Collected Writings of John Maynard Keynes* (London: Macmillan, 1973), 156.

3. Quoted from Poundstone, *Prisoner's Dilemma*, 220.

4. Readers who want a little more formal detail on each of these games will find

useful articles at http://en.wikipedia.org/wiki/Game_theory and www.game theory.net.

5. Gambit, which is useful for drawing and solving trees, also has a module for setting up and solving game tables. See chapter 2, note 3, for more information.

6. At a higher level of analysis, the two are seen to be equivalent in two-player games if mixed strategies are allowed for each player; see Avinash Dixit and Susan Skeath, *Games of Strategy*, 2nd ed. (New York: W. W. Norton, 2004), 207.

7. For readers with some mathematical background, here are a few steps in the calculation. The formula for the quantity sold by BB can be written as:

quantity sold by BB = 2800 – 100 × BB's price + 80 × RE's price.

On each unit, BB makes a profit equal to its price minus 20, its cost. Therefore BB's total profit is

BB's profit = (2800 – 100 × BB's price + 80 × RE's price)
× (BB's price – 20).

If BB sets its price equal to its cost, namely 20, it makes zero profit. If it sets its price equal to

(2800 + 80 × RE's price)/100 = 28 + 0.8 × RE's price,

it makes zero sales and therefore zero profit. BB's profit is maximized by choosing a price somewhere between these two extremes, and in fact for our linear demand formula this occurs at a price exactly half way between the extremes. Therefore

BB's best response price = $\frac{1}{2}$(20 + 28 + 0.8 × RE's price)
= 24 + 0.4 × RE's price.

Similarly, RE's best response price = 24 + 0.4 × BB's price.

When RE's price is $40, BB's best response price is 24 + 0.4 × 40 = 24 × 16 = 40, and vice versa. This confirms that in the Nash equilibrium outcome each firm charges $40. For more details of such calculations, see Dixit and Skeath, *Games of Strategy*, 124–28.

8. For readers interested in pursuing this topic, we recommend the survey by Peter C. Reiss and Frank A. Wolak, "Structural Econometric Modeling: Rationales and Examples from Industrial Organization," in *Handbook of Econometrics, Volume 6B*, ed. James Heckman and Edward Leamer (Amsterdam: North-Holland, 2008).

9. This research is surveyed by Susan Athey and Philip A. Haile: "Empirical Models of Auctions," in *Advances in Economic Theory and Econometrics, Theory and Applications, Ninth World Congress, Volume II*, ed. Richard Blundell, Whitney K. Newey, and Torsten Persson (Cambridge: Cambridge University Press, 2006), 1–45.

10. Richard McKelvey and Thomas Palfrey, "Quantal Response Equilibria for Normal Form Games," *Games and Economic Behavior* 10, no. 1 (July 1995): 6–38.

11. Charles A. Holt and Alvin E. Roth, "The Nash Equilibrium: A Perspective," *Proceedings of the National Academy of Sciences* 101, no. 12 (March 23, 2004): 3999–4002.

Chapter 5

1. The research contributions include Pierre-Andre Chiappori, Steven Levitt, and Timothy Groseclose, "Testing Mixed-Strategy Equilibria When Players Are Heterogeneous: The Case of Penalty Kicks in Soccer," *American Economic Review* 92, no. 4 (September 2002): 1138–51; and Ignacio Palacios-Huerta, "Professionals Play Minimax," *Review of Economic Studies* 70, no. 2 (April 2003): 395–415. Coverage in the popular media includes Daniel Altman, "On the Spot from Soccer's Penalty Area," *New York Times,* June 18, 2006.

2. The book was published by Princeton University Press in 1944.

3. Some numbers differ slightly from Palacios-Huerta's because he uses data accurate to two decimal places, while we have chosen to round them for neater exposition.

4. Mark Walker and John Wooders, "Minimax Play at Wimbledon," *American Economic Review* 91, no. 5 (December 2001): 1521–38.

5. Douglas D. Davis and Charles A. Holt, *Experimental Economics* (Princeton, NJ: Princeton University Press, 1993): 99.

6. Stanley Milgram, *Obedience to Authority: An Experimental View* (New York: Harper and Row, 1974).

7. The articles cited in note 1, above, cite and discuss these experiments in some detail.

8. E-mail from Graham Walker at the World RPS Society, July 13, 2006.

9. Rajiv Lal, "Price Promotions: Limiting Competitive Encroachment," *Marketing Science* 9, no. 3 (Summer 1990): 247–62, examines this and other related cases.

10. John McDonald, *Strategy in Poker, Business, and War* (New York: W. W. Norton, 1950), 126.

11. Many programs of this kind are available, including Gambit (see chapter 2, note 3) and ComLabGames. The latter enables experimentation with and analysis of games and their outcomes over the Internet; it is downloadable at www.comlabgames.com.

12. For a few more details, see Dixit and Skeath, *Games of Strategy*, chapter 7. A really thorough treatment is in R. Duncan Luce and Howard Raiffa, *Games and Decisions* (New York: Wiley, 1957), chapter 4 and appendices 2–6.

Chapter 6

1. See www.firstgov.gov/Citizen/Topics/New_Years_Resolutions.shtml.

2. Available at www.cnn.com/2004/HEALTH/diet.fitness/02/02/sprj.nyr.reso lutions/index.html.

3. See chapter 1, note 7.

4. An excellent, and still highly useful, exposition of the theory as it existed in the mid-1950s can be found in Luce and Raiffa, *Games and Decisions*.

5. Thomas C. Schelling, *The Strategy of Conflict* (Cambridge, MA: Harvard University Press); and Schelling, *Arms and Influence* (New Haven, CT: Yale University Press).

6. Thomas Schelling coined this term as a part of his pioneering analysis of the concept. See William Safire's *On Language* column in the *New York Times Magazine*, May 16, 1993.

7. James Ellroy, *L.A. Confidential* (Warner Books, 1990), 135–36, in the 1997 trade paperback edition.

8. Schelling, *Arms and Influence*, 97–98, 99.

9. For a detailed account of the crisis, see Elie Abel, *The Missile Crisis* (New York: J. B. Lippincott, 1966). Graham Allison offers a wonderful game-theoretic analysis in his book *Essence of Decision: Explaining the Cuban Missile Crisis* (Boston: Little, Brown, 1971).

10. This evidence is in Allison's *Essence of Decision*, 129–30.

Chapter 7

1. All Bible quotations are taken from the New International Version unless noted otherwise.

2. *Bartlett's Familiar Quotations* (Boston: Little, Brown, 1968), 967.

3. Dashiell Hammett, *The Maltese Falcon* (New York: Knopf, 1930); quotation taken from 1992 Random House Vintage Crime ed., 174.

4. Thomas Hobbes, *Leviathan* (London: J. M. Dent & Sons, 1973), 71.

5. *Wall Street Journal,* January 2, 1990.

6. This example is from his commencement address to the Rand Graduate School, later published as "Strategy and Self-Command," *Negotiation Journal*, October 1989, 343–47.

7. Paul Milgrom, Douglass North, and Barry R. Weingast, "The Role of Institutions in the Revival of Trade: The Law Merchant, Private Judges, and the Champagne Fairs," *Economics and Politics* 2, no. 1 (March 1990): 1–23.

8. Diego Gambetta, *The Sicilian Mafia: The Business of Private Protection* (Cambridge, MA: Harvard University Press, 1993), 15.

9. Lisa Bernstein, "Opting Out of the Legal System: Extralegal Contractual

Relations in the Diamond Industry," *Journal of Legal Studies* 21 (1992): 115–57.

10. Gambetta, *Sicilian Mafia*, 44. In the original opera, available at http://opera.stanford.edu/Verdi/Rigoletto/III.html, Sparafucile sings:

> Uccider quel gobbo! . . .
> che diavol dicesti!
> Un ladro son forse? . . .
> Son forse un bandito? . . .
> Qual altro cliente
> da me fu tradito? . . .
> Mi paga quest'uomo . . .
> fedele m'avrà

11. Ibid., 45.

12. Many of Kennedy's most famous speeches have been collected in a book and on CD, with accompanying explanation and commentary: Robert Dallek and Terry Golway, *Let Every Nation Know* (Naperville, IL: Sourcebooks, Inc., 2006). The quotation from the inaugural address is on p. 83; that from the Cuban missile crisis address to the nation is on p. 183. The quotation from the Berlin speech is on the CD but not in the printed book. A printed reference to this speech is in Fred Ikle, *How Nations Negotiate* (New York: Harper and Row, 1964), 67.

13. The quote, and others from the same movie used later in this chapter, are taken from www.filmsite.org/drst.html, which gives a detailed outline and analysis of the movie.

14. According to the *Guardian*: "Donald Rumsfeld can be criticised for a lot of things. But the US defence secretary's use of English is not one of them. . . . 'Reports that say something hasn't happened are always interesting to me,' Mr Rumsfeld said, 'because, as we know, there are known knowns, there are things we know we know. We also know there are known unknowns; that is to say, we know there are some things we do not know. But there are also unknown unknowns—the ones we don't know we don't know.' This is indeed a complex, almost Kantian, thought. It needs a little concentration to follow it. Yet it is anything but foolish. It is also perfectly clear. It is expressed in admirably plain English, with not a word of jargon or gobbledygook in it." See www.guardian.co.uk/usa/story/0,12271,1098 489,00.html.

15. See Schelling's "Strategic Analysis and Social Problems," in his *Choice and Consequence* (Cambridge, MA: Harvard University Press, 1984).

16. William H. Prescott, *History of the Conquest of Mexico,* vol. 1, chapter 8. The book was first published in 1843 and is now available in the Barnes & Noble Library of Essential Readings series, 2004. We recognize that this interpretation of Cortés's action is not universally accepted by modern historians.

17. This description and quote come from Michael Porter, *Cases in Competitive Strategy* (New York: Free Press, 1983), 75.

18. Schelling, *Arms and Influence*, 39.

19. For a fascinating account of incentives used to motivate soldiers, see John Keegan's *The Face of Battle* (New York: Viking Press, 1976).

20. The Sun Tzu translation is from Lionel Giles, *Sun Tzu on the Art of War* (London and New York: Viking Penguin, 2002).

21. Schelling, *Arms and Influence*, 66–67.

22. Convincing evidence that students anticipate textbook revisions comes from Judith Chevalier and Austan Goolsbee, "Are Durable Goods Consumers Forward Looking? Evidence from College Textbooks," NBER Working Paper No. 11421, 2006.

23. Professor Michael Granof is an early advocate of the textbook license; see his proposal at www.mccombs.utexas.edu/news/mentions/arts/2004/11.26_chron_Granof.asp.

Epilogue to Part II

1. "Secrets and the Prize," *The Economist,* October 12, 1996.

Chapter 8

1. C. P. Snow's *The Affair* (London: Penguin, 1962), 69.

2. Michael Spence pioneered the concept of signaling and developed it in an important and highly readable book, *Market Signaling* (Cambridge, MA: Harvard University Press, 1974).

3. George A. Akerlof, "The Market for 'Lemons': Quality Uncertainty and the Market Mechanism," *Quarterly Journal of Economics* 84, no. 3 (August 1970): 488–500.

4. Peter Kerr, "Vast Amount of Fraud Discovered In Workers' Compensation System," *New York Times*, December 29, 1991.

5. This point is developed in Albert L. Nichols and Richard J. Zeckhauser, "Targeting Transfers through Restrictions on Recipients," *American Economic Review* 72, no. 2 (May 1982): 372–77.

6. Nick Feltovich, Richmond Harbaugh, and Ted To, "Too Cool for School? Signaling and Countersignaling," *Rand Journal of Economics* 33 (2002): 630–49.

7. Nasar, *A Beautiful Mind*, 144.

8. Rick Harbaugh and Theodore To, "False Modesty: When Disclosing Good News Looks Bad," working paper, 2007.

9. Taken from Sigmund Freud's *Jokes and Their Relationship to the Unconscious* (New York: W. W. Norton, 1963).

10. This story is based on Howard Blum's op-ed "Who Killed Ashraf Marwan?" *New York Times*, July 13, 2007. Blum is the author of *The Eve of Destruction: The Untold Story of the Yom Kippur War* (New York: HarperCollins, 2003), which identifies Marwan as a potential Israeli agent and may have led to his assassination.

11. McDonald, *Strategy in Poker, Business, and War*, 30.

12. This strategy is explored in Raymond J. Deneckere and R. Preston McAfee, "Damaged Goods," *Journal of Economics & Management Strategy* 5 (1996): 149–74. The example of the IBM printer comes from their paper and M. Jones, "Low-Cost IBM LaserPrinter E Beats HP LaserJet IIP on Performance and Features," *PC Magazine*, May 29, 1990, 33–36. Deneckere and McAfee offer a series of damaged-goods examples, from chips and calculators to disk drives and chemicals.

13. We learned about this story from McAfee, "Pricing Damaged Goods," Economics Discussion Papers, no. 2007–2, available at www.economics-ejournal.org/economics/discussionpapers/2007-2. McAfee's paper provides a general theory of when firms will want to take such actions.

14. Many examples are entertainingly explained in Tim Harford, *The Undercover Economist* (New York: Oxford University Press, 2006); see chapter 2 and also parts of chapter 3. An excellent discussion of the principles and applications from the information industries can be found in Carl Shapiro and Hal Varian, *Information Rules* (Boston: Harvard Business School Press, 1999), chapter 3. A thorough treatment of the theories, with focus on regulation, is in Jean-Jacques Laffont and Jean Tirole, *A Theory of Incentives in Procurement and Regulation* (Cambridge, MA: MIT Press, 1993).

Chapter 9

1. This estimate for the advantage of DSK over QWERTY is found in Donald Norman and David Rumelhart, "Studies of Typing from the LNR Research Group," in *Cognitive Aspects of Skilled Typewriting*, ed. William E. Cooper (New York: Springer-Verlag, 1983).

2. The sad facts of this story come from Stanford economist W. Brian Arthur, "Competing Technologies and Economic Prediction," *Options*, International Institute for Applied Systems Analysis, Laxenburg, Austria, April 1984. Additional information is provided by Paul David, an economic historian at Stanford, in "Clio and the Economics of QWERTY," *American Economic Review* 75 (May 1985): 332–37.

3. See S. J. Liebowitz and Stephen Margolis, "The Fable of the Keys," *Journal of Law & Economics* 33 (April 1990): 1–25.

4. See W. Brian Arthur, Yuri Ermoliev, and Yuri Kaniovski, "On Generalized Urn Schemes of the Polya Kind." Originally published in the Soviet journal

Kibernetika, it was translated and reprinted in *Cybernetics* 19 (1983): 61–71. Similar results were shown, through the use of different mathematical techniques, by Bruce Hill, D. Lane, and William Sudderth, "A Strong Law for Some Generalized Urn Processes," *Annals of Probability* 8 (1980): 214–26.

5. Arthur, "Competing Technologies and Economic Prediction," 10–13.

6. See R. Burton, "Recent Advances in Vehicular Steam Efficiency," Society of Automotive Engineers Preprint 760340 (1976); and W. Strack, "Condensers and Boilers for Steam-powered Cars," NASA Technical Note, TN D-5813 (Washington, D.C., 1970). While the overall superiority may be in dispute among engineers, an unambiguous advantage of steam- or electric-powered cars is the reduction in tailpipe emissions.

7. These comparisons are catalogued in Robin Cowen's "Nuclear Power Reactors: A Study in Technological Lock-in," *Journal of Economic History* 50 (1990): 541–67. The engineering sources for these conclusions include Hugh McIntyre, "Natural-Uranium Heavy-Water Reactors," *Scientific American*, October 1975; Harold Agnew, "Gas-Cooled Nuclear Power Reactors," *Scientific American*, June 1981; and Eliot Marshall, "The Gas Reactor Makes a Comeback," *Science*, n.s., 224 (May 1984): 699–701.

8. The quote is from M. Hertsgaard, *The Men and Money Behind Nuclear Energy* (New York: Pantheon, 1983). Murray used the words "power-hungry" rather than "energy-poor," but of course he meant power in the electrical sense.

9. Charles Lave of the University of California, Irvine, finds strong statistical evidence to support this. See his "Speeding, Coordination and the 55 MPH Limit," *American Economic Review* 75 (December 1985): 1159–64.

10. Cyrus C. Y. Chu, an economist at National Taiwan University, develops this idea into a mathematical justification for the cyclic behavior of crackdowns followed by lax enforcement in his paper, "Oscillatory vs. Stationary Enforcement of Law," *International Review of Law and Economics* 13, no. 3 (1993): 303–15.

11. James Surowiecki laid out this argument in *The New Yorker*: see "Fuel for Thought," July 23, 2007.

12. Milton Friedman, *Capitalism and Freedom* (Chicago: University of Chicago Press, 1962), 191.

13. See his book *Micromotives and Macrobehavior* (New York: W. W. Norton, 1978), chapter 4. Software that lets you experiment with tipping in various conditions of heterogeneity and crowding of populations is available on the web. Two such programs are at http://ccl.northwestern.edu/netlogo/mod els/Segregation and www.econ.iastate.edu/tesfatsi/demos/schelling/schellhp .htm.

14. See his article "Stability in Competition," *Economic Journal* 39 (March 1929): 41–57.

Chapter 10

1. See Peter Cramton, "Spectrum Auctions," in *Handbook of Telecommunications Economics*, ed. Martin Cave, Sumit Majumdar, and Ingo Vogelsang (Amsterdam: Elsevier Science B.V., 2002), 605–39; and Cramton, "Lessons Learned from the UK 3G Spectrum Auction," in U.K. National Audit Office Report, The Auction of Radio Spectrum for the Third Generation of Mobile Telephones, Appendix 3, October 2001.

Chapter 11

1. The generalization to bargaining without procedures is based on work by economists Motty Perry and Philip Reny.

2. Roger Fisher and William Ury, *Getting to Yes: Negotiating Agreement without Giving In* (New York: Penguin Books, 1983).

3. See Adam Brandenburger, Harborne Stuart Jr., and Barry Nalebuff, "A Bankruptcy Problem from the Talmud," Harvard Business School Publishing case 9-795-087; and Barry O'Neill, "A Problem of Rights Arbitration from the Talmud," *Mathematical Social Science*s 2 (1982): 345–71.

4. This case is discussed by Larry DeBrock and Alvin Roth in "Strike Two: Labor-Management Negotiations in Major League Baseball," *Bell Journal of Economics* 12, no. 2 (Autumn 1981): 413–25.

5. This argument is developed more formally in M. Keith Chen's paper "Agenda in Multi-Issue Bargaining," available at www.som.yale.edu/fac ulty/keith.chen/papers/rubbarg.pdf. Howard Raiffa's *The Art and Science of Negotiation* is an excellent source for strategy in multiple-issue bargaining.

6. The idea of a virtual strike was proposed by Harvard negotiation gurus Howard Raiffa and David Lax as a tool to resolve the 1982 NFL strike. See also Ian Ayres and Barry Nalebuff, "The Virtues of a Virtual Strike," in *Forbes*, November 25, 2002.

7. The solution is described in most game theory textbooks. The original article is Ariel Rubinstein, "Perfect Equilibrium in a Bargaining Model," *Econometrica* 50 (1982): 97–100.

Chapter 12

1. This deep result is due to Stanford University professor and Nobel Laureate Kenneth Arrow. His famous "impossibility" theorem shows that any system for aggregating unrestricted preferences over three or more alternatives into a group decision cannot simultaneously satisfy the following minimally desirable properties: (i) transitivity, (ii) unanimity, (iii) independence of irrelevant alternatives, (iv) nondictatorship. Transitivity requires that if A is chosen over B and B is chosen over C, then A must be chosen over C. Unanimity requires A to be chosen over B when A is unanimously preferred to B. Inde-

pendence of irrelevant alternatives requires that the choice between A and B does not depend on whether some other alternative C is available. Nondictatorship requires that there is no individual who always gets his way and thus has dictatorial powers. See Kenneth Arrow, *Social Choice and Individual Values*, 2nd ed. (New Haven, CT: Yale University Press, 1970).

2. In Colorado, Clinton beat Bush 40 to 36, but Perot's 23 percent of the vote could have swung Colorado's 8 electoral votes the other way. Clinton won Georgia's 13 electoral votes with 43 percent of the vote. Bush had 43 percent as well (though fewer total). Perot's 13 percent would surely have swung the election. Kentucky is a Republican stronghold, with two Republican senators. Clinton had a 4-point lead over Bush, but Perot's 14 percent could have swung the election. Other states where Perot's influence was likely felt include Montana, New Hampshire, and Nevada. See www.fairvote.org/plurality/perot.htm.

3. Arrow's short monograph *Social Choice and Individual Values* explains this remarkable result. The issue of strategic manipulability of social choice mechanisms was the focus of Alan Gibbard, "Manipulation of Voting Schemes: A General Result," *Econometrica* 41, no. 4 (July 1973): 587–601; and Mark Satterthwaite, "Strategy-Proofness and Arrow's Conditions," *Journal of Economic Theory* 10, no. 2 (April 1975): 187–217.

4. Similar results hold even when there are many more outcomes.

5. The story of Pliny the Younger was first told from the strategic viewpoint in Robin Farquharson's 1957 Oxford University doctoral thesis, which was later published as *Theory of Voting* (New Haven, CT: Yale University Press, 1969). William Riker's *The Art of Political Manipulation* (New Haven, CT: Yale University Press, 1986) provides much more detail and forms the basis for this modern retelling. Riker's book is filled with compelling historical examples of sophisticated voting strategies ranging from the Constitutional Convention to attempts to pass the Equal Rights Amendment.

6. The idea of using the smallest super-majority rule that ensures the existence of a stable outcome is known as the Simpson-Kramer minmax rule. Here that majority size is no more than 64 percent. See Paul B. Simpson, "On Defining Areas of Voter Choice: Professor Tullock On Stable Voting," *Quarterly Journal of Economics* 83, no. 3 (1969): 478–87, and Gerald H. Kramer, "A Dynamic Model of Political Equilibrium," *Journal of Economic Theory* 16, no. 2 (1977): 538–48.

7. The original papers can be found at www.som.yale.edu/Faculty/bn1/. See "On 64%-Majority Rule," *Econometrica* 56 (July 1988): 787–815, and then the generalization in "Aggregation and Social Choice: A Mean Voter Theorem," *Econometrica* 59 (January 1991): 1–24.

8. The arguments are presented in their book *Approval Voting* (Boston: Birkhauser, 1983).

9. This topic is addressed in Hal Varian, "A Solution to the Problem of Externalities When Agents Are Well-Informed," *The American Economic Review* 84, no. 5 (December 1994): 1278–93.

Chapter 13

1. Canice Prendergast, "The Provision of Incentives in Firms," *Journal of Economic Literature* 37, no. 1 (March 1999): 7–63, is an excellent review discussing numerous applications in relation to the theories. A more theory-based review is Robert Gibbons, "Incentives and Careers in Organizations," in *Advances in Economics and Econometrics, Volume III*, ed. D. M. Kreps and K. F. Wallis (Cambridge: Cambridge University Press, 1997), 1–37. The pioneering analysis of multitask incentive problems is Bengt Holmstrom and Paul Milgrom, "Multitask Principal-Agent Analysis: Incentive Contracts, Asset Ownership, and Job Design," *Journal of Law, Economics, and Organization* 7 (Special Issue, 1991): 24–52. Incentive problems take somewhat different forms and need different solutions in public sector firms and bureaucracies; these are reviewed in Avinash Dixit, "Incentives and Organizations in the Public Sector," *Journal of Human Resources* 37, no. 4 (Fall 2002): 696–727.

2. See Uri Gneezy and Aldo Rustichini, "Pay Enough or Don't Pay At All," *Quarterly Journal of Economics* 115 (August 2000): 791–810.

3. Matthew 6:24 in the King James Version.

Chapter 14

1. A raider who gains control of the company has a right to take the company private and thus buy out all remaining shareholders. By law, these shareholders must be given a "fair market" price for their stock. Typically, the lower tier of a two-tiered bid is still in the range of what might be accepted as fair market value.

2. More on this problem, including a historical perspective, can be found in Paul Hoffman's informative and entertaining *Archimedes' Revenge* (New York: W. W. Norton, 1988).

3. For more on this topic see Barry Nalebuff and Ian Ayres, "In Praise of Honest Pricing," *MIT Sloan Management Review* 45, no. 1 (2003): 24–28, and Xavier Gabaix and David Laibson, "Shrouded Attributes, Consumer Myopia, and Information Suppression in Competitive Markets," *Quarterly Journal of Economics* 121, no. 2 (2006): 505–40.

4. For a full discussion of this problem, see John Moore, "Implementation, Contracts, and Renegotiation," in *Advances in Economic Theory*, vol. 1, ed. Jean-Jacques Laffont (Cambridge: Cambridge University Press, 1992): 184–85 and 190–94.

5. Martin Shubik, "The Dollar Auction Game: A Paradox in Noncooperative Behavior and Escalation," *Journal of Conflict Resolution* 15 (1971): 109–11.

6. This idea of using a fixed budget and then applying backward logic is based on research by Barry O'Neill, "International Escalation and the Dollar Auction," *Journal of Conflict Resolution* 30, no. 1 (1986): 33–50.

7. A summary of the arguments appears in F. M. Scherer, *Industrial Market Structure and Economic Performance* (Chicago: Rand McNally, 1980).

INDEX

Page numbers in *italics* refer to illustrations.
Page numbers beginning with 457 refer to endnotes.